Magnonics
Interface Transmission Tutorial Book Series

Interface Transmission Tutorial Book Series

Series editor: Leonard Dobrzyński

In this Series:

Phononics
Magnonics

Magnonics

Interface Transmission Tutorial Book Series

Abdellatif Akjouj

Leonard Dobrzyński

Housni Al-Wahsh

El Houssaine El Boudouti

Gaëtan Lévêque

Yan Pennec

Bahram Djafari-Rouhani

Series Editor
Leonard Dobrzyński

ELSEVIER

Elsevier
Radarweg 29, PO Box 211, 1000 AE Amsterdam, Netherlands
The Boulevard, Langford Lane, Kidlington, Oxford OX5 1GB, United Kingdom
50 Hampshire Street, 5th Floor, Cambridge, MA 02139, United States

Notices
Knowledge and best practice in this field are constantly changing. As new research and experience broaden our understanding, changes in research methods, professional practices, or medical treatment may become necessary.

Practitioners and researchers must always rely on their own experience and knowledge in evaluating and using any information, methods, compounds, or experiments described herein. In using such information or methods they should be mindful of their own safety and the safety of others, including parties for whom they have a professional responsibility.

To the fullest extent of the law, neither the Publisher nor the authors, contributors, or editors, assume any liability for any injury and/or damage to persons or property as a matter of products liability, negligence or otherwise, or from any use or operation of any methods, products, instructions, or ideas contained in the material herein.

Library of Congress Cataloging-in-Publication Data
A catalog record for this book is available from the Library of Congress

British Library Cataloguing-in-Publication Data
A catalogue record for this book is available from the British Library

ISBN: 978-0-12-813366-8

For information on all Academic Press publications visit
our website at https://www.elsevier.com/books-and-journals

Working together
to grow libraries in
developing countries

www.elsevier.com • www.bookaid.org

Publisher: Susan Dennis
Acquisition Editor: Anita Koch
Editorial Project Manager: Sara Pianavilla
Production Project Manager: Prem Kumar Kaliamoorthi
Cover Designer: Matthew Limbert

Typeset by SPi Global, India

Search and you shall find.
Matthew 7:8

Contents

Contributors   xi
Preface   xiii
Acknowledgments   xv

1.   **Centered System Magnons**   1

Leonard Dobrzyński, Abdellatif Akjouj, Gaëtan Lévêque,
El Houssaine El Boudouti and Housni Al-Wahsh

1.1 Introduction   1
1.2 Heisenberg Exchange Interaction Model   3
     1.2.1 Two Interacting Objects   4
     1.2.2 Three Interacting Objects   7
     1.2.3 Four Interacting Objects   10
     1.2.4 Five Interacting Objects   14
     1.2.5 N Interacting Objects, $N \geq 2$ 18
1.3 Long-Wavelength Limit of the Heisenberg Exchange Approach   21
     1.3.1 Introduction   21
     1.3.2 Two Interacting Objects   22
     1.3.3 Three Interacting Objects   25
     1.3.4 Four Interacting Objects   29
     1.3.5 Five Interacting Objects   33
     1.3.6 N Interacting Objects   37
     1.3.7 Tuning Considerations   40
1.4 Centered System Phonons   41
     1.4.1 Discrete Phonon Model   42
     1.4.2 Continuous Phonon Model   44
1.5 Bound in Continuum States   47
     1.5.1 Discrete Systems   48
     1.5.2 Continuous Systems   48
     1.5.3 Discrete-Continuous Systems   49
     1.5.4 General Theorem for Bound in Continuum States   49
     1.5.5 Fano and Induced Transparency Resonances   49
1.6 Some General Considerations and Perspectives   50
     References   50

2.   Magnonic Circuits: Comb Structures   53
Abdellatif Akjouj, Housni Al-Wahsh, Leonard Dobrzyński,
Bahram Djafari-Rouhani and El Houssaine El Boudouti

2.1   Introduction   54
2.2   Motivations and Outline   55
2.3   Interface Response Theory   56
   2.3.1   Discrete Theory   57
   2.3.2   Continuous Theory   58
2.4   Inverse Surface Green's Functions of the Elementary
    Constituents   59
   2.4.1   Green's Function for an Infinite Ferromagnetic Medium   59
   2.4.2   Inverse Surface Green's Functions of the Semiinfinite
       Medium   61
   2.4.3   Inverse Surface Green's Functions of a Segment
       (or Finite Wire)   61
2.5   Comb Structures   62
   2.5.1   Propagation of Spin Waves in an Infinite Backbone With
       One Grafted Finite Segment   62
   2.5.2   One-Dimensional Infinite Backbone With Periodic Array
       of Finite Segments: Infinite Comb   64
   2.5.3   Transmission Coefficient of the Finite Comb-Like
       Structures   66
   2.5.4   Magnonic Stop Bands and Transmission Spectrum   68
   2.5.5   Defect Modes   72
   2.5.6   Summary and Concluding Remarks   76
2.6   Effects of Pinning Fields in Comb Structures   78
   2.6.1   Dispersion Relations and Transmission Coefficients   78
   2.6.2   Numerical Results and Discussion   79
   2.6.3   Summary   82
2.7   Fano Resonances in a Simple Comb Structure   83
   2.7.1   Theoretical Discussion   84
   2.7.2   Transmission Gaps and Fano Resonances   86
   2.7.3   Summary   94
2.8   Magnonic Analog of Electromagnetic Induced Transparency in
    Detuned Magnetic Circuit   96
   2.8.1   Motivations   96
   2.8.2   Transmission and Reflection Coefficients   98
   2.8.3   Analytical Results Around the MIT Resonances   101
   2.8.4   Summary   102
2.9   Conclusion and Perspectives   103
   References   104
   Further Reading   110

3.   Magnon Mono-Mode Circuits: Serial Loop Structures   111
Housni Al-Wahsh, Abdellatif Akjouj, Leonard Dobrzyński
and Bahram Djafari-Rouhani

3.1   Introduction   111

3.2 Symmetric Serial Loop Structures   114
   3.2.1 Theoretical Model   114
   3.2.2 Numerical Results and Discussion   119
   3.2.3 Summary   127
3.3 Asymmetric Serial Loop Structures   128
3.4 Stop Bands and Defect Modes in a Magnonic Chain
    of Cells Showing Single-Cell Spectral Gaps   131
3.5 General Conclusions and Prospectives   138
   Acknowledgments   138
   References   138

4. Magnons in Nanometric Discrete Structures   143

Housni Al-Wahsh, Abdellatif Akjouj, Leonard Dobrzyński,
Bahram Djafari-Rouhani and El Houssaine El Boudouti

4.1 Introduction   143
4.2 Interface Response Theory   146
4.3 Effects of Coupling Infinite Linear Chain of Nano-Particles
    to Three Local Resonators   148
   4.3.1 Model and Calculations   148
   4.3.2 Results and Discussion   155
4.4 Quasibox Structures   160
   4.4.1 Model and Calculations   160
   4.4.2 Applications and Discussion of the Results   163
   4.4.3 Summary   167
4.5 Magnon Nanometric Multiplexer in Cluster Chains   167
   4.5.1 Introduction   167
   4.5.2 Calculations   169
   4.5.3 Applications and Discussion of the Results   171
   4.5.4 Summary   177
4.6 General Conclusions and Prospectives   177
   Acknowledgments   178
   References   178

5. Surface, Interface, and Confined Slab Magnons   185

Leonard Dobrzyński, Abdellatif Akjouj, Housni Al-Wahsh and
Bahram Djafari-Rouhani

5.1 Introduction   185
5.2 Bulk Heisenberg Model   186
5.3 Bulk Response Function   187
5.4 Planar Defect Magnons   188
5.5 Surface Magnons   189
   5.5.1 Model With First Nearest-Neighbor Interactions   189
   5.5.2 Model With First and Second Nearest-Neighbor
       Interactions   190
5.6 Interface Magnons   190
5.7 Confined Slab Magnons   192
   5.7.1 Introduction to Slabs   192

	5.7.2	The Surface Response Operators	192
	5.7.3	The Response Function	193
	5.7.4	The Localized Modes	198
	5.7.5	Discussion	199
5.8	Surface Reconstruction and Soft Surface Magnons		200
5.9	The Effect of Surface-Pinning Fields on the Thermodynamic Properties of a Ferromagnet		203
	5.9.1	Motivations	203
	5.9.2	Elementary Considerations	205
	5.9.3	Surface-Specific Heat in Presence of Finite Surface-Pinning Field	213
5.10	Summary		218
	References		219

6.   One-Dimensional Magnonic Crystals   221

Leonard Dobrzyński, Abdellatif Akjouj, Housni Al-Wahsh and Bahram Djafari-Rouhani

6.1	Introduction	221
6.2	Bulk Magnetic Response Function for a Two-Slab 1D Crystal	223
6.3	Surface Ferromagnetic Response Function for a Two-Slab 1D Crystal	226
6.4	Bulk and Surface Magnons in a Two-Slab 1D Crystal	227
6.5	Bulk and Surface Magnons in a Three-Slab 1D Crystal	229
6.6	Discussion	231
	References	231

7.   Two-Dimensional Magnonic Crystals   233

Leonard Dobrzyński, Housni Al-Wahsh, Abdellatif Akjouj, Yan Pennec and Bahram Djafari-Rouhani

7.1	Introduction	233
7.2	Method of Calculation	234
7.3	Magnon Band Structures	239
7.4	Summary	245
	References	247

| Index | 251 |

Contributors

Numbers in parentheses indicate the pages on which the authors' contributions begin.

Abdellatif Akjouj (1, 53, 111, 143, 185, 221, 233), Department of Physics, Faculty of Sciences and Technologies, Institute of Electronics, Microelectronics and Nanotechnology, UMR CNRS 8520, Lille University, Villeneuve d'Ascq Cedex, France

Housni Al-Wahsh (1, 53, 111, 143, 185, 221, 233), Faculty of Engineering, Benha University, Cairo, Egypt; Department of Physics, Faculty of Sciences and Technologies, Institute of Electronics, Microelectronics and Nanotechnology, UMR CNRS 8520, Lille University, Villeneuve d'Ascq Cedex, France

El Houssaine El Boudouti (1, 53, 143), LPMR, Department of Physics, Faculty of Sciences, University Mohammed I, Oujda, Morocco; Department of Physics, Faculty of Sciences and Technologies, Institute of Electronics, Microelectronics and Nanotechnology UMR CNRS 8520, Lille University, Villeneuve d'Ascq Cedex, France

Bahram Djafari-Rouhani (53, 111, 143, 185, 221, 233), Department of Physics, Faculty of Sciences and Technologies, Institute of Electronics, Microelectronics and Nanotechnology, UMR CNRS 8520, Lille University, Villeneuve d'Ascq Cedex, France

Leonard Dobrzyński (1, 53, 111, 143, 185, 221, 233), Department of Physics, Faculty of Sciences and Technologies, Institute of Electronics, Microelectronics and Nanotechnology, UMR CNRS 8520, Lille University, Villeneuve d'Ascq Cedex, France

Gaëtan Lévêque (1), Department of Physics, Faculty of Sciences and Technologies, Institute of Electronics, Microelectronics and Nanotechnology, UMR CNRS 8520, Lille University, Villeneuve d'Ascq Cedex, France

Yan Pennec (233), Department of Physics, Faculty of Sciences and Technologies, Institute of Electronics, Microelectronics and Nanotechnology, UMR CNRS 8520, Lille University, Villeneuve d'Ascq Cedex, France

Preface

The Interface Transmission Tutorial Book Series reviews a few simple examples of interface transmission in phononics, magnonics, plasmonics, photonics, electronics, and polaritonics. It gives an unified synthesis of reexamined previous publications of a team of authors, keeping mostly what seems to have some tutorial value. Such review may bring also new ideas and open new research paths. This is intended also to help to go from one field to other ones and transmit like that general analysis methods, physical concepts. An example of such field transposition, from magnon to phonon, is given at the end of Chapter 1. This series may be also considered as an introduction to novel physics domains. Nevertheless this book series cannot be an exhaustive review of interface transmission in any of the physics fields introduced here. This would be an endless task. It leaves space for other books. The volume on Magnonics provides a few simple model examples of interface transmission of magnons in composite materials. A magnon is a particle and a wave associated with a spin vibration. Interface magnon simple transmission models contribute to the foundation of a novel science called Magnonics. All the examples given in this book are aimed to help the interested reader to understand and use for new modern magnonic problems some of well-known theoretical methods. These simple examples have also some pedagogical values for other science domains. Some examples may even be easily transposed from one field to another one and be the starting points for new research directions. In Chapter 1 are present some new simple examples of magnons in finite centered systems. In finite systems, only discrete magnon states exist. Some of these discrete states are subsystem states. Subsystem states are states localized in a part of a system. Finite systems help to understand better some physical effects like trapping, bound in continuum states, resonances, etc. Chapter 2 deals with simple magnonic circuits made out comb structures: transmission gaps, pinning field effects, resonances, magnonic induced transparencies, etc. Chapter 3 addresses magnonic circuits made out of loop structures and stress gaps, defect modes, etc. Chapter 4 is about magnon propagation in nanometric discrete structures. Chapter 5 presents planar surface, interface and confined slab magnons, as well as surface reconstruction and surface pinning field effects. Chapter 6 describes a simple example of magnons in one-dimensional magnonic crystals, called also superlattices. Chapter 7 introduces two-dimensional magnonic crystals. The examples presented in this Magnonics book are often similar to some presented in the other volumes of this book series. Applications, for example, filters, multiplexers, traps, insulators, may benefit from such simple examples.

Leonard Dobrzyński
Villeneuve d'Ascq, France
October 29, 2018

Acknowledgments

We wish to acknowledge Douglas Mills, Jérôme Vasseur, Henryk Puszkarski, Piotr Zielinski, Abdallah Mir, Pierre Deymier, Gregorio Hernandez-Cocoletzi, and Abdelkader Mouadili for their collaboration to some of the original papers reviewed in this book.

It is a pleasure to thank Susan Dennis (publisher), Anita Koch (acquisition editor), Sara Pianavilla (editorial project manager), Prem Kumar Kaliamoorthi (production project manager), Matthew Limbert (cover designer), and their Elsevier teams for an excellent book publication.

We dedicate this book to our families in partial compensation for taking so much of our time away from them. One of us (Leonard Dobrzyński) thanks his children Laetitia, Marie-Laure, François, and Coralie for their support in this project. He acknowledges also François for stimulating discussions about this tutorial book series title and Coralie for her help with the book series project document.

Chapter 1

Centered System Magnons

**Leonard Dobrzyński*, Abdellatif Akjouj*, Gaëtan Lévêque*,
El Houssaine El Boudouti† and Housni Al-Wahsh‡**

**Department of Physics, Faculty of Sciences and Technologies, Institute of Electronics,
Microelectronics and Nanotechnology, UMR CNRS 8520, Lille University, Villeneuve d'Ascq
Cedex, France †LPMR, Department of Physics, Faculty of Sciences, University Mohammed I,
Oujda, Morocco ‡Faculty of Engineering, Benha University, Cairo, Egypt*

Chapter Outline

1.1 **Introduction**	**1**	
1.2 **Heisenberg Exchange Interaction**		
Model	**3**	
1.2.1 Two Interacting Objects	4	
1.2.2 Three Interacting Objects	7	
1.2.3 Four Interacting Objects	10	
1.2.4 Five Interacting Objects	14	
1.2.5 N Interacting Objects,		
$N \geq 2$	18	
1.3 **Long-Wavelength Limit of the**		
Heisenberg Exchange Approach	**21**	
1.3.1 Introduction	21	
1.3.2 Two Interacting Objects	22	
1.3.3 Three Interacting Objects	25	
1.3.4 Four Interacting Objects	29	
1.3.5 Five Interacting Objects	33	
1.3.6 N Interacting Objects	37	
1.3.7 Tuning Considerations	40	

1.4 **Centered System Phonons**	**41**
1.4.1 Discrete Phonon Model	42
1.4.2 Continuous Phonon	
Model	44
1.5 **Bound in Continuum States**	**47**
1.5.1 Discrete Systems	48
1.5.2 Continuous Systems	48
1.5.3 Discrete-Continuous	
Systems	49
1.5.4 General Theorem for	
Bound in Continuum	
States	49
1.5.5 Fano and Induced	
Transparency Resonances	49
1.6 **Some General Considerations**	
and Perspectives	**50**
References	**50**

1.1 INTRODUCTION

This chapter presents first two simple models of magnons in centered systems. The first model represents N, with $N \geq 2$, interacting electron spins, called here objects. One of these objects is in the center of a circle. The others are on the circle and are constituting the surface of this system, see Fig. 1.1.

The first model uses the Heisenberg interaction Hamiltonian between the central spin and each of the surface ones. The surface objects have all the

Magnonics. https://doi.org/10.1016/B978-0-12-813366-8.00001-7

1

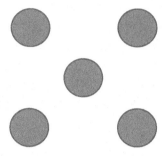

FIG. 1.1 Sketch of the geometry of the considered first kind of objects for $N = 5$.

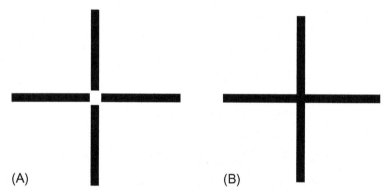

(A)   (B)

FIG. 1.2 Sketch of the geometry of the considered second kind of objects for $N = 5$, (A) for the reference system of four disconnected guides and (B) for the final system with the connected guides.

same interaction with the central object. All other interactions are excluded, by appropriate isolation techniques. The simplicity of this model is of tutorial value.

The second model uses the long-wavelength approximation of the Heisenberg Hamiltonian. Consider $N - 1$ identical magnonic waveguides. Connect together one of their free ends and put this connection in the center of a circle and the remaining unconnected ends on the circle, see Fig. 1.2. It is possible also to consider the free ends of these connected guides and the centered connected site to be the objects of investigation.

All the objects have a spin pointing in the direction perpendicular to the plane in which they are situated. This direction is imposed by a static magnetic field. The procession of these spins around their respective static positions and the propagation of this procession creates the magnons [1]. These two models present degenerate resonant subsurface magnon states, localized each on two of the $N - 1$ surface sites. They show also trapping and isolation effects.

Then a transposition of these examples to centered system phonons is presented, as an example of how it is possible to go from magnon to phonon models and vice versa. Such transpositions are one of the aim of this interface transmission book series.

A final prospective discussion shows other applications of such models: bound in continuum states (e.g., [2]), breaking symmetry effects, possible transposition to other waves (plasmons, electrons, photons, polaritons), and so on.

1.2   HEISENBERG EXCHANGE INTERACTION MODEL

Let us define the Heisenberg interaction J between electron spins $\mathbf{S}_i$ and $\mathbf{S}_j$ pointing in the same space direction and situated at two different space positions

$$\mathbf{H} = -\sum_{ik} J \mathbf{S}_i \cdot \mathbf{S}_k. \tag{1.1}$$

The procession of these magnetic moments around their static positions produces magnons [1]. Such excitations are the solutions of (e.g., [3])

$$\hbar \frac{d\mathbf{S}_i}{dt} = \sum_{k} J(\mathbf{S}_i \wedge \mathbf{S}_k). \tag{1.2}$$

One considers then the three space components of $\mathbf{S}_i$ to be S_i^x, S_i^y, and $S_i^z = S$. Define then

$$S_i^{+} = S_i^x + j S_i^y, \tag{1.3}$$

where $j = \sqrt{-1}$, enables to obtain

$$j\hbar \frac{dS_i^{+}}{dt} = S \sum_{k} J(S_i^{+} - S_k^{+}). \tag{1.4}$$

Define for all these operators

$$S_i^{+} = u_i \exp(-j\omega t), \tag{1.5}$$

where ω is the magnon frequency and t the time.

The energy of these magnons is $\hbar\omega$, where $\hbar$ is the Planck constant and ω the magnon frequency. Then define the dimensionless E to be

$$E = \frac{\hbar\omega}{SJ}. \tag{1.6}$$

With this model, one may undertake simple investigations of magnons in centered systems. These systems are formed out of interacting electron spins, called in what follows objects. The first of these objects is in the center of a circle. The others are on the circle and are constituting the surface of these systems, see Fig. 1.1.

Eq. (1.4), for a centered system with N sites, may be rewritten in the following matrix form

$$\mathbf{h}_N\mathbf{u} = 0, \tag{1.7}$$

where the components of the vector $\mathbf{u}$ are the u_i defined in Eq. (1.5). The eigenvalues and eigenvectors of the matrix $\mathbf{h}_N$ provide the system eigenstates.

The system may be submitted to an action due to a force $\mathbf{F}$ with components F_i on each site. In this case Eq. (1.7) becomes

$$\mathbf{h}_N\mathbf{u} = \mathbf{F}. \tag{1.8}$$

So the system response $\mathbf{u}$ to such an action $\mathbf{F}$ reads

$$\mathbf{u} = \mathbf{g}_N\mathbf{F}, \tag{1.9}$$

where

$$\mathbf{g}_N = [\mathbf{h}_N]^{-1}. \tag{1.10}$$

The matrix $\mathbf{g}_N$, inverse of $\mathbf{h}_N$, is the system response matrix.

1.2.1 Two Interacting Objects

1.2.1.1 Matrix

Consider the following dynamical matrix enabling to investigate two interacting objects as defined previously by Eq. (1.4)

$$\mathbf{h}_2 = \begin{pmatrix} E - 1 & 1 \\ 1 & E - 1 \end{pmatrix}. \tag{1.11}$$

This model has one central object and one surface one. For $N = 2$, there is an in-determination between which is the central object and which is the surface one.

1.2.1.2 Determinant

The determinant of $\mathbf{h}_2$ is

$$|\mathbf{h}_2| = E(E - 2). \tag{1.12}$$

1.2.1.3 Eigenvalues and Eigenvectors

The first eigenvalue and eigenvector values are

$$E_1 = 0 \tag{1.13}$$

and

$$\mathbf{e}_1 = \frac{1}{\sqrt{2}}\begin{pmatrix} 1 \\ 1 \end{pmatrix}. \tag{1.14}$$

The second eigenvalue and eigenvector values are

$$E_2 = 2 \tag{1.15}$$

FIG. 1.3 The spin procession for the eigenvector corresponding to the eigenvalue $E_2 = 2$ and $N = 2$.

and

$$\mathbf{e}_2 = \frac{1}{\sqrt{2}} \begin{pmatrix} 1 \\ -1 \end{pmatrix}. \tag{1.16}$$

Fig. 1.3 shows the spin procession for this $\mathbf{e}_2$ eigenvector.

1.2.1.4 Response Matrix

The matrix inverse of $\mathbf{h}_2$ is

$$\mathbf{g}_2 = \frac{1}{E(E-2)} \begin{pmatrix} E-1 & -1 \\ -1 & E-1 \end{pmatrix}. \tag{1.17}$$

This is the response matrix as defined in the interface response theory [4]. It enables in particular to obtain the system response to any action.

1.2.1.5 System Responses

Consider, for example, the response of this system to the action represented by the following force (magnetic local fields)

$$\mathbf{F} = \begin{pmatrix} F_1 \\ F_2 \end{pmatrix}. \tag{1.18}$$

The system response $\mathbf{u}$ to the action $\mathbf{F}$ [4] is given by

$$\mathbf{u} = \begin{pmatrix} u_1 \\ u_2 \end{pmatrix} = \mathbf{g}_2 \mathbf{F}. \tag{1.19}$$

More explicitly $\mathbf{u}$ is given by

$$\begin{pmatrix} u_1 \\ u_2 \end{pmatrix} = \frac{1}{E(E-2)} \begin{pmatrix} (E-1)F_1 - F_2 \\ -F_1 + (E-1)F_2 \end{pmatrix}. \tag{1.20}$$

1.2.1.6 The Inverse Problem

It is possible also to consider the inverse problem: namely find an action $\mathbf{F}$ such that the system has a specific response $\mathbf{u}$, of interest for some applications. Then as $\mathbf{g}_2$ is the inverse matrix of $\mathbf{h}_2$, it is simpler to use the following equation

$$\mathbf{F} = \mathbf{h}_2 \mathbf{u}. \tag{1.21}$$

More explicitly

$$\begin{pmatrix} F_1 \\ F_2 \end{pmatrix} = \begin{pmatrix} (E-1)u_1 + u_2 \\ u_1 + (E-1)u_2 \end{pmatrix}. \tag{1.22}$$

1.2.1.7 Resonant Responses

The system responds for all values of E, once an action $\mathbf{F}$ is applied to it. However, the resonant responses for each of the eigenvalues, namely here $E_1 = 0$ and $E_2 = 2$, are the only ones for which the system motion diverges, see Eq. (1.20).

1.2.1.8 Forced Trapping and Isolation

A special response of the system may be achieved, with $F_1 \neq \pm F_2$, when

$$E = 1 + \frac{F_1}{F_2}. \tag{1.23}$$

Then

$$\mathbf{u} = \begin{pmatrix} F_2 \\ 0 \end{pmatrix}. \tag{1.24}$$

Another similar special response is obtained, also with $F_1 \neq \pm F_2$, for

$$E = 1 + \frac{F_2}{F_1}. \tag{1.25}$$

Then

$$\mathbf{u} = \begin{pmatrix} 0 \\ F_1 \end{pmatrix}. \tag{1.26}$$

These conditions for getting forced trapping of one of the two spins are shown in Fig. 1.4. The dotted lines correspond to $u_1 = 0$ for ($F_1/F_2 = 1/(E-1)$) on figure (A), and to $u_2 = 0$ for ($F_1/F_2 = E-1$) on figure (B). Figure (C) gives the absolute values of u_1 and u_2 in function of E, for $F_1/F_2 = 0.95$. Note that for negative values of F_1/F_2, one may also obtain forced trapping near the $E = 0$ resonance. However, it is not possible to obtain forced trapping with $F_1 = \pm F_2$ as such forces excite only the eigenvectors e_1 and e_2 of the resonant eigenvalues E_1 and E_2.

Note that in these last two cases, one of the two objects is not responding to the corresponding excitations, for which there is a specific relation, given earlier, between E, F_1, and F_2. These are trapping and isolation effects. These forced object motions may be obtained for all values of E by tuning the force amplitudes F_1 and F_2 as indicated here earlier.

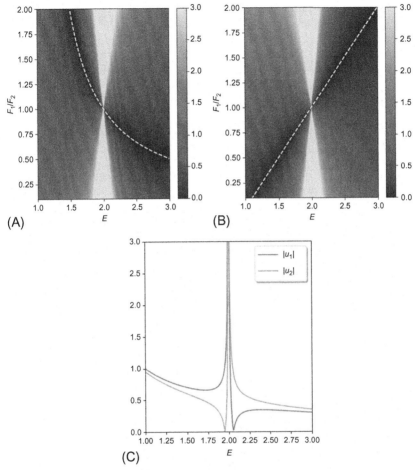

FIG. 1.4 Forced trapping of one of the two spins. The *dotted lines* correspond to $u_1 = 0$ for $(F_1/F_2 = 1/(E-1))$ on figure (A), and to $u_2 = 0$ for $(F_1/F_2 = E - 1)$ on figure (B). Figure (C) gives the absolute values of u_1 and u_2 in function of E, for $F_1/F_2 = 0.95$.

1.2.2 Three Interacting Objects

1.2.2.1 Matrix

Consider the following matrix

$$\mathbf{h}_3 = \begin{pmatrix} E - 2 & 1 & 1 \\ 1 & E - 1 & 0 \\ 1 & 0 & E - 1 \end{pmatrix}. \tag{1.27}$$

One has here one central and two surface objects.

1.2.2.2 Determinant

Its determinant is

$$|\mathbf{h}_3| = E(E-1)(E-3). \tag{1.28}$$

1.2.2.3 Eigenvalues and Eigenvectors

The first eigenvalue and eigenvector values are

$$E_1 = 0 \tag{1.29}$$

and

$$e_1 = \frac{1}{\sqrt{3}}\begin{pmatrix} 1 \\ 1 \\ 1 \end{pmatrix}. \tag{1.30}$$

This eigenvalue shows that this is a state inducing motion of all the system objects.

The second eigenvalue and eigenvector values are

$$E_2 = 1 \tag{1.31}$$

and

$$\mathbf{e}_2 = \frac{1}{\sqrt{2}}\begin{pmatrix} 0 \\ 1 \\ -1 \end{pmatrix}. \tag{1.32}$$

Fig. 1.5 shows the spin procession for this $\mathbf{e}_2$ eigenvector corresponding to the $E_2 = 1$ eigenvalue for $N = 3$.

FIG. 1.5 The spin procession for this $\mathbf{e}_2$ eigenvector corresponding to the $E_2 = 1$ eigenvalue for $N = 3$.

This eigenvector corresponds to a surface state localized on the two surface objects of this three-object system. This state traps and immobilizes the central object.

The third eigenvalue and eigenvector values are

$$E_3 = 3 \tag{1.33}$$

and

$$\mathbf{e}_3 = \frac{1}{\sqrt{6}} \begin{pmatrix} -2 \\ 1 \\ 1 \end{pmatrix}. \tag{1.34}$$

This eigenvalue shows that this is a state spread on all the system objects.

1.2.2.4 Response Matrix

The matrix inverse of h_3 is

$$\mathbf{g}_3 = \frac{1}{E(E-1)(E-3)} \begin{pmatrix} (E-1)^2 & -(E-1) & -(E-1) \\ -(E-1) & E^2 - 3E + 1 & 1 \\ -(E-1) & 1 & E^2 - 3E + 1 \end{pmatrix}. \tag{1.35}$$

This is the response matrix as defined in the interface response theory [4].

1.2.2.5 System Responses

Consider the following action

$$\mathbf{F} = \begin{pmatrix} F_1 \\ F_2 \\ F_3 \end{pmatrix}. \tag{1.36}$$

The system response [4] to the previous action is

$$\mathbf{u} = \mathbf{g}_3 \mathbf{F}. \tag{1.37}$$

More explicitly the response reads

$$\begin{pmatrix} u_1 \\ u_2 \\ u_3 \end{pmatrix} = \frac{1}{E(E-1)(E-3)} \begin{pmatrix} (E-1)((E-1)F_1 - F_2 - F_3) \\ -(E-1)F_1 + (E^2 - 3E + 1)F_2 + F_3 \\ -(E-1)F_1 + F_2 + (E^2 - 3E + 1)F_3 \end{pmatrix}. \tag{1.38}$$

1.2.2.6 The Inverse Problem

It is possible also to consider the inverse problem: namely find an action **F** such that the system has a specific response **u**, of interest for some applications. Then as $\mathbf{g}_3$ is the inverse matrix of $\mathbf{h}_3$, it is simpler to use the following equation

$$\begin{pmatrix} F_1 \\ F_2 \\ F_3 \end{pmatrix} = \begin{pmatrix} (E-2)u_1 + u_2 + u_3 \\ u_1 + (E-1)u_2 \\ u_1 + (E-1)u_3 \end{pmatrix}. \tag{1.39}$$

1.2.2.7 Resonant Responses

The resonant responses for each of the eigenvalues, namely here $E_1 = 0$, $E_2 = 1$, and $E_3 = 3$, are the only values of E for which the system motion diverges. The $E_2 = 1$ state is a surface state, localized on the subsystem formed by the two surface objects of this three-object system. This response matrix confirms also that the central object is trapped and cannot be excited for this value of E. Indeed for $E = 1$, only the surface elements of $\mathbf{g}_3$ diverge. For the two other eigenstates all elements of $\mathbf{g}_3$ diverge.

1.2.2.8 Forced Trapping and Isolation

For $E = \frac{3 \pm \sqrt{5}}{2}$, $g_3(22) = g_3(33) = 0$, it is possible, with the help of Eq. (1.37) or its inverse

$$\mathbf{F} = \mathbf{h}_3 \mathbf{u}, \tag{1.40}$$

to show that simple solutions exist enabling to trap each of the three system objects. For example,

$$\mathbf{u} = \begin{pmatrix} 0 \\ 1 \\ 1 \end{pmatrix}, \tag{1.41}$$

may be obtained for specific values of F_1, F_2, and F_3.

Similarly, simple solutions exist also for trapping any two of these three objects. For example,

$$\mathbf{u} = \begin{pmatrix} 1 \\ 0 \\ 0 \end{pmatrix}. \tag{1.42}$$

1.2.3 Four Interacting Objects

1.2.3.1 Matrix

Consider the following matrix

$$\mathbf{h}_4 = \begin{pmatrix} E-3 & 1 & 1 & 1 \\ 1 & E-1 & 0 & 0 \\ 1 & 0 & E-1 & 0 \\ 1 & 0 & 0 & E-1 \end{pmatrix}. \tag{1.43}$$

With the previous definitions one has here one central object and three surface ones.

1.2.3.2 Determinant

The determinant is

$$|\mathbf{h}_4| = E(E-1)^2(E-4). \tag{1.44}$$

1.2.3.3 Eigenvalues and Eigenvectors

The first eigenvalue and eigenvector values are

$$E_1 = 0 \tag{1.45}$$

and

$$\mathbf{e}_1 = \frac{1}{2}\begin{pmatrix} 1 \\ 1 \\ 1 \\ 1 \end{pmatrix}. \tag{1.46}$$

The second eigenvalue and eigenvector values are

$$E_2 = 1 \tag{1.47}$$

and

$$\mathbf{e}_2 = \frac{1}{\sqrt{2}}\begin{pmatrix} 0 \\ 1 \\ -1 \\ 0 \end{pmatrix}. \tag{1.48}$$

Fig. 1.6 shows the spin procession for this $\mathbf{e}_2$ eigenvector corresponding to $E_2 = 1$ and $N = 4$.

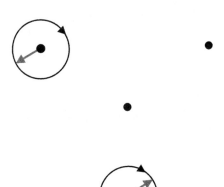

FIG. 1.6 The spin procession for this $\mathbf{e}_2$ eigenvector corresponding to $E_2 = 1$ and $N = 4$.

The third eigenvalue and eigenvector values are

$$E_3 = 1 \qquad (1.49)$$

and

$$\mathbf{e}_3 = \frac{1}{\sqrt{2}} \begin{pmatrix} 0 \\ 0 \\ 1 \\ -1 \end{pmatrix}. \qquad (1.50)$$

Here exist for $E = 1$, two degenerate surface eigenvalues, E_2 and E_3. Their elementary corresponding eigenvectors e_2 and e_3 are localized, respectively, on only two of the three surface objects. However, any linear combination of these two eigenvectors is also a valid surface solution.

The fourth eigenvalue and eigenvector values are

$$E_1 = 4 \qquad (1.51)$$

and

$$\mathbf{e}_1 = \frac{1}{2\sqrt{3}} \begin{pmatrix} -3 \\ 1 \\ 1 \\ 1 \end{pmatrix}. \qquad (1.52)$$

This eigenvalue shows that this state induces motion of all the system objects.

1.2.3.4 Response Matrix

The matrix inverse of h_4 is

$$\mathbf{g}_4 = \frac{1}{E(E-1)(E-4)}$$
$$\begin{pmatrix} (E-1)^2 & -(E-1) & -(E-1) & -(E-1) \\ -(E-1) & E^2 - 4E + 1 & 1 & 1 \\ -(E-1) & 1 & E^2 - 4E + 1 & 1 \\ -(E-1) & 1 & 1 & E^2 - 4E + 1 \end{pmatrix}. \qquad (1.53)$$

This is the response matrix as defined in the interface response theory [4]. Note that for the two time degenerate surface eigenvalues $E = 1$, there is one and the same pole in $\mathbf{g}_4$.

1.2.3.5 System Responses

A response of the system may be excited, for any value of E. More explicitly the response reads

$$\begin{pmatrix} u_1 \\ u_2 \\ u_3 \\ u_4 \end{pmatrix} = \frac{1}{E(E-1)(E-4)} \begin{pmatrix} (E-1)((E-1)F_1 - F_2 - F_3 - F_4) \\ -(E-1)F_1 + (E^2 - 4E + 1)F_2 + F_3 + F_4 \\ -(E-1)F_1 + F_2 + (E^2 - 4E + 1)F_3 + F_4 \\ -(E-1)F_1 + F_2 + F_3 + (E^2 - 4E + 1)F_4 \end{pmatrix}.$$

$$(1.54)$$

1.2.3.6   The Inverse Problem

It is possible also to consider the inverse problem: namely find an action $\mathbf{F}$ such that the system has a specific response $\mathbf{u}$, of interest for some applications. Then as $\mathbf{g}_4$ is the inverse matrix of $\mathbf{h}_4$, it is simpler to use the following equation

$$\begin{pmatrix} F_1 \\ F_2 \\ F_3 \\ F_4 \end{pmatrix} = \begin{pmatrix} (E-3)u_1 + u_2 + u_3 + u_4 \\ u_1 + (E-1)u_2 \\ u_1 + (E-1)u_3 \\ u_1 + (E-1)u_4 \end{pmatrix}.$$

$$(1.55)$$

1.2.3.7   Resonant Responses

Each of the two $E = 1$ resonant surface states is localized within a subsurface formed by two of the three surface objects of this system. Indeed for $E = 1$, only the surface elements of $\mathbf{g}_4$ diverge. By linear combination of these two elementary eigenvectors, it is possible to produce other excitations of the three surface objects. This confirms also that the central object is trapped and cannot be excited for $E = 1$.

The two other resonant states $E_1 = 0$ and $E_4 = 4$ induce motion of the whole system.

1.2.3.8   Forced Trapping and Isolation

For $E = 2 \pm \sqrt{3}$, $g_3(22) = g_3(33) = g_3(44) = 0$, it is possible, with the help of Eq. (1.55) or its inverse

$$\mathbf{F} = \mathbf{h}_4\mathbf{u}, \qquad (1.56)$$

to show that simple solutions exist enabling to trap any one, two, or three of the four system objects. For example,

$$\mathbf{u} = \begin{pmatrix} 0 \\ 1 \\ 1 \\ 1 \end{pmatrix} \qquad (1.57)$$

or

$$\mathbf{u} = \begin{pmatrix} 1 \\ 0 \\ 0 \\ 0 \end{pmatrix}. \qquad (1.58)$$

1.2.4 Five Interacting Objects

1.2.4.1 Matrix

Consider the following matrix

$$\mathbf{h}_5 = \begin{pmatrix} E-4 & 1 & 1 & 1 & 1 \\ 1 & E-1 & 0 & 0 & 0 \\ 1 & 0 & E-1 & 0 & 0 \\ 1 & 0 & 0 & E-1 & 0 \\ 1 & 0 & 0 & 0 & E-1 \end{pmatrix}. \tag{1.59}$$

With the previous definitions one has here one central object and four surface ones.

1.2.4.2 Determinant

Its determinant is

$$|\mathbf{h}_5| = E(E-1)^3(E-5). \tag{1.60}$$

1.2.4.3 Eigenvalues and Eigenvectors

The first eigenvalue and eigenvector values are

$$E_1 = 0 \tag{1.61}$$

and

$$\mathbf{e}_1 = \frac{1}{\sqrt{5}} \begin{pmatrix} 1 \\ 1 \\ 1 \\ 1 \\ 1 \end{pmatrix}. \tag{1.62}$$

This state is spread on the whole system.

The second eigenvalue and eigenvector values are

$$E_2 = 1 \tag{1.63}$$

and

$$\mathbf{e}_2 = \frac{1}{\sqrt{2}} \begin{pmatrix} 0 \\ 1 \\ -1 \\ 0 \\ 0 \end{pmatrix}. \tag{1.64}$$

Fig. 1.7 shows the spin procession for this $\mathbf{e}_2$ eigenvector corresponding to $E_2 = 1$ and $N = 5$.

The third eigenvalue and eigenvector values are

$$E_3 = 1 \tag{1.65}$$

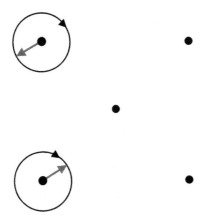

FIG. 1.7 The spin procession for this $\mathbf{e}_2$ eigenvector corresponding to $E_2 = 1$ and $N = 5$.

and

$$\mathbf{e}_3 = \frac{1}{\sqrt{2}} \begin{pmatrix} 0 \\ 0 \\ 1 \\ -1 \\ 0 \end{pmatrix}. \tag{1.66}$$

The fourth eigenvalue and eigenvector values are

$$E_4 = 1 \tag{1.67}$$

and

$$\mathbf{e}_4 = \frac{1}{\sqrt{2}} \begin{pmatrix} 0 \\ 0 \\ 0 \\ 1 \\ -1 \end{pmatrix}. \tag{1.68}$$

These last three states are surface states localized on two of the four surface objects. Any linear combination of these three eigenvectors is also a valid surface eigenvector.

The fifth eigenvalue and eigenvector values are

$$E_5 = 5 \tag{1.69}$$

and

$$\mathbf{e}_5 = \frac{1}{2\sqrt{5}} \begin{pmatrix} -4 \\ 1 \\ 1 \\ 1 \\ 1 \end{pmatrix}. \tag{1.70}$$

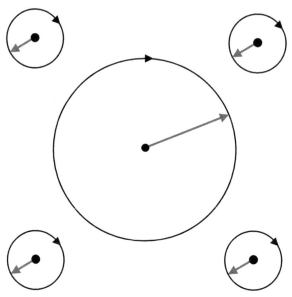

FIG. 1.8 The spin procession for the $\mathbf{e}_5$ eigenvector corresponding to $E_5 = 5$ and $N = 5$.

Fig. 1.8 shows the spin procession for this $\mathbf{e}_5$ eigenvector corresponding to $E_5 = 5$ and $N = 5$.

This state is spread on the whole system.

1.2.4.4 Response Matrix

The matrix inverse of $\mathbf{h}_5$ is

$$\mathbf{g}_5 = \frac{1}{E(E-1)(E-5)}$$

$$\begin{pmatrix} (E-1)^2 & -(E-1) & -(E-1) & -(E-1) & -(E-1) \\ -(E-1) & E^2 - 5E + 1 & 1 & 1 & 1 \\ -(E-1) & 1 & E^2 - 5E + 1 & 1 & 1 \\ -(E-1) & 1 & 1 & E^2 - 5E + 1 & 1 \\ -(E-1) & 1 & 1 & 1 & E^2 - 5E + 1 \end{pmatrix}.$$

$$(1.71)$$

This is the response matrix as defined in the interface response theory [4]. Note that the three times degenerate eigenvalues have one and the same pole at $E = 1$ in $\mathbf{g}_5$.

1.2.4.5 System Responses

A response of the system may be excited, for any value of E. More explicitly the response reads

$$\begin{pmatrix} u_1 \\ u_2 \\ u_3 \\ u_4 \\ u_5 \end{pmatrix} = \frac{1}{E(E-1)(E-5)}$$

$$\begin{pmatrix} (E-1)((E-1)F_1 - F_2 - F_3 - F_4 - F_5) \\ -(E-1)F_1 + (E^2 - 5E + 1)F_2 + F_3 + F_4 + F_5 \\ -(E-1)F_1 + F_2 + (E^2 - 5E + 1)F_3 + F_4 + F_5 \\ -(E-1)F_1 + F_2 + F_3 + (E^2 - 5E + 1)F_4 + F_5 \\ -(E-1)F_1 + F_2 + F_3 + F_4 + (E^2 - 5E + 1)F_5 \end{pmatrix}. \qquad (1.72)$$

1.2.4.6 The Inverse Problem

It is possible also to consider the inverse problem: namely find an action **F** such that the system has a specific response **u**, of interest for some applications. Then as g_5 is the inverse matrix of h_5, it is simpler to use the following equation

$$\begin{pmatrix} F_1 \\ F_2 \\ F_3 \\ F_4 \\ F_5 \end{pmatrix} = \begin{pmatrix} (E-4)u_1 + u_2 + u_3 + u_4 + u_5 \\ u_1 + (E-1)u_2 \\ u_1 + (E-1)u_3 \\ u_1 + (E-1)u_4 \\ u_1 + (E-1)u_5 \end{pmatrix}. \qquad (1.73)$$

1.2.4.7 Resonant Responses

Each of the three $E = 1$ resonant surface states is localized within a subsurface formed by two of the four surface objects of this system. Indeed for $E = 1$, only the surface elements of g_5 diverge. By linear combination of these three elementary eigenvectors, it is possible to produce other excitations of the four surface objects. This confirms also that the central object is trapped and cannot be excited for $E = 1$.

The two other resonant states $E_1 = 0$ and $E_5 = 5$ induce motion of the whole system.

1.2.4.8 Forced Trapping and Isolation

For $E = \frac{5 \pm \sqrt{21}}{2}$, $g_3(22) = g_3(33) = g_3(44) = g_3(55) = 0$, it is possible, with the help of Eq. (1.73) or its inverse

$$\mathbf{F} = \mathbf{h_5 u}, \qquad (1.74)$$

to show that simple solutions exist enabling to trap each of the five system objects. For example,

$$\mathbf{u} = \begin{pmatrix} 0 \\ 1 \\ 1 \\ 1 \\ 1 \end{pmatrix}. \tag{1.75}$$

Similar solutions exist for forced trapping one, two, three, or four of these five objects.

1.2.5 N Interacting Objects, $N \geq 2$

It is easy to generalize the earlier results to N objects ($N \geq 2$) by mathematical induction.

1.2.5.1 Matrix

The $\mathbf{h}_N$ matrix generalization from $\mathbf{h}_5$ is straightforward. Replace just $E - 4$ by $E - N + 1$

$$\mathbf{h}_N = \begin{pmatrix} E - N + 1 & 1 & 1 & \cdots & 1 & 1 & 1 \\ 1 & E - 1 & 0 & \cdots & 0 & 0 & 0 \\ 1 & 0 & E - 1 & \cdots & 0 & 0 & 0 \\ \cdot & \cdot & \cdot & \cdots & \cdot & \cdot & \cdot \\ \cdot & \cdot & \cdot & \cdots & \cdot & \cdot & \cdot \\ \cdot & \cdot & \cdot & \cdots & \cdot & \cdot & \cdot \\ 1 & 0 & 0 & \cdots & E - 1 & 0 & 0 \\ 1 & 0 & 0 & \cdots & 0 & E - 1 & 0 \\ 1 & 0 & 0 & \cdots & 0 & 0 & E - 1 \end{pmatrix}. \tag{1.76}$$

1.2.5.2 Determinant

The determinant of $|\mathbf{h}_N|$ is

$$|\mathbf{h}_N| = E(E - 1)^{N-2}(E - N). \tag{1.77}$$

1.2.5.3 Eigenvalues and Eigenvectors

The first eigenvalue and eigenvector values are

$$E_1 = 0 \tag{1.78}$$

and

$$\mathbf{e}_1 = \frac{1}{\sqrt{N}} \begin{pmatrix} 1 \\ 1 \\ \vdots \\ 1 \\ 1 \end{pmatrix}, \tag{1.79}$$

where the earlier stands for an N column vector.

The second eigenvalue and eigenvector values are

$$E_2 = 1 \tag{1.80}$$

and

$$e_2 = \frac{1}{\sqrt{2}} \begin{pmatrix} 0 \\ 1 \\ -1 \\ 0 \\ \vdots \\ 0 \\ 0 \end{pmatrix}. \tag{1.81}$$

The third eigenvalue and eigenvector values are

$$E_3 = 1 \tag{1.82}$$

and

$$e_3 = \frac{1}{\sqrt{2}} \begin{pmatrix} 0 \\ 0 \\ 1 \\ -1 \\ 0 \\ \vdots \\ 0 \\ 0 \end{pmatrix}. \tag{1.83}$$

The $(N-1)$th eigenvalue and eigenvector values are

$$E_{N-1} = 1 \tag{1.84}$$

and

$$e_{N-1} = \frac{1}{\sqrt{2}} \begin{pmatrix} 0 \\ 0 \\ 0 \\ \vdots \\ 0 \\ 1 \\ -1 \end{pmatrix}. \tag{1.85}$$

Here exist for $E = 1$, $N - 2$ degenerate surface eigenvalues, $E_2, E_3, \ldots,$ $E_N - 1$. Their elementary corresponding eigenvectors $e_2, e_3, \ldots, e_{N-1}$ are localized, respectively, on only two of the $N - 1$ surface objects. However, any linear combination of these two eigenvectors is also a valid solution.

The Nth eigenvalue and eigenvector values are

$$E_N = N \tag{1.86}$$

and

$$
\mathbf{e}_N = \frac{1}{\sqrt{N(N-1)}} \begin{pmatrix} -(N-1) \\ 1 \\ 1 \\ \vdots \\ 1 \\ 1 \end{pmatrix}. \tag{1.87}
$$

1.2.5.4 Response Matrix

The response matrix $\mathbf{g}_N$ inverse of $\mathbf{h}_N$ is obtained also by mathematical induction to be as given here after

$$
\mathbf{g}_N = \frac{1}{E(E-1)(E-N)}
\begin{pmatrix}
(E-1)^2 & 1-E & \cdots & 1-E & 1-E \\
1-E & E^2-NE+1 & \cdots & 1 & 1 \\
\vdots & \vdots & \vdots & \vdots & \vdots \\
1-E & 1 & \cdots & E^2-NE+1 & 1 \\
1-E & 1 & \cdots & 1 & E^2-NE+1
\end{pmatrix}. \tag{1.88}
$$

The denominators of all its terms, but those of its first line and first column, are $E(E-1)(E-N)$. Note that although the $E = 1$ state is $N-2$ times degenerate, only $(E-1)$ appears in these response matrix denominators. The numerators of all the diagonal terms, but the first one, are $E^2 - NE + 1$. The numerators of all the remaining terms are 1.

Note that $g_N(11) = \frac{E-1}{E(E-N)}$. So for the resonant eigenstate $E = 1$, this confirms that the central object is trapped. The system cannot be excited for $E = 1$ by any force acting on this central object.

1.2.5.5 System Responses

Examination of the earlier response matrix confirms that each of the $(N-2)$ surface states $(E = 1)$ is localized within a subsurface formed by two of the $(N-1)$ surface objects of this system. Indeed for $E = 1$, only the surface elements of $\mathbf{g}_N$ diverge. By linear combination of these $N-2$ elementary eigenvectors, it is possible to produce other excitations of the $(N-1)$ surface objects. This confirms also that the central object is trapped and cannot be excited for $E_2 = 1$.

When $E^2 - NE + 1 = 0$, all the surface site elements of $\mathbf{g}_N$ are zero. That means that all these surface sites are trapped when $E = \frac{N \pm \sqrt{N^2 - 4}}{2}$. So for these values of E the system cannot be forced through its surface sites.

The generalization to N objects of the resonant and forced responses obtained earlier for $N \leq 5$ is straightforward.

1.3 LONG-WAVELENGTH LIMIT OF THE HEISENBERG EXCHANGE APPROACH

1.3.1 Introduction

Consider a magnonic waveguide (e.g., [5]). This is one-dimensional continuous ferromagnet. In the long-wavelength approximation the magnons in such a waveguide have the following dispersion relation

$$\omega = Dk^2 + \gamma H_0, \tag{1.89}$$

where ω is the magnon frequency, D a constant, k the propagation vector, γ the gyromagnetic ratio, and H_0 a static magnetic field perpendicular to the waveguide.

The differential equation enabling to obtain magnons in such an infinite waveguide is

$$\frac{F}{k} \left(\frac{\partial^2}{\partial x^2} + k^2 \right) G(x - x') = \delta(x - x'), \tag{1.90}$$

with

$$F = \frac{Dk}{\gamma M_0}, \tag{1.91}$$

M_0 the static magnetization, x is the spatial position along the guide, $\delta(x - x')$ the Dirac delta function, and $G(x - x')$ the response function

$$G(x - x') = \frac{1}{2jF} \exp jk \left| x - x' \right|. \tag{1.92}$$

The magnon propagation vector k is such that

$$k = \frac{2\pi}{\lambda}, \tag{1.93}$$

where λ is the magnon wavelength.

In what follows, the reference system consists of $N - 1$ ($N \geq 2$) noninteracting finite waveguides of length L. Their lateral dimensions are supposed small compared to L and to the wavelengths λ. The final system is an N object centered system constructed out of the $N - 1$ finite waveguides. In this final system the central object is formed by the connection of one free end of all the $N - 1$ guides. Its other $N - 1$ objects are the unconnected free ends of the guides. All these objects are assumed to be in the same plane. An external magnetic field perpendicular to this plane makes the guide magnetic moments point in the direction perpendicular to the plane containing the final system. In the following tutorial presentation, are given first the physical results for $N = 2, 3, 4, 5$ and then the general result for any value of $N \geq 2$.

1.3.2 Two Interacting Objects

1.3.2.1 Matrix

In order to find the response function of a finite guide of length L, one needs within the interface response theory [4] to define first the surface cleavage operator

$$V(x) = \frac{F}{k} \frac{\partial}{\partial x}, \tag{1.94}$$

the surface response operator

$$A_s(x, x') = |V(x'')G(x'' - x')|_{x''=x}, \tag{1.95}$$

and

$$\Delta(M_2 M_2) = \mathbf{I}(M_2 M_2) + \mathbf{A}_s(M_2 M_2), \tag{1.96}$$

where $M_2 = (1, 2)$ is the interface space. The site 1 of the finite guide is at $x = 0$ and the site 2 at $x = L$.

This enables to obtain the response function $\mathbf{g}_2$ of the finite guide and first

$$[\mathbf{g}_2(M_2 M_2)]^{-1} = \Delta(M_2 M_2)[\mathbf{G}(M_2 M_2)]^{-1} \tag{1.97}$$

to be

$$[\mathbf{g}_2(M_2 M_2)]^{-1} = \frac{F}{S} \begin{pmatrix} -C & 1 \\ 1 & -C \end{pmatrix}. \tag{1.98}$$

This is the (2×2) matrix for the two interacting objects (the free ends of the finite guide),

where

$$S = \sin(kL) \tag{1.99}$$

and

$$C = \cos(kL). \tag{1.100}$$

1.3.2.2 Determinant

The determinant of the earlier matrix is equal to

$$\left| [\mathbf{g}_2(M_2 M_2)]^{-1} \right| = -(F)^2. \tag{1.101}$$

1.3.2.3 Eigenvalue and Eigenvector Values at the Guide Ends

From the inverse of Eq. (1.98), one sees that the eigenvalues of this waveguide are the values of kL for which $S = 0$ or equivalently those for which $C = \pm 1$.

So the first eigenvalue is

$$C = 1,$$
$$kL = 2n\pi, \tag{1.102}$$

and the corresponding guide end eigenvector

$$\mathbf{e}_1 = \frac{1}{\sqrt{2}} \begin{pmatrix} 1 \\ 1 \end{pmatrix}. \tag{1.103}$$

The second eigenvalue is

$$C = -1,$$
$$kL = (1 + 2n)\pi, \tag{1.104}$$

and the corresponding guide end eigenvector

$$\mathbf{e}_2 = \frac{1}{\sqrt{2}} \begin{pmatrix} 1 \\ -1 \end{pmatrix}. \tag{1.105}$$

1.3.2.4 Response Function

The matrix inverse of $[\mathbf{g}_2(M_2M_2)]^{-1}$ is the interface response matrix for the two interacting objects

$$\mathbf{g}_2(M_2M_2) = \frac{1}{FS} \begin{pmatrix} C & 1 \\ 1 & C \end{pmatrix}. \tag{1.106}$$

It is possible also to obtain the response function between any two points x and x' belonging to the space D of this finite wire. This may be done with the help of the general equation [4]

$$\mathbf{g}(DD) = \mathbf{G}(DD) - \mathbf{G}(DM)[\mathbf{G}(MM)]^{-1}\mathbf{G}(MD)$$
$$+ \mathbf{G}(DM)[\mathbf{G}(MM)]^{-1}\mathbf{g}(MM)[\mathbf{G}(MM)]^{-1}\mathbf{G}(MD), \tag{1.107}$$

where $\mathbf{G}$ is the response function of the infinite wire given by Eq. (1.92), written here in its matrix form.

It is helpful to recall the following general identities of the interface response theory [4], easily derived in this context with the help of Eq. (1.97):

$$\mathbf{G}(DM)[\mathbf{G}(MM)]^{-1} = \mathbf{g}(DM)[\mathbf{g}(MM)]^{-1} \tag{1.108}$$

and

$$\mathbf{G}(DD) - \mathbf{G}(DM)[\mathbf{G}(MM)]^{-1}\mathbf{G}(MD) = \mathbf{g}(DD) - \mathbf{g}(DM)\mathbf{g}(MM)\mathbf{g}(MD). \tag{1.109}$$

These identities between expressions of the reference and final systems show in this context that these expressions have the same values, no matter if they are calculated from the infinite bulk response function (1.92) or from the finite waveguide one. The explicit first expression is here a (1×2) matrix

$$\mathbf{G}(D_2 M_2)[\mathbf{G}(M_2 M_2)]^{-1} = \frac{1}{S}\left(\sin[k(L-x)] \quad \sin(kx)\right), \tag{1.110}$$

where the index 2 recalls that x is inside the finite waveguide of length L and $M_2 = (0, L)$.

The explicit second expression is within D_2, for $x \le x'$

$$\mathbf{G}(D_2 D_2) - \mathbf{G}(D_2 M_2)[\mathbf{G}(M_2 M_2)]^{-1}\mathbf{G}(M_2 D_2) = -\frac{1}{FS}\sin(kx)\sin[k(L-x')]. \tag{1.111}$$

For $x \ge x'$, just exchange x and x' in the previous expression.

The explicit form of the finite guide response function is for $x \le x'$

$$g_2(x, x') = \frac{1}{FS}\cos(kx)\cos[k(L-x')]. \tag{1.112}$$

For $x \ge x'$, just exchange x and x' in the previous expression.

1.3.2.5 Complete Eigenfunctions

With the help of Eq. (1.112), applying an unity excitation [4] on $x' = L$, one obtains the complete eigenfunctions, for $kL = n\pi$ ($S = 0$), in agreement with its earlier corresponding guide end eigenvectors (Eqs. 1.103, 1.105) to be

$$\mathbf{e}(x) = \sqrt{\frac{2}{L}}\cos\left(\frac{n\pi x}{L}\right). \tag{1.113}$$

1.3.2.6 System Responses

Consider, for example, the response of this system to the action represented by the following force

$$\mathbf{F} = \begin{pmatrix} F_1 \\ F_2 \end{pmatrix}. \tag{1.114}$$

The system response $\mathbf{u}$ to the action $\mathbf{F}$ [4] is given by

$$\mathbf{u} = \mathbf{g}_2(M_2 M_2)\mathbf{F}. \tag{1.115}$$

More explicitly

$$\begin{pmatrix} u_1 \\ u_2 \end{pmatrix} = \frac{1}{FS}\begin{pmatrix} CF_1 + F_2 \\ F_1 + CF_2 \end{pmatrix}. \tag{1.116}$$

1.3.2.7 The Inverse Problem

It is possible also to consider the inverse problem; namely find an action $\mathbf{F}$ such that the system has a specific response $\mathbf{u}$, of interest for some applications. It is simpler to use the following equation

$$\mathbf{F} = [\mathbf{g}_2(M_2 M_2)]^{-1}\mathbf{u}. \tag{1.117}$$

More explicitly

$$\begin{pmatrix} F_1 \\ F_2 \end{pmatrix} = \frac{F}{S} \begin{pmatrix} -Cu_1 + u_2 \\ u_1 - Cu_2 \end{pmatrix}. \tag{1.118}$$

1.3.2.8 Resonant Responses

The system responds for all values of kL, once an action $\mathbf{F}$ is applied to it. However, the resonant responses for each of the eigenvalues, namely here those for which $S = 0$, are the only ones for which the system motion diverges.

1.3.2.9 Forced Responses and Trapping

For values of kL for which there is no resonance, this guide may have forced responses and in particular forced trapping effects. Such behavior is obtained by tuning the force amplitudes F_1 and F_2 as explained in the previous section, see Eqs. (1.23)–(1.26).

1.3.3 Three Interacting Objects

The three interacting object system is formed by two connected waveguides of length L. It is therefore equivalent to the previous two interacting object systems but formed by one waveguide of length $2L$. So all the previous results are valid here once L is replaced by $2L$. However, for an easier generalization to the N interacting object system, is given in what follows a presentation of this three interacting object systems.

1.3.3.1 Matrix

Consider the following matrix representing three interacting objects

$$[\mathbf{g}_3(M_3 M_3)]^{-1} = \frac{F}{S} \begin{pmatrix} -2C & 1 & 1 \\ 1 & -C & 0 \\ 1 & 0 & -C \end{pmatrix}, \tag{1.119}$$

where $M_3 = (1, 2, 3)$. The connection site is 1. The free end sites are 2 and 3.

It is obtained by linear superposition [4] of two $[\mathbf{g}_2(M_2 M_2)]^{-1}$, which represent two guides with free ends. These disconnected two guides constitute the reference system. One has here in the final system one central and two surface objects.

1.3.3.2 Determinant

The determinant of the earlier matrix is

$$\left| [\mathbf{g}_3(M_3 M_3)]^{-1} \right| = 2(F)^3 \frac{C}{S}. \tag{1.120}$$

1.3.3.3 State Conservation

The final system is obtained by connecting together two guides by one of their antinode end. The other end of each guide remains open. The eigenstates and eigenfunctions are shifted as compared to those of the disconnected guides. This state phase shift [4] between the final and the reference system is obtained from Eqs. (1.101), (1.120) to be

$$\eta(kL) = \arg \det[\mathbf{g}_3(M_3M_3)]^{-1} - \arg \det[[\mathbf{g}_2(M_2M_2)]^{-1}]^2 = \arg\left(\frac{C}{S}\right).$$
(1.121)

It is just equal to $-\pi$ for the lost eigenstates and $+\pi$ for the new eigenstates. In the reference system, one has two guides and then two degenerate states for each value of kL such that $S - \sin(kL) = 0$. The earlier state phase shift shows that one of these two degenerate states is replaced by new eigenstates for the values of kL such that $C = \cos(kL) = 0$. So in the final system, the eigenstates are those given by $S = 0$ (or equivalently by $C = 1$ and $C = -1$) and $C = 0$.

1.3.3.4 Eigenvalues and Eigenvector Values at the Guide Ends

First,

$$\begin{aligned} C &= 1, \\ kL &= 2n\pi \end{aligned}$$
(1.122)

and

$$\mathbf{e}_1 = \frac{1}{\sqrt{3}}\begin{pmatrix} 1 \\ 1 \\ 1 \end{pmatrix}.$$
(1.123)

These eigenvalues show that these states induce motion of all the system objects.

Second,

$$\begin{aligned} C &= 0, \\ kL &= (1 + 2n)\frac{\pi}{2} \end{aligned}$$
(1.124)

and

$$\mathbf{e}_2 = \frac{1}{\sqrt{2}}\begin{pmatrix} 0 \\ 1 \\ -1 \end{pmatrix}.$$
(1.125)

This eigenvector corresponds to a surface state localized on the two surface objects of this three-object system. This magnon has a wave function node in the center of this guide of length $2L$ and antinodes at the guide ends. This state traps the central object.

Third,

$$C = -1,$$
$$kL = (1 + 2n)\pi$$

(1.126)

and

$$\mathbf{e}_3 = \frac{1}{\sqrt{3}} \begin{pmatrix} -1 \\ 1 \\ 1 \end{pmatrix}.$$

(1.127)

This eigenvalue shows that this state is spread on all the system objects.

Note that these results are equivalent to those obtained earlier for $N = 2$ but here for one guide of length $2L$. Here, however, as the center of the guide of length $2L$ is inside the interface space, the $C = 0$ surface state is directly visible.

1.3.3.5 Response Matrix

The matrix inverse of $[\mathbf{g}_3(M_3 M_3)]^{-1}$ is [4]

$$\mathbf{g}_3(M_3 M_3) = \frac{1}{2FSC} \begin{pmatrix} C^2 & C & C \\ C & 2C^2 - 1 & 1 \\ C & 1 & 2C^2 - 1 \end{pmatrix}.$$

(1.128)

The response function between any two points x and x' belonging to the space D_3 of this system may be obtained with the help of the general equation (1.107) and the four equations which follow it. The variables x and x' are chosen to be positive, their value 0 being in the center of the system. This choice enables an easier generalization to the cases of higher number of connected guides.

The explicit form of this response function for x within one waveguide of length L and x' within the other waveguide of length L is

$$\mathbf{g}_3(x, x') = \frac{1}{2FSC} \cos[k(L - x')] \cos[k(L - x)].$$

(1.129)

The explicit form of this response function for x and x' within the same waveguide of length L is for $x \le x'$

$$\mathbf{g}_3(x, x') = \frac{1}{2F} \cos[k(L - x')] \left(\frac{\cos(kx)}{S} - \frac{\sin(kx)}{C} \right).$$

(1.130)

For $x \ge x'$, just exchange x and x' in the earlier expression.

1.3.3.6 Complete Eigenfunctions

With the help of Eq. (1.130), one obtains within each of the two branches of this system the complete eigenfunctions to be for $kL = n\pi$ ($S = 0$).

$$e(x) = \pm\sqrt{\frac{1}{L}}\cos\left(\frac{n\pi x}{L}\right), \tag{1.131}$$

with the $+$ sign when n is even and the $-$ sign when n is odd.

For $kL = (1 + 2n)\frac{\pi}{2}$ ($C = 0$), the complete eigenfunctions are found to be

$$e(x) = \pm\sqrt{\frac{1}{L}}\sin\left(\frac{(1 + 2n)\pi x}{2L}\right), \tag{1.132}$$

with the $+$ sign when x is within the first branch of the system and the $-$ sign when x is in the second branch.

These results are in agreement with its earlier corresponding guide end eigenvectors (Eqs. 1.123, 1.125, 1.127).

1.3.3.7 System Responses

Consider the following action

$$\mathbf{F} = \begin{pmatrix} F_1 \\ F_2 \\ F_3 \end{pmatrix}. \tag{1.133}$$

The system response [4] to the earlier action is

$$\mathbf{u} = \mathbf{g}_3(M_3 M_3)\mathbf{F}. \tag{1.134}$$

More explicitly the response reads

$$\begin{pmatrix} u_1 \\ u_2 \\ u_3 \end{pmatrix} = \frac{1}{2FSC} \begin{pmatrix} C(CF_1 + F_2 + F_3) \\ CF_1 + (2C^2 - 1)F_2 + F_3 \\ CF_1 + F_2 + (2C^2 - 1)F_3 \end{pmatrix}. \tag{1.135}$$

1.3.3.8 The Inverse Problem

It is possible also to consider the inverse problem: namely find an action $\mathbf{F}$ such that the system has a specific response $\mathbf{u}$, of interest for some applications. Then it is simpler to use the following equation

$$\begin{pmatrix} F_1 \\ F_2 \\ F_3 \end{pmatrix} = \frac{F}{S} \begin{pmatrix} -2Cu_1 + u_2 + u_3 \\ u_1 - Cu_2 \\ u_1 - Cu_3 \end{pmatrix}. \tag{1.136}$$

1.3.3.9 Resonant Responses

The resonant responses for each of the eigenvalues are the only values of kL for which the system response diverges. This response matrix confirms also that the central object is trapped and cannot be excited for $C = 0$. Indeed for this value of C, only the surface elements of the response function diverge. For the two other eigenstates, all elements of the response function diverge.

1.3.3.10 Forced Trapping

When $2C^2 = 1$, the (22) and (33) elements of the response function vanish. It is possible, with the help of the response function or its inverse to show that simple solutions exist enabling to trap each of the three system objects. Simple solutions exist also for trapping any two of these three objects.

1.3.4 Four Interacting Objects

1.3.4.1 Matrix

Consider the following (4×4) matrix representing four interacting objects, obtained by linear superposition of the three Eq. (1.98) for each of the three wires connected at the central site

$$[\mathbf{g}_4(M_4M_4)]^{-1} = \frac{F}{S}\begin{pmatrix} -3C & 1 & 1 & 1 \\ 1 & -C & 0 & 0 \\ 1 & 0 & -C & 0 \\ 1 & 0 & 0 & -C \end{pmatrix}, \tag{1.137}$$

where $M_4 = (1, 2, 3, 4)$ is the corresponding interface space.

With the earlier definitions one has here one central object and three surface ones.

1.3.4.2 Determinant

The determinant of the earlier matrix is

$$\left|[\mathbf{g}_4(M_4M_4)]^{-1}\right| = -3(F)^4 \frac{C^2}{S^2}. \tag{1.138}$$

1.3.4.3 State Conservation

The final system is obtained by connecting together three guides by one of their antinode end. The other end of each guide remains open. The eigenstates and eigenfunctions are shifted as compared to those of the disconnected guides. This state phase shift [4] between the final and the reference system (Eqs. 1.101, 1.138) is

$$\eta(kL) = \arg\det[\mathbf{g}_4(M_4M_4)]^{-1} - \arg\det[[\mathbf{g}_2(M_2M_2)]^{-1}]^3 = \arg\left(\frac{C^2}{S^2}\right). \tag{1.139}$$

It is just equal to $-\pi$ for the lost eigenstates and $+\pi$ for the new eigenstates. In the reference system, one has three guides and so three degenerate eigenstates for each value of kL such that $S = \sin(kL) = 0$. The previous state phase shift shows that two of these three degenerate states are replaced by new eigenstates for the values of kL such that $C = \cos(kL) = 0$. So in the final system, the eigenstates are those given by $S = 0$ (or equivalently by $C = 1$ and $C = -1$) and $C = 0$, two time degenerate.

1.3.4.4 Eigenvalues and Eigenvector Values at the Guide Ends

The first eigenvalue and eigenvector values are

$$C = 1,$$
$$kL = 2n\pi \tag{1.140}$$

and

$$e_1 = \frac{1}{2} \begin{pmatrix} 1 \\ 1 \\ 1 \\ 1 \end{pmatrix}. \tag{1.141}$$

This eigenvalue shows that this state induces motion of all the system objects.

The second eigenvalue and eigenvector values are

$$C = 0,$$
$$kL = (1 + 2n)\frac{\pi}{2} \tag{1.142}$$

and

$$e_2 = \frac{1}{\sqrt{2}} \begin{pmatrix} 0 \\ 1 \\ -1 \\ 0 \end{pmatrix}. \tag{1.143}$$

The third eigenvalue and eigenvector values are

$$C = 0,$$
$$kL = (1 + 2n)\frac{\pi}{2} \tag{1.144}$$

and

$$e_3 = \frac{1}{\sqrt{2}} \begin{pmatrix} 0 \\ 0 \\ 1 \\ -1 \end{pmatrix}. \tag{1.145}$$

Here, exist for $C = 0$, two degenerate surface eigenvalues. Their elementary corresponding eigenvectors e_2 and e_3 are localized, respectively, on only two of the three surface objects. However, any linear combination of these two eigenvectors is also a valid surface solution.

The fourth eigenvalue and eigenvector values are

$$C = -1,$$
$$kL = (1 + 2n)\pi \tag{1.146}$$

and

$$\mathbf{e}_1 = \frac{1}{2}\begin{pmatrix} -1 \\ 1 \\ 1 \\ 1 \end{pmatrix}.$$ (1.147)

This eigenvalue shows that this state induces motion of all the system objects.

1.3.4.5 Response Matrix

The matrix inverse of $[\mathbf{g}_4(M_4 M_4)]^{-1}$ is

$$\mathbf{g}_4 = \frac{1}{3FSC}\begin{pmatrix} C^2 & C & C & C \\ C & 3C^2-2 & 1 & 1 \\ C & 1 & 3C^2-2 & 1 \\ C & 1 & 1 & 3C^2-2 \end{pmatrix}.$$ (1.148)

Note that for the two time degenerate surface eigenvalues $C = 0$, there is one and the same pole in this response matrix.

The response function between any two points x and x' belonging to the space D_4 of this system may be obtained with the help of the general equation (1.107) and the four equations which follow it. The variables x and x' are chosen to be positive, their value 0 being in the center of the system.

The explicit form of this response function for x within one waveguide of length L and x' within the other waveguide of length L is

$$\mathbf{g}_4(x, x') = \frac{1}{3FSC}\cos[k(L - x)]\cos[k(L - x')].$$ (1.149)

The explicit form of this response function for x and x' within the same waveguide of length L is for $x \le x'$

$$\mathbf{g}_4(x, x') = \frac{1}{3F}\cos[k(L - x')]\left(\frac{\cos(kx)}{S} - \frac{2\sin(kx)}{C}\right).$$ (1.150)

For $x \ge x'$, just exchange x and x' in the earlier expression.

1.3.4.6 Complete Eigenfunctions

With the help of Eq. (1.150), one obtains within each of the two branches of this system the complete eigenfunctions to be for $kL = n\pi$ $(S = 0)$.

$$\mathbf{e}(x) = \pm\sqrt{\frac{2}{3L}}\cos\left(\frac{n\pi x}{L}\right),$$ (1.151)

with the + sign when n is even and the − sign when n is odd.

For $kL = (1 + 2n)\frac{\pi}{2}$ $(C = 0)$, the complete eigenfunctions are found to be

$$\mathbf{e}(x) = \pm\sqrt{\frac{1}{L}}\sin\left(\frac{(1 + 2n)\pi x}{2L}\right),$$ (1.152)

with the $+$ sign when x is within the first excited branch of the system and the $-$ sign when x is in the second excited branch. Within the nonexcited branch $\mathbf{e}(x) = 0$.

These results are in agreement with its earlier corresponding guide end eigenvectors (Eqs. 1.141, 1.143, 1.145, 1.147).

1.3.4.7 System Responses

Consider the following action

$$\mathbf{F} = \begin{pmatrix} F_1 \\ F_2 \\ F_3 \\ F_4 \end{pmatrix}. \tag{1.153}$$

The system response [4] to the previous action is

$$\mathbf{u} = \mathbf{g}_4 (M_4 M_4) \mathbf{F}. \tag{1.154}$$

More explicitly the response reads

$$\begin{pmatrix} u_1 \\ u_2 \\ u_3 \\ u_4 \end{pmatrix} = \frac{1}{2FSC} \begin{pmatrix} C(CF_1 + F_2 + F_3 + F_4) \\ CF_1 + (3C^2 - 2)F_2 + F_3 + F_4 \\ CF_1 + F_2 + (3C^2 - 2)F_3 + F_4 \\ CF_1 + F_2 + F_3 + (3C^2 - 2)F_4 \end{pmatrix}. \tag{1.155}$$

1.3.4.8 The Inverse Problem

It is possible also to consider the inverse problem; namely find an action $\mathbf{F}$ such that the system has a specific response $\mathbf{u}$, of interest for some applications. Then it is simpler to use the following equation

$$\begin{pmatrix} F_1 \\ F_2 \\ F_3 \\ F_4 \end{pmatrix} = \frac{F}{S} \begin{pmatrix} -3Cu_1 + u_2 + u_3 + u_4 \\ u_1 - Cu_2 \\ u_1 - Cu_3 \\ u_1 - Cu_4 \end{pmatrix}. \tag{1.156}$$

1.3.4.9 Resonant Responses

Each of the two $C = 0$ resonant states is localized within a subsystem formed by two of the three branches of this system. By linear combination of these two elementary eigenvectors, it is possible to produce other excitations of the three branches. This confirms also that the central object is trapped and cannot be excited for $C = 0$.

The two other resonant states induce motion of the whole system.

1.3.4.10 Forced Trapping and Isolation

When $3C^2 = 2$, the (22), (33), and (44) elements of the response matrix vanish. It is possible, with the help of Eq. (1.148) or its inverse to show that simple solutions exist enabling to trap each of the four system objects. Simple solutions exist also for trapping any two or three of these four objects.

1.3.5   Five Interacting Objects

1.3.5.1   Matrix

Consider the following (5×5) matrix representing five interacting objects, obtained by linear superposition of the four Eq. (1.98) for each of the four wires connected at the site 1

$$[\mathbf{g}_5(M_5 M_5)]^{-1} = \frac{F}{S} \begin{pmatrix} -4C & 1 & 1 & 1 & 1 \\ 1 & -C & 0 & 0 & 0 \\ 1 & 0 & -C & 0 & 0 \\ 1 & 0 & 0 & -C & 0 \\ 1 & 0 & 0 & 0 & -C \end{pmatrix}. \tag{1.157}$$

With the previous definitions one has here one central object and four surface ones.

1.3.5.2   Determinant

The determinant of the previous matrix is

$$\left| [\mathbf{g}_5(M_5 M_5)]^{-1} \right| = 4(F)^5 \frac{C^3}{S^3}. \tag{1.158}$$

1.3.5.3   State Conservation

The final system is obtained by connecting together four guides by one of their antinode end. The other end of each guide remains open. The eigenstates and eigenfunctions are shifted as compared to those of the disconnected wires. This state phase shift [4] between the final and the reference system, Eqs. (1.101), (1.158), is

$$\eta(kL) = \arg \det[\mathbf{g}_5(M_5 M_5)]^{-1} - \arg \det[[\mathbf{g}_2(M_2 M_2)]^{-1}]^4 = \arg \left(\frac{C^3}{S^3} \right). \tag{1.159}$$

It is just equal to $-\pi$ for the lost eigenstates and $+\pi$ for the new eigenstates. In the reference system, one has four guides and so four degenerate eigenstates for each value of kL such that $S = \sin(kL) = 0$. The previous state phase shift shows that three of these four degenerate states are replaced by new eigenstates for the values of kL such that $C = \cos(kL) = 0$. So in the final system, the eigenstates are those given by $S = 0$ (or equivalently by $C = 1$ and $C = -1$) and $C = 0$, three times degenerate.

1.3.5.4 Eigenvalues and Eigenvector Values at the Guide Ends

The first eigenvalue and eigenvector values are

$$C = 1,$$
$$kL = 2n\pi \tag{1.160}$$

and

$$\mathbf{e}_1 = \frac{1}{\sqrt{5}} \begin{pmatrix} 1 \\ 1 \\ 1 \\ 1 \\ 1 \end{pmatrix}. \tag{1.161}$$

This state is spread on the whole system.

The second eigenvalue and eigenvector values are

$$C = 0,$$
$$kL = (1 + 2n)\frac{\pi}{2} \tag{1.162}$$

and

$$\mathbf{e}_2 = \frac{1}{\sqrt{2}} \begin{pmatrix} 0 \\ 1 \\ -1 \\ 0 \\ 0 \end{pmatrix}. \tag{1.163}$$

The third eigenvalue and eigenvector values are

$$C = 0,$$
$$kL = (1 + 2n)\frac{\pi}{2} \tag{1.164}$$

and

$$\mathbf{e}_3 = \frac{1}{\sqrt{2}} \begin{pmatrix} 0 \\ 0 \\ 1 \\ -1 \\ 0 \end{pmatrix}. \tag{1.165}$$

The fourth eigenvalue and eigenvector values are

$$C = 0,$$
$$kL = (1 + 2n)\frac{\pi}{2} \tag{1.166}$$

and

$$\mathbf{e}_4 = \frac{1}{\sqrt{2}} \begin{pmatrix} 0 \\ 0 \\ 0 \\ 1 \\ -1 \end{pmatrix}. \tag{1.167}$$

These last three states are localized on two of the four system branches. Any linear combination of these three eigenvectors is also a valid surface eigenvector.

The fifth eigenvalue and eigenvector values are

$$C = -1,$$
$$kL = (1 + 2n)\pi$$

(1.168)

and

$$\mathbf{e}_5 = \frac{1}{\sqrt{5}} \begin{pmatrix} -1 \\ 1 \\ 1 \\ 1 \\ 1 \end{pmatrix}.$$

(1.169)

This fifth eigenvector is spread on all the five system objects.

1.3.5.5 Response Matrix

The matrix inverse of $[\mathbf{g}_5(M_5 M_5)]^{-1}$ is

$$\mathbf{g}_5(M_5 M_5) = \frac{1}{4FSC} \begin{pmatrix} C^2 & C & C & C & C \\ C & 4C^2 - 3 & 1 & 1 & 1 \\ C & 1 & 4C^2 - 3 & 1 & 1 \\ C & 1 & 1 & 4C^2 - 3 & 1 \\ C & 1 & 1 & 1 & 4C^2 - 3 \end{pmatrix}.$$

(1.170)

This is the response matrix as defined in the interface response theory [4]. Note that the three times degenerate eigenvalues have one and the same pole at $C = 0$ in this response matrix.

The response function between any two points x and x' belonging to the space D_5 of this system may be obtained, with the help of the general equation (1.107) and the four equations which follow it. The variables x and x' are chosen to be positive, their value 0 being in the center of the system.

The explicit form of this response function for x within one waveguide of length L and x' within the other waveguide of length L is

$$\mathbf{g}_5(x, x') = \frac{1}{4FSC} \cos[k(L - x)] \cos[k(L - x')].$$

(1.171)

The explicit form of this response function for x and x' within the same waveguide of length L is for $x \leq x'$

$$\mathbf{g}_5(x, x') = \frac{1}{4F} \cos[k(L - x')] \left(\frac{\cos(kx)}{S} - \frac{3\sin(kx)}{C} \right).$$

(1.172)

For $x \geq x'$, just exchange x and x' in the earlier expression.

1.3.5.6 Complete Eigenfunctions

With the help of Eq. (1.172), one obtains within each of the two branches of this system the complete eigenfunctions to be for $kL = n\pi$ $(S = 0)$.

$$\mathbf{e}(x) = \pm\sqrt{\frac{1}{2L}} \cos\left(\frac{n\pi x}{L}\right), \tag{1.173}$$

with the $+$ sign when n is even and the $-$ sign when n is odd. This fits in each of the four connected branches the right numbers of $\frac{\lambda}{2}$ oscillations of the previous eigenfunction.

For $kL = (1 + 2n)\frac{\pi}{2}$ $(C = 0)$, the complete eigenfunctions are found to be

$$\mathbf{e}(x) = \pm\sqrt{\frac{1}{L}} \sin\left(\frac{(1 + 2n)\pi x}{2L}\right), \tag{1.174}$$

with the $+$ sign when x is within the first excited branch of the system and the $-$ sign when x is in the second excited branch. This fits in each of the two excited branches the right numbers of $\frac{\lambda}{4}$ oscillations of the previous eigenfunction. Within the two nonexcited branches $\mathbf{e}(x) = 0$.

These results are in agreement with its earlier corresponding guide end eigenvectors (Eqs. 1.161, 1.163, 1.165, 1.167, 1.169).

1.3.5.7 System Responses

The system responses read

$$\begin{pmatrix} u_1 \\ u_2 \\ u_3 \\ u_4 \\ u_5 \end{pmatrix} = \frac{1}{4FSC} \begin{pmatrix} C(CF_1 + F_2 + F_3 + F_4 + F_5) \\ CF_1 + (4C^2 - 3)F_2 + F_3 + F_4 + F_5 \\ CF_1 + F_2 + (4C^2 - 3)F_3 + F_4 + F_5 \\ CF_1 + F_2 + F_3 + (4C^2 - 3)F_4 + F_5 \\ CF_1 + F_2 + F_3 + F_4 + (4C^2 - 3)F_5 \end{pmatrix}. \tag{1.175}$$

1.3.5.8 The Inverse Problem

It is possible also to consider the inverse problem; namely find an action $\mathbf{F}$ such that the system has a specific response $\mathbf{u}$, of interest for some applications. Then it is simpler to use the following equation

$$\begin{pmatrix} F_1 \\ F_2 \\ F_3 \\ F_4 \\ F_5 \end{pmatrix} = \frac{F}{S} \begin{pmatrix} -4Cu_1 + u_2 + u_3 + u_4 + u_5 \\ u_1 - Cu_2 \\ u_1 - Cu_3 \\ u_1 - Cu_4 \\ u_1 - Cu_5 \end{pmatrix}. \tag{1.176}$$

1.3.5.9 Resonant Responses

Each of the three $C = 0$ resonant states is localized within a subsurface formed by two of the four branches of this system. By linear combination of these three elementary eigenvectors, it is possible to produce other excitations of the four branches. This confirms also that the central object is trapped and cannot be excited for $C = 0$.

The two other resonant states induce motion of the whole system.

1.3.5.10 Forced Trapping and Isolation

When $4C^2 = 3$, all but the (11) diagonal element of the response matrix vanish. It is possible, with the help of the response matrix or its inverse to show that simple solutions exist enabling to trap each of the five system objects. Simple solutions exist also for trapping any two, three, or four of these five objects.

1.3.6 N Interacting Objects

It is easy to generalize the earlier results to N objects ($N \geq 2$) by mathematical induction. The matrix $[\mathbf{g}_N(M_N M_N)]^{-1}$ is obtained by linear superposition [4] of $(N - 1)$ $[\mathbf{g}_2(M_2 M_2)]^{-1}$ (Eq. 1.98) connected at the site 1. This matrix has the same form as that for $N = 5$, just replace the $-4C$ term by $-(N - 1)C$. With the previous definitions one has here one central object and $N - 1$ surface ones.

1.3.6.1 Determinant

The determinant of $[\mathbf{g}_N(M_N M_N)]^{-1}$ is

$$|[\mathbf{g}_N(M_N M_N)]^{-1}| = (-1)^{N-1}(F)^N(N - 1)\frac{C^{N-2}}{S^{N-2}}. \qquad (1.177)$$

1.3.6.2 State Conservation

The final system is obtained by connecting together $N - 1$ guides by one of their antinode end. The other end of each guide remains open. The eigenstates and eigenfunctions are shifted as compared to those of the disconnected guides. This state phase shift [4] between the final and the reference system (Eqs. 1.101, 1.177) is

$$\eta(kL) = \arg \det[\mathbf{g}_N(M_N M_N)]^{-1} - \arg \det[[\mathbf{g}_2(M_2 M_2)]^{-1}]^{N-1} = \arg\left(\frac{C^{N-2}}{S^{N-2}}\right). \qquad (1.178)$$

It is just equal to $-\pi$ for the lost eigenstates and $+\pi$ for the new eigenstates. In the reference system, one has $N-1$ guides and so $N-1$ degenerate eigenstates for each value of kL such that $S = \sin(kL) = 0$. The earlier state phase shift shows that $N-2$ of these $N-1$ degenerate states are replaced by new eigenstates for the values of kL such that $C = \cos(kL) = 0$. So in the final system, the eigenstates are those given by $S = 0$ (or equivalently by $C = 1$ and $C = -1$) and $C = 0$, $N - 2$ time degenerate.

1.3.6.3 Eigenvalue and Eigenvector Values at the Guide Ends

The first eigenvalue and eigenvector values are

$$C = 1,$$
$$kL = 2n\pi \tag{1.179}$$

and

$$\mathbf{e}_1 = \frac{1}{\sqrt{N}} \begin{pmatrix} 1 \\ 1 \\ 1 \\ \vdots \\ 1 \\ 1 \end{pmatrix}, \tag{1.180}$$

where the earlier stands for an N column vector.

The second eigenvalue and eigenvector values are

$$C = 0,$$
$$kL = (1 + 2n)\frac{\pi}{2} \tag{1.181}$$

and

$$\mathbf{e}_2 = \frac{1}{\sqrt{2}} \begin{pmatrix} 0 \\ 1 \\ -1 \\ 0 \\ \vdots \\ 0 \\ 0 \end{pmatrix}. \tag{1.182}$$

The third eigenvalue and eigenvector values are

$$C = 0,$$
$$kL = (1 + 2n)\frac{\pi}{2} \tag{1.183}$$

and

$$\mathbf{e}_3 = \frac{1}{\sqrt{2}} \begin{pmatrix} 0 \\ 0 \\ 1 \\ -1 \\ 0 \\ \vdots \\ 0 \\ 0 \end{pmatrix}. \tag{1.184}$$

The $(N-1)$th eigenvalue and eigenvector values are

$$C = 0,$$
$$kL = (1+2n)\frac{\pi}{2} \tag{1.185}$$

and

$$\mathbf{e}_{N-1} = \frac{1}{\sqrt{2}} \begin{pmatrix} 0 \\ 0 \\ 0 \\ \vdots \\ 0 \\ 1 \\ -1 \end{pmatrix}. \tag{1.186}$$

Here exist for $C = 0$, $N - 2$ degenerate eigenvalues. Their elementary corresponding eigenvectors $\mathbf{e}_2, \mathbf{e}_3, \ldots, \mathbf{e}_{N-1}$ are localized, respectively, on only two of the $(N - 1)$ system branches. However, any linear combination of these two eigenvectors is also a valid solution.

The Nth eigenvalue and eigenvector values are

$$C = -1,$$
$$kL = (1+2n)\pi \tag{1.187}$$

and

$$\mathbf{e}_N = \frac{1}{\sqrt{N}} \begin{pmatrix} -1 \\ 1 \\ 1 \\ \vdots \\ 1 \\ 1 \end{pmatrix}. \tag{1.188}$$

1.3.6.4 Response Matrix

The response matrix within the system interface space is the inverse of $[\mathbf{g}_N(M_N M_N)]^{-1}$

$$\mathbf{g}_N = \frac{1}{(N-1)FSC}$$
$$\begin{pmatrix} C^2 & C & \cdots & C & C \\ C & (N-1)C^2 - N + 2 & \cdots & 1 & 1 \\ \vdots & \vdots & \vdots & \vdots & \vdots \\ C & 1 & \cdots & (N-1)C^2 - N + 2 & 1 \\ C & 1 & \cdots & 1 & (N-1)C^2 - N + 2 \end{pmatrix}. \tag{1.189}$$

This is the response matrix as defined in the interface response theory [4]. Note that the $N - 2$ times degenerate eigenvalues have one and the same pole at $C = 0$ in this response matrix.

The response function between any two points x and x' belonging to the space D_N of this system may be obtained, with the help of the general equation (1.107) and the four equations which follow it. The variables x and x' are chosen to be positive, their value 0 being in the center of the system.

The explicit form of this response function for x within one waveguide of length L and x' within the other waveguide of length L is

$$\mathbf{g}_N(x, x') = \frac{1}{(N-1)FSC} \cos[k(L-x)] \cos[k(L-x')]. \qquad (1.190)$$

The explicit form of this response function for x and x' within the same waveguide of length L is for $x \leq x'$

$$\mathbf{g}_N(x, x') = \frac{1}{(N-1)F} \cos[k(L-x')] \left(\frac{\cos(kx)}{S} - \frac{(N-2)\sin(kx)}{C} \right). \qquad (1.191)$$

For $x \geq x'$, just exchange x and x' in the earlier expression.

1.3.6.5 Complete Eigenfunctions

With the help of Eq. (1.191), one obtains within each of the two branches of this system the complete eigenfunctions to be for $kL = n\pi$ $(S = 0)$.

$$\mathbf{e}(x) = \pm \sqrt{\frac{2}{(N-1)L}} \cos\left(\frac{n\pi x}{L} \right), \qquad (1.192)$$

with the $+$ sign when n is even and the $-$ sign when n is odd. This fits in each of the four connected branches the right numbers of $\frac{\lambda}{2}$ oscillations of the earlier eigenfunction.

For $kL = (1 + 2n)\frac{\pi}{2}$ $(C = 0)$, the complete eigenfunctions are found to be

$$\mathbf{e}(x) = \pm \sqrt{\frac{1}{L}} \sin\left(\frac{(1+2n)\pi x}{2L} \right), \qquad (1.193)$$

with the $+$ sign when x is within the first excited branch of the system and the $-$ sign when x is in the second excited branch. This fits in each of the two excited branches the right numbers of $\frac{\lambda}{4}$ oscillations of the previous eigenfunction. Within the two nonexcited branches $e(x) = 0$.

These results are in agreement with its earlier corresponding guide end eigenvectors.

1.3.7 Tuning Considerations

In the continuous systems discussed previously, it is important to stress that the eigenstates, called also resonances, are obtained by adjusting the system to a given resonant wavelength.

The continuous systems presented here are build with the help of $(N - 1)$ finite one-dimensional open-ended magnonic waveguides of length L. The ensemble of these independent guides is the reference system.

Each independent guide has stationary eigenstates. These eigenstate wavelengths λ are such that

$$L = n\frac{\lambda}{2}, \tag{1.194}$$

where n is a positive integer.

So the system eigenfunctions fit an integer number of half-wavelength oscillations into each independent waveguide of length L.

In agreement with the open boundary conditions at the guide ends, the eigenstates have antinodes at the wire ends.

The final system is obtained by connecting together all these guides by one of their antinode end. The other end of each guide remains open.

Due to this connection, the state phase shift is just equal to $-\pi$ for the lost eigenstates and $+\pi$ for the new eigenstates. It shows, for each value of the integer n, annihilation of $N - 2$ states of the disconnected four guides and creation of $N - 2$ shifted states with a shifted wavelength value such that

$$L = (2n + 1)\frac{\lambda}{4}. \tag{1.195}$$

So the corresponding system elementary eigenfunctions fit an odd number of quarter-wavelength oscillations into two branches of the final system. These new states have a wave function node at the system connecting point and antinodes with opposite signs at the two other guide ends. The final system keeps one state unshifted. This state has a wave function antinode at the connecting point and at all surface ends. Recall also that these unshifted eigenfunctions fit an integer number of half-wavelength oscillations into all system branches.

1.3.7.1 System Responses

Examination of the earlier response matrix confirms that each of the $(N - 2)$ degenerate states for $(C = 0)$ is localized within a subsystem formed by two of the $(N - 1)$ branches of this system. By linear combination of these $N - 2$ elementary eigenvectors, it is possible to produce many other excitations of the $(N - 1)$ branches. This confirms also that the central object is trapped and cannot be excited for $C = 0$. The generalization to N objects of the resonant, trapped, isolated, and forced responses obtained earlier for $N \leq 5$ is straightforward.

1.4 CENTERED SYSTEM PHONONS

This section presents two simple models of phonons in centered systems. This is an example of transposition from magnon to phonon models and vice versa. Such transpositions are one of the aim of this interface transmission book series. For those who may be interested only in the phonon model and not in the magnon one, are given here a summary of the same mathematical considerations. As earlier for magnons, the present models represent N, with $N \geq 2$, interacting objects. The first of these objects is in the center of a circle. The others are on the circle. They may be considered to be the surface of such centered systems. Here only transverse phonons, polarized perpendicular to the plane containing the N objects, are investigated. Fig. 1.1 used for magnons may also represent the structure of the first phonon model for $N = 5$. The second phonon model is build out of $N - 1$ identical waveguides. These guides are connected together by one of their free ends. This connection in the center of a circle and the remaining unconnected ends on the circle, see Fig. 1.2. These models have both $(N-2)$ degenerate states, with no motion of the central object. So at this state frequency the central object is trapped.

1.4.1 Discrete Phonon Model

Let us investigate the first phonon model. In this model all the objects have the same mass m. The central object interacts directly, with a spring interaction β, with all the other objects. Any other interactions are excluded. Call ω their phonon frequency. In order to work with entities without dimensions, it is helpful to define

$$E = \frac{m\omega^2}{\beta}. \tag{1.196}$$

Consider the following $N \times N$ dynamical matrix representing the transverse vibrations of the N objects defined earlier

$$\mathbf{h}_N = \begin{pmatrix} E - N + 1 & 1 & 1 & \cdots & 1 & 1 & 1 \\ 1 & E - 1 & 0 & \cdots & 0 & 0 & 0 \\ 1 & 0 & E - 1 & \cdots & 0 & 0 & 0 \\ \vdots & \vdots & \vdots & \vdots & \vdots & \vdots & \vdots \\ 1 & 0 & 0 & \cdots & E - 1 & 0 & 0 \\ 1 & 0 & 0 & \cdots & 0 & E - 1 & 0 \\ 1 & 0 & 0 & \cdots & 0 & 0 & E - 1 \end{pmatrix}. \tag{1.197}$$

Its determinant is

$$|\mathbf{h}_N| = E(E - 1)^{N-2}(E - N). \tag{1.198}$$

The first eigenvalue and eigenvector values are

$$E = 0 \tag{1.199}$$

and

$$\mathbf{e}_1 = \frac{1}{\sqrt{N}} \begin{pmatrix} 1 \\ \cdot \\ 1 \end{pmatrix},$$ (1.200)

where the earlier stands for an N column vector.

The nth eigenvalue, with $2 \leq n \leq (N-1)$, is $N-2$ times degenerate for

$$E = 1.$$ (1.201)

The corresponding $N-2$ eigenvectors have the following generic form

$$\mathbf{e}_n = \frac{1}{\sqrt{2}} \begin{pmatrix} 0 \\ 0 \\ 1 \\ -1 \\ 0 \\ \vdots \\ 0 \end{pmatrix}.$$ (1.202)

Each eigenvector has just two components $\frac{1}{\sqrt{2}}$ and $\frac{-1}{\sqrt{2}}$ on two of the $N-1$ surface objects. All these eigenvectors have a 0 component on the central object. That means that this central object is trapped and isolated for the resonant eigenvalue $E = 1$.

Note that the other possible forms for the eigenvectors of the $N-2$ degenerate eigenvalues are linear combinations of the elementary ones given here earlier.

The Nth eigenvalue and eigenvector values are

$$E = N$$ (1.203)

and

$$\mathbf{e}_N = \frac{1}{\sqrt{N(N-1)}} \begin{pmatrix} -(N-1) \\ 1 \\ \vdots \\ 1 \end{pmatrix}.$$ (1.204)

The response matrix $\mathbf{g}_N$ inverse of $\mathbf{h}_N$ is

$$\mathbf{g}_N = \frac{1}{E(E-1)(E-N)}$$
$$\begin{pmatrix} (E-1)^2 & 1-E & \cdots & 1-E & 1-E \\ 1-E & E^2-NE+1 & \cdots & 1 & 1 \\ \vdots & \vdots & \vdots & \vdots & \vdots \\ 1-E & 1 & \cdots & E^2-NE+1 & 1 \\ 1-E & 1 & \cdots & 1 & E^2-NE+1 \end{pmatrix}.$$
(1.205)

The denominators of all its terms, but those of its first row and first column, are $E(E-1)(E-N)$. Note that although the $E = 1$ state is $N-2$ times degenerate, only $E - 1$ appears in these response matrix denominators. The numerators of all the diagonal terms, but the first one, are $E^2 - NE + 1$. The numerators of all the remaining terms are 1.

Note that $g_N(11) = \frac{E-1}{E(E-N)}$. So for the resonant eigenstate $E = 1$, this confirms that the central object is trapped. The system cannot be excited for $E = 1$ by any force acting on this central object.

When $E^2 - NE + 1 = 0$, all the surface site elements of $\mathbf{g}_N$ are zero. That means that all these surface sites are trapped when $E = \frac{N \pm \sqrt{N^2-4}}{2}$. So for these values of E the system cannot be forced through its surface sites.

1.4.2 Continuous Phonon Model

Let us consider now the second model, it is build out of $N-1$ identical phononic waveguides.

For each of this finite guides, with open-ended boundary conditions, of length L such that $0 \leq x \leq L$, the surface elements of the response function for transverse vibrations, within the interface space $M_2 = (1, 2)$, where 1 and 2 are the free end sites, are given by (e.g., [6])

$$[\mathbf{g}_2(M_2M_2)]^{-1} = \frac{F}{S}\begin{pmatrix} -C & 1 \\ 1 & -C \end{pmatrix}, \tag{1.206}$$

where

$$S = \sin(kL), \tag{1.207}$$

$$C = \cos(kL), \tag{1.208}$$

$$F = C_{44}k, \tag{1.209}$$

C_{44} is the usual elastic constant, $k = 2\pi/\lambda$ and λ is the transverse phonon wavelength.

Note also that this expression of $[g_2(M_2M_2)]^{-1}$ enables to obtain the phonon eigenstates of this finite wire to be given by the zeros of S.

So these eigenstates are such that

$$kL = n\pi \tag{1.210}$$

or

$$\lambda = \frac{2L}{n}, \tag{1.211}$$

where $n = 0, 1, 2, 3, \ldots, \infty$.

In order to construct a centered system, let us connect together $N - 1$, with $N \geq 2$, such guides by one of their free ends. Put this connection in the center of a circle and the remaining unconnected ends on the circle, see Fig. 1.2. With the previous definitions one has here one central object and $N - 1$ surface ones. The $(N \times N)$ matrix $[\mathbf{g}_N(M_N M_N)]^{-1}$ is obtained by linear superposition (e.g., [6]) of $(N - 1)$ $[\mathbf{g}_2(M_2 M_2)]^{-1}$ (Eq. 1.206) connected at the site 1 and has the following expression

$$[\mathbf{g}_N(M_N M_N)]^{-1} = \frac{F}{S} \begin{pmatrix} -(N-1)C & 1 & 1 & \cdots & 1 & 1 & 1 \\ 1 & -C & 0 & \cdots & 0 & 0 & 0 \\ 1 & 0 & -C & \cdots & 0 & 0 & 0 \\ \vdots & \vdots & \vdots & \vdots & \vdots & \vdots & \vdots \\ 1 & 0 & 0 & \cdots & -C & 0 & 0 \\ 1 & 0 & 0 & \cdots & 0 & -C & 0 \\ 1 & 0 & 0 & \cdots & 0 & 0 & -C \end{pmatrix}.$$
(1.212)

The determinant of $[\mathbf{g}_N(M_N M_N)]^{-1}$ is

$$\left| [\mathbf{g}_N(M_N M_N)]^{-1} \right| = (-1)^{N-1} (F)^N (N-1) \frac{C^{N-2}}{S^{N-2}}.$$
(1.213)

The final system is obtained by connecting together $N - 1$ guides by one of their antinode end. The other end of each guide remains open. The eigenstates and eigenfunctions are shifted as compared to those of the disconnected guides. This state phase shift (e.g., [6]) between the final and the reference system is

$$\eta(kL) = \arg \det[\mathbf{g}_N(M_N M_N)]^{-1} - \arg \det[[\mathbf{g}_2(M_2 M_2)]^{-1}]^{N-1} = \arg \left(\frac{C^{N-2}}{S^{N-2}} \right).$$
(1.214)

It is just equal to $-\pi$ for the lost eigenstates and $+\pi$ for the new eigenstates. In the reference system, one has $(N - 1)$ guides and so $(N - 1)$ degenerate eigenstates for each value of kL such that $S = \sin(kL) = 0$. The earlier state phase shift shows that $(N - 2)$ of these $(N - 1)$ degenerate states are replaced by new eigenstates for the values of kL such that $C = \cos(kL) = 0$. So in the final system, the eigenstates are those given by $S = 0$ (or equivalently by $C = 1$ and $C = -1$) and $C = 0$, $(N - 2)$ times degenerate.

The first eigenvalues are $C = 1$ or equivalently $kL = 2\pi n$, with $n = 0, 1, 2, 3, \ldots, \infty$. The corresponding eigenvector values within the interface space formed by the N guide ends of the centered system are

$$\mathbf{e}_1 = \frac{1}{\sqrt{N}} \begin{pmatrix} 1 \\ \vdots \\ 1 \end{pmatrix}.$$
(1.215)

The mth eigenvalues, with $2 \leq m \leq (N - 1)$, are $N - 2$ times degenerate and are given by $C = 0$ or equivalently by $kL = \frac{(2n+1)\pi}{2}$.

The corresponding $N - 2$ eigenvectors have the following generic form

$$
\mathbf{e}_3 = \frac{1}{\sqrt{2}} \begin{pmatrix} 0 \\ 0 \\ 1 \\ -1 \\ 0 \\ \vdots \\ 0 \end{pmatrix}. \tag{1.216}
$$

Each eigenvector has just two components $\frac{1}{\sqrt{2}}$ and $\frac{-1}{\sqrt{2}}$ on two of the $(N-1)$ surface guide ends. All these eigenvectors have a 0 component on the central object. That means that this central object is trapped for the resonant eigenvalue $C = 0$. Note that the other possible forms for the eigenvectors of the $N - 2$ degenerate eigenvalues are linear combinations of the elementary ones given here earlier.

The Nth eigenvalues are $C = -1$ or equivalently $kL = \pi(2n + 1)$. The corresponding eigenvector values within the interface space formed by the N guide ends of the centered system are

$$
\mathbf{e}_N = \frac{1}{\sqrt{N}} \begin{pmatrix} -1 \\ 1 \\ \vdots \\ 1 \end{pmatrix}. \tag{1.217}
$$

The response function within the system interface space is the inverse of $[\mathbf{g}_N(M_N M_N)]^{-1}$

$$
\mathbf{g}_N = \frac{1}{(N-1)FSC}
\begin{pmatrix}
C^2 & C & \cdots & C & C \\
C & (N-1)C^2 - N + 2 & \cdots & 1 & 1 \\
\vdots & \vdots & \vdots & \vdots & \vdots \\
C & 1 & \cdots & (N-1)C^2 - N + 2 & 1 \\
C & 1 & \cdots & 1 & (N-1)C^2 - N + 2
\end{pmatrix}. \tag{1.218}
$$

The response function between any two points x and x' belonging to the space D_N of this system may be obtained, with the help of the general equation (1.107) and the four equations which follow it. The variables x and x' are chosen to be positive, their value 0 being in the center of the system.

The explicit form of this response function for x within one waveguide of length L and x' within the other waveguide of length L is

$$
\mathbf{g}_N(x, x') = \frac{1}{(N-1)FSC} \cos[k(L-x)] \cos[k(L-x')]. \tag{1.219}
$$

The explicit form of this response function for x and x' within the same waveguide of length L is for $x \leq x'$

$$g_N(x, x') = \frac{1}{(N-1)F} \cos[k(L - x')] \left(\frac{\cos(kx)}{S} - \frac{(N-2)\sin(kx)}{C} \right). \quad (1.220)$$

For $x \geq x'$, just exchange x and x' in the earlier expression.

With the help of Eq. (1.220), one obtains within each of the two branches of this system the complete eigenfunctions to be for $kL = n\pi$ $(S = 0)$.

$$e(x) = \pm\sqrt{\frac{2}{(N-1)L}} \cos\left(\frac{n\pi x}{L}\right), \quad (1.221)$$

with the $+$ sign when n is even and the $-$ sign when n is odd. This fits in each of the four connected branches the right numbers of $\frac{\lambda}{2}$ oscillations of the earlier eigenfunction.

For $kL = (1 + 2n)\frac{\pi}{2}$ $(C = 0)$, the complete eigenfunctions are found to be

$$e(x) = \pm\sqrt{\frac{1}{L}} \sin\left(\frac{(1 + 2n)\pi x}{2L}\right), \quad (1.222)$$

with the $+$ sign when x is within the first excited branch of the system and the $-$ sign when x is in the second excited branch. This fits in each of the two excited branches the right numbers of $\frac{\lambda}{4}$ oscillations of the earlier eigenfunction. Within the two nonexcited branches $e(x) = 0$.

These results are in agreement with its earlier corresponding guide end eigenvectors.

Examination of the previous response matrix confirms that each of the $(N-2)$ degenerate states for $(C = 0)$ is localized within a subsystem formed by two of the $(N-1)$ guides. Indeed for $C = 0$, only within these two excited guides, the elements of the response matrix diverge. By linear combination of these $N-2$ elementary eigenvectors, it is possible to produce many other excitations of the $(N-1)$ connected guides. This confirms also that the central object is trapped and cannot be excited for $C = 0$.

Experimental checks of the predictions of this continuum phonon model are easy. Some were done (P. Journée, L. Dobrzynski, 2017, unpublished) with one string touching another one at one point. That means with the previous definitions four guides, see Fig. 1.2.

1.5 BOUND IN CONTINUUM STATES

In the first chapter of this book are presented magnons and phonons of some finite centered systems. All finite systems have discrete eigenstates. Infinite and semiinfinite systems have continuous bulk band states and eventually localized states within gaps between their bulk bands. In present section is addressed the question of what happens when one couples a finite system with a semiinfinite or an infinite one. One may expect that the finite system discrete states will couple with the bulk band states or remain localized within the gaps. There exists another possibility for these discrete states, namely to remain localized even if their frequencies fall within a bulk band. The conditions of existence of these bound in continuum states are given in what follows. These very special localized states called also bound in continuum states are of interest for many applications, as well for discrete as continuous systems.

1.5.1 Discrete Systems

Consider the coupling of a finite system S_1 to another semiinfinite or infinite system S_2 through an interaction $\mathbf{V}_I(M'M')$ between one or several objects of the space M_1 of S_1 and one or several sites of the space M_2 of S_2, where M' is the space containing M_1 and M_2.

Let the coupled system dynamical matrix be $\mathbf{h}'$ and the uncoupled system one $\mathbf{h}$ such that

$$\mathbf{h}'(M'M') = \mathbf{h}(M'M') + \mathbf{V}_I(M'M'), \tag{1.223}$$

with

$$\mathbf{h}(M'M') = \begin{pmatrix} \mathbf{h}(M_1M_1) & 0 \\ 0 & \mathbf{h}(M_2M_2) \end{pmatrix}. \tag{1.224}$$

Consider now an eigenvector $\mathbf{e}(M_1)$ of the finite system S_1,

$$\mathbf{h}(M_1M_1)\mathbf{e}(M_1) = 0. \tag{1.225}$$

This eigenvector components are 0 on all sites of M_2 and may be also 0 on some sites of M_1. When the interface interaction $\mathbf{V}_I(M'M')$ affects in M_1 only the sites on which this eigenvector components are 0, then this eigenvector remains a valid eigenvector of the final system, although localized within the space M_1 of the finite system. If the value of its eigenstate energy falls within the bulk band of the semiinfinite or infinite system, it becomes a bound in continuum state.

This can be checked with the help of any eigenvector having at least one component with value 0 presented here earlier in this chapter.

1.5.2 Continuous Systems

Note that for any finite discrete system S_1

$$\mathbf{h}(M_1M_1)\mathbf{g}(M_1M_1) = \mathbf{I}(M_1M_1), \tag{1.226}$$

where $\mathbf{g}(M_1M_1)$ is the response function associated with $\mathbf{h}(M_1M_1)$.

This enables to repeat for continuous systems the earlier demonstration for discrete ones.

For continuous systems (e.g., [6]), the $[\mathbf{g}'(M'M')]^{-1}$ of the final system is obtained by linear superposition of the interface elements of $[\mathbf{g}(M_1M_1)]^{-1}$ and $[\mathbf{g}(M_2M_2)]^{-1}$.

For example, when the semiinfinite system S_2 is just a one-dimensional lead attached to one single site of S_1, this adds only an iF to the corresponding element of the $[\mathbf{g}(M_1 M_1)]^{-1}$, where F has the values defined earlier for magnons and phonons.

Consider now an eigenvector $e(M_1)$ of the finite system S_1,

$$[\mathbf{g}(M_1 M_1)]^{-1} \, \mathbf{e}(M_1) = 0. \tag{1.227}$$

If this eigenvector has components with a 0 value on the sites on which the interaction with the S_2 system acts, because of these 0 values, such an eigenvector is not perturbed by this interaction.

This eigenvector remains therefore a valid eigenvector of the final system, but localized within the space M_1 of the finite system. If the value of its eigenstate falls within the bulk band of the semiinfinite or infinite system, it becomes a bound in continuum state.

This can be checked with the help of any eigenvector having at least one component with value 0 presented here earlier in this chapter.

1.5.3 Discrete-Continuous Systems

The previous demonstrations may also be repeated, with the help of the response functions for discrete-continuous systems (e.g., [6]). Considering the general mathematical demonstrations given in this section, it is possible to present the following general theorem.

1.5.4 General Theorem for Bound in Continuum States

A finite system can have eigenstates whose eigenvectors are 0 on some sites of the system. Each of such eigenstates becomes a bound in continuum state, once this finite system is interfaced with an infinite or semiinfinite system only through one or several of these 0 sites and its energy falls within a bulk band.

1.5.5 Fano and Induced Transparency Resonances

Among the possible applications of the earlier bound in continuum states, this section stress how these states may used to obtain high-quality Fano [7] and induced transparency resonances [8]. Start with the simple example of two connected waveguides of length L (the $N = 3$ case discussed earlier for magnons and phonons).

Connect two semiinfinite leads to the middle of the earlier $N = 3$ finite structure of length $2L$. The states, such that their wavelengths λ are given by

$$L = (2n + 1)\frac{\lambda}{4}, \tag{1.228}$$

where n is an integer, are bound in the continuum states of the infinite guide.

When the length L of one of the two finite guides is out of tune by a small amount δ, each of the bound in continuum states becomes Fano resonances, with a transmission zero on one side of the asymmetric resonances.

When the length L of one finite guide is out of tune by a small amount δ and the length of the other by $-\delta$, then the bound in continuum states becomes induced transparency resonances, with a transmission zero on each side of the symmetric resonances.

Such simple examples are developed in the next chapter of this book.

1.6 SOME GENERAL CONSIDERATIONS AND PERSPECTIVES

Magnon and phonon trapping effects, for any value of the wavelength λ, may be achieved within some of the tuned waveguides of the centered systems described in this chapter by fitting the different possible actions F_n on the system sites. When such a localization is obtained within some branches, there is no transmission between these branches and the remaining ones. Other simpler manner to get such effects is to break some system symmetries by changing the length of some branches. Such procedures are pretty straightforward as they are used by music artists on cord instruments when they tune them before playing. As an example consider variations of the interactions between the discrete objects. Another example of broken symmetry may be obtained by changing the lengths of the $(N - 1)$ guides in the continuous models. It is possible to obtain like that new localization effects, high-quality resonances, multiplexers, memories, and so on.

Among the possible applications of all the results presented in this chapter for magnons and for phonons are also localized states in a bulk band, called also bound in continuum states. Such states are easily obtained with the discrete and continuous models presented in this chapter. One manner to get such states consists to add a finite or a semiinfinite one-dimensional medium, bringing a continuum band of states, to any of the previous two finite systems. This may be achieved through a connection to the object of these centered systems on which an eigenvector component vanishes. In this case the corresponding eigenstate becomes a bound in continuum state.

Note that in this chapter, all damping effects are neglected. The damping effects may be compensated in some real systems by appropriate amplifying devices. Hope also that more realistic models, simulations, and experiments can shortly appear for such centered systems.

Similar centered system models may be obtained for other kind of waves (i.e., plasmons, electrons, photons, and polaritons).

REFERENCES

[1] F. Bloch, Zur Theorie des Ferromagnetismus, Z. Phys. 61 (1930) 206.
[2] C.W. Hsu, B. Zhen, A.D. Stone, J.D. Joannopoulos, M. Soljacic, Bound states in the continuum, Nat. Rev. Mater. 1.9 (2016) 16048.

[3] M.G. Cottam, D.R. Tilley, Introduction to Surface and Superlattice Excitations, Cambridge University Press, Cambridge, 1989, p. 127.

[4] L. Dobrzynski, A. Akjouj, E. El Boudouti, et al., Interface Response Theory, in: Interface Transmission Tutorial, Book Series: Phononics, Elsevier, 2017, p. 1 (In this reference, on p. 16, Eq. (93) should read $\mathbf{u}(D) \propto \mathbf{G}(DM)(G(MM))^{-1}\mathbf{g}(MM)(G(MM))^{-1}\mathbf{U}(M)$).

[5] H. Al-Wahsh, A. Akjouj, B. Djafari-Rouhani, L. Dobrzynski, Magnonic circuits and crystals, Surf. Sci. Rep. 66 (2011) 29.

[6] L. Dobrzynski, Interface response theory of continuous composite systems, Surf. Sci. Rep. 11 (1990) 164.

[7] U. Fano, Effects of configuration interaction on intensities and phase shifts, Phys. Rev. 124 (1961) 1866.

[8] K.J. Boller, A. Imamoglu, S.E. Harris, Observation of electromagnetically induced transparency, Phys. Rev. Lett. 66 (1991) 2593.

Chapter 2

Magnonic Circuits: Comb Structures

Abdellatif Akjouj*, Housni Al-Wahsh†, Leonard Dobrzyński*,
Bahram Djafari-Rouhani* and El Houssaine El Boudouti‡
*Department of Physics, Faculty of Sciences and Technologies, Institute of Electronics,
Microelectronics and Nanotechnology, UMR CNRS 8520, Lille University, Villeneuve d'Ascq
Cedex, France †Faculty of Engineering, Benha University, Cairo, Egypt ‡LPMR, Department of
Physics, Faculty of Sciences, University Mohammed I, Oujda, Morocco

Chapter Outline

2.1 **Introduction**   54
2.2 **Motivations and Outline**   55
2.3 **Interface Response Theory**   56
    2.3.1 Discrete Theory   57
    2.3.2 Continuous Theory   58
2.4 **Inverse Surface Green's Functions**
    of the Elementary Constituents   59
    2.4.1 Green's Function for an Infinite
       Ferromagnetic Medium   59
    2.4.2 Inverse Surface Green's
       Functions of the Semiinfinite
       Medium   61
    2.4.3 Inverse Surface Green's
       Functions of a Segment
       (or Finite Wire)   61
2.5 **Comb Structures**   62
    2.5.1 Propagation of Spin Waves in
       an Infinite Backbone With One
       Grafted Finite Segment   62
    2.5.2 One-Dimensional Infinite
       Backbone With Periodic Array
       of Finite Segments: Infinite
       Comb   64
    2.5.3 Transmission Coefficient of the
       Finite Comb-Like Structures   66
    2.5.4 Magnonic Stop Bands and
       Transmission Spectrum   68
    2.5.5 Defect Modes   72

    2.5.6 Summary and Concluding
       Remarks   76
2.6 **Effects of Pinning Fields in Comb**
    Structures   78
    2.6.1 Dispersion Relations and
       Transmission Coefficients   78
    2.6.2 Numerical Results and
       Discussion   79
    2.6.3 Summary   82
2.7 **Fano Resonances in a Simple**
    Comb Structure   83
    2.7.1 Theoretical Discussion   84
    2.7.2 Transmission Gaps and Fano
       Resonances   86
    2.7.3 Summary   94
2.8 **Magnonic Analog of Electromagnetic**
    Induced Transparency in Detuned
    Magnetic Circuit   96
    2.8.1 Motivations   96
    2.8.2 Transmission and Reflection
       Coefficients   98
    2.8.3 Analytical Results Around
       the MIT Resonances   101
    2.8.4 Summary   102
2.9 **Conclusion and Perspectives**   103
References   104
Further Reading   110

Magnonics. https://doi.org/10.1016/B978-0-12-813366-8.00002-9

2.1 INTRODUCTION

An exciting development in material science is the appearance of new samples of alternating thin layers of two different materials, with the thickness and composition of each element subject to precise control. The resulting entity may possess new physical properties. Most particularly, by means of sputtering technique, specimens from two metals were prepared and each of them is present as a layer with thickness from a few angstroms to several hundred angstroms [1–6].

The nature of the spin wave spectrum of such systems is studied theoretically [2, 3, 7–10] and experimentally by light scattering [11]. Since these early works on magnetic superlattices, very exciting developments followed. Our aim is not to review them here. Let us mostly cite those who lead to giant magnetoresistance and to Albert Fert [12] and Peter Grunberg [13] Nobel prize. Recent experimental developments [14–19] that enable to master the scattering and transmission of surface spin waves, start paving the ways for the use of magnons for circuits.

Low-dimensional spin systems—that is, magnetic configuration with a dimensionality less than three—have attracted enthusiastic attention [20–23]. This was related both to the fundamental interest and to the potential applications of spin-transport devices supported by the advanced progress in nanofabrication technology [24, 25]. Arrays of very long ferromagnetic nanowires of, for example, Ni, permalloy, and Co, with diameters in the range of 30–500 nm have been created [26, 27]. These are very uniform in cross-section, with lengths in the range of 20 μm. Thus they are realizations of nanowires which can be reasonably considered as infinite in length, to excellent approximation. In addition to the static and spin-transport properties of magnetic nanowire arrays, the dynamical properties of magnetic nanostructures are also of considerable interest in both fundamental research as well as applied research [27]. The study of spin waves is a powerful tool for probing the dynamical properties of magnetic media in general and those of laterally patterned magnetic structures in particular [28]. On the other hand, due to the possible use of electron spin for storage and information transfer in quantum computers [29, 30], there have been many recent studies of spin transport in semiconductor nanostructures [31].

Magnonic systems, which have a regular distribution of scattering centers, have been seen to possess a distinct and interesting array of magnonic properties. Frequency band gaps within which magnons cannot propagate through the structure, called magnonic band gaps [32–40], are one of these properties. The interest in these band gaps is related to a number of advantages that magnonic crystals have in comparison with photonic crystals. The wavelength of a spin wave and, hence, the properties of such crystals depend on the external magnetic field and can be controlled by this field. The wavelength of propagating spin waves for a wide class of ferromagnetic materials in the microwave range is on the order of tens or even hundreds of micrometers. The phase and group velocities of spin waves are also functions of the structure size and the applied external field and may vary over a wide range [35]. As a rule, the velocity of spin waves is several orders of magnitude smaller than the velocity of electromagnetic waves in a given medium. Thus, it is possible to obtain crystals with a magnon band gap whose width is on the order of millimeters, in unit of wavelength. Such crystals may have a planar geometry, which can be extremely important for designing integrated devices such as narrow-frequency optical or microwave filters and high-speed switches.

In addition to the band gaps, a great interest has been paid to the so-called Fano resonances that may be introduced in such gaps. Some analytical models in phononic crystals have been proposed to explain the origin and the behavior of such resonances [41–44]. These resonances were first theoretically described by Fano [45] when he studied the inelastic auto ionizing resonances in atoms. The asymmetry (Fano profile) was explained as the result of the interference between the discrete resonance with the smooth continuum background in which the former is embedded. The symmetric and asymmetric line shapes have also been reported in the commonly studied Aharonov-Bohm systems [46–55]. Mainly, the subject of these studies was to use these interferometric systems to show the conditions for the existence and the collapse of Fano resonances as function of the applied current-voltage and magnetic flux. These studies are also related to the investigation of the electronic states of quantum dots [46–51] as well as to the understanding [56–59] of the transmission phase jumps by π between two adjacent resonances in relation with the experiments of Yacoby et al. [60, 61]. The analogy between scattering properties of electrons, phonons, and magnons suggests that this type of feature can also appear in other vibrational systems [62].

One aim of the present chapter is to give a comprehensive review of a simple theoretical approach to the propagation properties of magnons in one-dimensional (1D) mono-mode circuits; a short review of this can be found in Al-Wahsh et al. [63]. We will give here all the details necessary for graduate students to understand this new and fast-growing field. The second aim of this report is to review the magnonic crystals based on magnonic circuits and finally we show that a simple magnetic circuit made of two dangling stubs placed at the same or different places may induce either Fano or magnetic transparency resonances.

2.2 MOTIVATIONS AND OUTLINE

One general feature of most of the magnon circuits described in this chapter is that they are made out of mono-mode wires. By mono-mode wires, we mean wires such that their lateral dimensions are small as compared with the wavelength of the propagating spin waves. Only one type of mode can propagate in them and this mode behaves as that of a 1D structure. This simplifies greatly the understanding of the propagation of magnon excitations in mono-mode circuits, enables to study analytically most of their properties and helps to design microdevices which can then be realized with the help of current numerical simulations and surface technologies. This allows applications as selective frequency filters and efficient waveguides.

In this chapter, we review first the appealing possibility of designing a 1D comb-like structure waveguide exhibiting stop bands [64]. This structure is composed of a backbone (or substrate) waveguide along which finite side branches are grafted periodically. This study is conducted within the frame of the Dobrzynski's interface response theory (IRT) [65–68] which we recall briefly in Section 2.3. This theory allows the calculation of the Green's functions of a network structure in terms of the Green's functions of its elementary constituents (Section 2.4). Three network structures are then considered (Section 2.5), namely a single-side branch on an infinite 1D waveguide, an infinite periodic comb-like structure, and a finite comb with two semiinfinite leads. The first one is shown in Section 2.5.1; this structure gives rise to well-defined zeros of transmission due to destructive interference between the side branch and the backbone. These resonances are enlarged to absolute gaps in the limit of an infinite periodic comb. In Section 2.5.2, we show that additional gaps arise as a result of the periodicity. In Section 2.5.3, we calculate the transmission coefficient for spin waves of a finite comb. These theoretical results are illustrated in Section 2.5.4 through numerical examples of dispersion curves and transmission spectra. Despite its finite size, this device retains most of the features of the infinite periodic one. This work [64] demonstrates the possibility of designing simple homogeneous networks of 1D waveguides with absolute band gaps.

In Section 2.5.5, we address defect modes in 1D comb structures. In Section 2.6, we discuss the results obtained with boundary condition at the ends of the dangling branches, namely, the effect of the pinning field [69] (acting at the end of the resonators) on the spin wave band gaps and transmission spectra of this simple comb-like structure [70]. In Sections 2.7 and 2.8 we show the possibility of existence of magnetic induced transparency (MIT) resonances in simple circuits made of two stubs placed at the same site or at two different sites along the backbone.

2.3 INTERFACE RESPONSE THEORY

We will address the propagation of spin waves in composite systems composed of 1D continuous segments (or branches) grafted on different substrates. This study is performed with the help of the IRT [65–68, 71] which permits to calculate the Green's function of any composite material. In what follows, we present the basic concepts and the fundamental equations of this theory.

Let us consider any composite material contained in its space of definition D and formed out of N different homogeneous pieces situated in their domains $D_i(1 \leq i \leq N)$. Each piece is bounded by an interface M_i, adjacent in general to J other pieces through subinterface domains $M_{ij}(1 \leq j \leq J)$. The ensemble of all these interface spaces M_i is called the interface space M of the composite material.

2.3.1 Discrete Theory

We consider first the discrete theory designed for problems using matrix formulations for linear operators H_i. The starting point is an infinite homogenous material i described by an infinite matrix $[(E6 + j\varepsilon)I - H_i]$, where E is the eigenenergy, I the identity matrix, $j = \sqrt{-1}$ and ε an infinitesimally positive small number. The inverse of this matrix is called the corresponding Green's function G_i and

$$[(E + j\varepsilon)I - H_i]G_i = I. \qquad (2.1)$$

One cuts out of this medium a finite one with free surfaces in its space D_i with the help of a cleavage operator V_i in the interface space M_i. One defines A_{si} as the truncated part within D_i of $A_i = V_i G_i$. In the same manner one constructs also G_{si} out of the truncated part of the G_i. One defines then block diagonal matrices G and A_s by juxtaposition of, respectively, all the G_{si} and A_{si} defined for N different homogenous materials i. A composite material is then constructed by assembling such finite media with the help of a coupling operator V_I defined in the whole interface space M. One defines then in the whole space D of the composite the matrices

$$A = A_s + V_I G \qquad (2.2)$$

and in the interface space M

$$\Delta(MM) = I(MM) + A(MM). \qquad (2.3)$$

The elements of the Green's function $g(DD)$ of any composite material can be obtained from [65–67]

$$g(DD) = G(DD) - G(DM)[\Delta(MM)]^{-1}A(MD). \qquad (2.4)$$

The new interface states can be calculated from [65, 66]

$$\det[\Delta(MM)] = 0 \qquad (2.5)$$

showing that if one is interested in calculating the interface states of a composite, one only needs to know $\Delta(MM)$ in the interface space M.

The density of states n_i corresponding to H_i can then be obtained from the imaginary part of the trace of G_i, namely

$$n_i(E) = -\frac{1}{\pi}Im Tr G_i(E). \qquad (2.6)$$

Moreover, if $U(D)$ [72] represents an eigenvector of the reference system formed by all the infinite materials i, Eq. (2.4) enables one to calculate the eigenvectors $u(D)$ of the composite material

$$u(D) = U(D) - U(M)[\Delta(MM)]^{-1}A(MD). \qquad (2.7)$$

In Eq. (2.7), $U(D)$, $U(M)$, and $u(D)$ are row-vectors. Eq. (2.7) enables also to calculate all the waves reflected and transmitted by the interfaces as well as the reflection and the transmission coefficients of the composite system. In this case, $U(D)$ must be replaced by a bulk wave launched in one homogeneous piece of the composite material [72].

2.3.2 Continuous Theory

We consider now the continuous theory designed for problems using differential formulations for linear Hamiltonians. The elements of the Green's function $g(DD)$ of any composite material can now be obtained from [65–67, 73]

$$g(DD) = G(DD) - G(DM)[G(MM)]^{-1}G(MD)$$
$$+ G(DM)[G(MM)]^{-1}g(MM)[G(MM)]^{-1}G(MD), \qquad (2.8)$$

where D and M are the whole space and the space of the interfaces in the composite materials, respectively. $G(DD)$ is the block diagonal Green's function of a reference continuous medium and $g(MM)$, the interface elements of the Green's function of the composite system. The inverse $[g(MM)]^{-1}$ of $g(MM)$ is obtained for any point within the space of the interface $M = \{\bigcup M_i\}$ as a superposition of the different $[g_i(M_i, M_i)]^{-1}$ [65–68]. $[g_i(M_i, M_i)]^{-1}$ being the inverse of $g_i(M_i, M_i)$ for each constituent i of the composite system. The latter quantities are given by the equation

$$[g_i(M_i, M_i)]^{-1} = \Delta_i(M_i, M_i)[G_i(M_i, M_i)]^{-1}, \qquad (2.9)$$

where

$$\Delta_i(M_i, M_i) = I(M_i, M_i) + A_i(M_i, M_i). \qquad (2.10)$$

With I being the unit matrix and

$$A_i(x, x') = V_{c_i}(x'')G_i(x'', x')|_{x''=x}, \qquad (2.11)$$

where $\{x, x''\} \in M_i$ and $x' \in D_i$.

In Eq. (2.11), the cleavage operator V_{c_i} acts only in the surface domain M_i of D_i and cuts the finite or semiinfinite size block out of the infinite homogeneous medium [65–68]. A_i is called the surface response operator of block i.

The interface states can be calculated from [65, 66]

$$\det[g(MM)]^{-1} = 0 \qquad (2.12)$$

showing that if one is interested in calculating the interface states of a composite, one only needs to know the inverse of the Green's function of each individual block in the space of their respective surfaces and/or interfaces.

Moreover, if $U(D)$ [72] represents an eigenvector of the reference system, Eq. (2.8) enables the calculation of the eigenvectors $u(D)$ of the composite material

$$u(D) = U(D) - U(M)[G(MM)]^{-1}G(MD)$$
$$+ U(M)[G(MM)]^{-1}g(MM)[G(MM)]^{-1}G(MD). \qquad (2.13)$$

In Eq. (2.13), $U(D)$, $U(M)$, and $u(D)$ are row-vectors. Eq. (2.13) provides a description of all the waves reflected and transmitted by the interfaces, as well as the reflection and the transmission coefficients of the composite system. In this case, $U(D)$ must be replaced by a bulk wave launched in one homogeneous piece of the composite material [72].

2.4 INVERSE SURFACE GREEN'S FUNCTIONS OF THE ELEMENTARY CONSTITUENTS

We report here the expression for the Green's function of a homogeneous infinite ferromagnetic medium. We give also the inverse of the surface Green's function for a semiinfinite medium with a free surface and for a slab of thickness d (or segment of length d) with free surfaces.

2.4.1 Green's Function for an Infinite Ferromagnetic Medium

Here we turn to the calculation of the magnetic Green's function for an infinite ferromagnetic medium. We use the Heisenberg model of a ferromagnet and we neglect the effects of dipole interactions compared with the exchange contribution to the Hamiltonian. Therefore, in evaluating the needed Green's function, it is convenient to use a continuum approximation. Such an approximation is valid provided that the relevant wavelengths are large compared with the lattice spacing. Therefore, we will deal only with long-wavelength excitations.

A medium denoted "i" and described in a Cartesian coordinate system (O, x_1, x_2, x_3) is assumed to have a simple cubic structure with lattice parameter a. We take the spontaneous magnetization M_0 to be in the x_1 direction. The equation of motion for the total magnetization $\mathbf{M}$ can be expressed in terms of the total effective magnetic field $\mathbf{H}$ as

$$\frac{d\mathbf{M}}{dt} = \gamma(\mathbf{M} \times \mathbf{H}) - \Gamma(\mathbf{M} - M_0\mathbf{i_1}), \qquad (2.14)$$

where γ is the gyromagnetic ratio and Γ is a phenomenological damping factor (considered to be a positive constant). The fields $\mathbf{M}$ and $\mathbf{H}$ are given by

$$\mathbf{M} = M_0\mathbf{i_1} + \mathbf{m}(\mathbf{r}, t) \qquad (2.15)$$

and

$$\mathbf{H} = H_0\mathbf{i_1} + \mathbf{h}_{ex}(\mathbf{r}, t) + \mathbf{H}_{ext}\, e^{j(\mathbf{k}\cdot\mathbf{r}-\omega t)}, \qquad (2.16)$$

where $j = \sqrt{-1}$. It is understood that $\mathbf{i_1}$ is a unit vector parallel to the static fields M_0 and H_0 in the x_1 direction and $\mathbf{m}(r, t)$ represents the instantaneous deviation from its average value M_0. The term proportional to $\mathbf{H}_{ext}$ in Eq. (2.16) represents an externally applied driving field of the wave vector $\mathbf{k}$ and frequency ω. Finally the term $\mathbf{h}_{ex}(r, t)$ in Eq. (2.16) is an effective field arising from the exchange interactions between neighboring magnetic moments. This exchange field $\mathbf{h}_{ex}(r, t)$ may be written as [74]

$$\mathbf{h}_{ex}(\mathbf{r}, t) = \frac{2}{(\gamma \hbar)^2} \sum_{\delta} J_{r,r+\delta} \mathbf{M}(\mathbf{r} + \delta, t), \qquad (2.17)$$

where $J_{r,r+\delta}$ is the exchange interaction between magnetic sites at $\mathbf{r}$ and $\mathbf{r} + \delta$. We assume that $J_{r,r+\delta}$ couples only nearest neighbors in the simple cubic lattice. By expanding $\mathbf{M}(\mathbf{r} + \delta, t)$ in terms of $\mathbf{M}(\mathbf{r}, t)$ and its derivatives using Taylor series and taking into account that for each site $\mathbf{r}$ there are six neighbors coupled by the exchange J, we obtain to the lowest order

$$\mathbf{h}_{ex}(\mathbf{r}, t) = \frac{2J}{(\gamma \hbar)^2} [6 + a^2 \nabla^2] \mathbf{M}(\mathbf{r}, t). \qquad (2.18)$$

As was mentioned earlier, we note that in doing the previous expansion we use a continuum representation of the ferromagnet, and thus restricting ourselves to long-wavelength excitations. Inserting Eqs. (2.15), (2.16), (2.18) into the torque Eq. (2.14), and making the usual linear spin wave approximation (i.e., neglecting small terms that are of the second order in $\mathbf{m}$, since $|\mathbf{m}| \ll M_0$ at low temperatures) we arrive at the following equation of motion for $\mathbf{m}$:

$$\frac{d\mathbf{m}}{dt} + \Gamma \mathbf{m} = i_1 \times \{\gamma M_0 \, \mathbf{H}_{ext} \, e^{j(\mathbf{k} \cdot \mathbf{r} - \omega t)} - (\gamma H_0 - D' \nabla^2) \mathbf{m}\}, \qquad (2.19)$$

where $D' = 2Ja^2 M_0 / \gamma \hbar^2$. From the property of translational invariance of the medium and on assuming a time dependence in the form $e^{-j\omega t}$, we may write

$$\mathbf{m}(\mathbf{r}, t) = \mathbf{m}(x_3) \, e^{j(\mathbf{k}_\parallel \cdot \mathbf{l} - \omega t)}, \qquad (2.20)$$

where $\mathbf{k}_\parallel \equiv (k_1, k_2)$ and $\mathbf{l} \equiv (x_1, x_2)$ are 2D wave vectors. If we now substitute Eq. (2.20) into Eq. (2.19) and after some algebraic manipulations we arrive at the following differential equation for $m^+(x_3)$

$$\frac{D'}{\gamma M_0} \left(\frac{\partial^2}{\partial x_3^2} - \mathbf{k}_\parallel^2 + \frac{\omega + j\Gamma - \gamma H_0}{D'} \right) m^+(x_3) = -(H_{ext}^{x_3} + jH_{ext}^{x_2}) \, e^{jk_3 x_3}, \quad (2.21)$$

where $m^+(x_3) = m_3(x_3) + jm_2(x_3)$ and $m_1(x_3) = 0$. k_3 is the x_3 component of the propagation vector $\mathbf{k} = (\mathbf{k}_\parallel, k_3)$. Using Eq. (2.21), the Fourier-transformed Green's function between two points (sites) $\mathbf{r}(x_1, x_2, x_3)$ and $\mathbf{r}'(x_1', x_2', x_3')$ of the considered infinite ferromagnetic medium i associated with the magnetization $m^+(x_3)$ satisfies the following equation:

$$\frac{F_i}{\alpha_i(\omega)} \left(\frac{\partial^2}{\partial x_3^2} - \alpha_i^2(\omega) \right) G_i(\mathbf{k}_\parallel, x_3, x_3') = \delta(x_3 - x_3') \qquad (2.22)$$

and can be expressed as

$$G_i(\mathbf{k}_\parallel, x_3, x_3') = -\frac{e^{-\alpha_i(\omega)|x_3 - x_3'|}}{2F_i}, \qquad (2.23)$$

where

$$F_i = \frac{D_i' \alpha_i(\omega)}{\gamma_i M_i} \tag{2.24}$$

and

$$\alpha_i(\omega) = \sqrt{\mathbf{k}_\parallel^2 - \frac{(\omega - \gamma_i H_0)}{D_i'}}. \tag{2.25}$$

Let us note that Eq. (2.23) may be generalized to other excitations, such as elastic waves in solids or liquids [75], electrons [76], and electromagnetic waves [77]. In Eq. (2.25) and in what follows, the damping constant Γ is considered to be zero. The Green's function for a 1D infinite waveguide is obtained by setting $\mathbf{k}_\parallel = 0$ in Eq. (2.25), that is,

$$\alpha_i(\omega) = j\sqrt{(\omega - \gamma_i H_0)/D_i'} = j\alpha_i'(\omega). \tag{2.26}$$

2.4.2 Inverse Surface Green's Functions of the Semiinfinite Medium

One considers a semiinfinite medium "i" with a "free surface" located at the position $x_3 = 0$ in the direction Ox_3 of the Cartesian coordinate system (O, x_1, x_2, x_3) and infinite in the other two directions. In this case

$$[g_i(MM)]^{-1} = [g_i(00)]^{-1} = -F_i. \tag{2.27}$$

2.4.3 Inverse Surface Green's Functions of a Segment (or Finite Wire)

One considers a finite wire of length d_i bounded by two free surfaces located at $x_3 = 0$ and $x_3 = d_i$ in the direction Ox_3 of the Cartesian coordinates system (O, x_1, x_2, x_3). It is known [68] that for such a case the inverse of the (2×2) matrix $g_i(M_i M_i)$, within the interface space $M_i = \{0, d_i\}$ takes the following form

$$[g_i(M_i M_i)]^{-1} = \begin{pmatrix} A_i & B_i \\ B_i & A_i \end{pmatrix} = \begin{pmatrix} g_i(0, 0) & g_i(0, d_i) \\ g_i(d_i, 0) & g_i(d_i, d_i) \end{pmatrix}^{-1}, \tag{2.28}$$

where

$$A_i = -\frac{F_i C_i}{S_i} \tag{2.29}$$

$$B_i = \frac{F_i}{S_i}, \tag{2.30}$$

F_i has the same meaning as earlier and

$$C_i = \cosh[\alpha_i(\omega)d_i] \tag{2.31}$$
$$S_i = \sinh[\alpha_i(\omega)d_i]. \tag{2.32}$$

To obtain the Green's function for 1D segments of waveguides, one needs only to take the limit $\mathbf{k}_{\parallel} \rightarrow 0$ in Eq. (2.28). In order to study elementary spin wave excitations, we will calculate the Green's function for different composite devices composed of finite segments grafted on a 1D waveguide.

2.5 COMB STRUCTURES

2.5.1 Propagation of Spin Waves in an Infinite Backbone With One Grafted Finite Segment

One considers a quasi-1D composite system formed out of a finite segment of length d grafted on an infinite waveguide line (see Fig. 2.1A). In order to calculate the corresponding interface Green's function, we construct this system out of two semiinfinite lines constituted of the same material (medium 1) and a segment of finite length d (medium 2). These three blocks are coupled at their ends (see Fig. 2.1B). For the two semiinfinite lines and for the finite segment, the interface domains correspond to site 0 and sites 0 and 1', respectively.

The inverse surface Green's functions $[g_i(M_iM_i)]^{-1}$ for the two semiinfinite lines and for the finite segment are given by Eqs. (2.27), (2.28) with $i = 1$ and 2, respectively ($d_2 = d$). Media 1 and 2 are 1D and the parameters α_i ($i = 1, 2$) are given by Eq. (2.26). In this case, the interface domain of the composite system can be reduced to site 0 and the finite segment contribution to the surface Green's

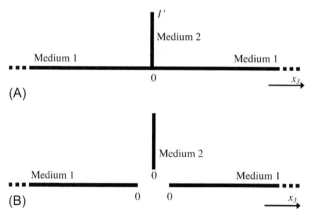

FIG. 2.1 (A) Waveguide with a single-grafted segment of length $d_2 = d$. (B) Elementary constituents of the waveguide with a single-grafted segment of length $d_2 = d$.

function of the composite system takes then the form

$$g_2(0,0) = -\frac{C_2}{F_2 S_2}. \tag{2.33}$$

Superposing these different linear contributions, one deduces [68] the inverse interface Green's function of the composite system as

$$[g(0,0)]^{-1} = -2F_1 - \frac{F_2 S_2}{C_2}. \tag{2.34}$$

Eq. (2.13) allows us to calculate the transmission coefficient of this composite system. Consider $U(x_3) = e^{-\alpha_1 x_3}$ a bulk propagating spin wave coming from $x_3 = -\infty$. Substituting this incident wave in Eq. (2.13), one obtains the transmitted wave $u(x_3')$, with $x_3' \geq 0$, as

$$u(x_3') = \frac{2F_1 C_2}{F_2 S_2 + 2F_1 C_2} e^{-\alpha_1 x_3'} \tag{2.35}$$

and the transmission coefficient as

$$T = \left| \frac{2F_1 C_2}{F_2 S_2 + 2F_1 C_2} \right|^2. \tag{2.36}$$

It is clear that this coefficient equals zero when $\cosh(\alpha_2 d_2) = 0$, that is,

$$\alpha_2' d_2 = \left(m + \frac{1}{2} \right) \pi, \quad m = 0, 1, 2, \ldots. \tag{2.37}$$

The corresponding frequency will be

$$\omega_g = \gamma_2 H_0 + D_2' \left(\left(m + \frac{1}{2} \right) \frac{\pi}{d_2} \right)^2 \tag{2.38}$$

or symbolically

$$\Omega_g = H_g + \left[\left(m + \frac{1}{2} \right) \pi \right]^2, \tag{2.39}$$

where m is a positive integer, $\Omega_g = \omega_g d_2^2 / D_2'$ is a reduced frequency, and $H_g = \gamma_2 H_0 d_2^2 / D_2'$. From the previous equation one can notice that for this composite system, there exists an infinite set of forbidden frequencies Ω_g corresponding to the eigenmodes of the side branch. This branch behaves as a resonator and consequently this composite system filters out the frequencies Ω_g. One also observes from Eq. (2.37) that T is equal to zero when $\alpha_2' d_2$ is an odd multiple of $\pi/2$ and reaches its maximum value of 1 when $\alpha_2' d_2$ is a multiple of π. This is illustrated in Fig. 2.2. If the number of dangling resonators N at the same site increases, the zeros of the transmission coefficient enlarge into gaps. It is worth mentioning that the existence of transmission zeros has also been demonstrated in waveguides with a resonantly coupled stub for electrons [78], phonons [79, 80], photons [81, 82], and acoustic waves [83]. This phenomenon is related to the resonances associated with the finite additional path offered to the wave propagation.

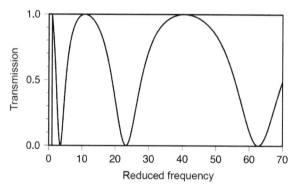

FIG. 2.2 Transmission coefficient versus the reduced frequency for a waveguide with one dangling resonator in the case of identical media 1 and 2. For convenience H_g is considered to be 1.

2.5.2 One-Dimensional Infinite Backbone With Periodic Array of Finite Segments: Infinite Comb

Here, we treat the case of a comb-like structure composed of an infinite 1D waveguide or backbone (medium 1), along which N' finite side branches (medium 2) of length d_2 are grafted periodically with spacing period d_1 at N sites, N and N' being integers (see Fig. 2.3). Let us first write the Green's function of this composite system. The infinite line can be modeled as an infinite number of finite segments of length d_1 in the direction x_3, each one being pasted to two neighbors. The interface domain is constituted of all the connection points between finite segments. Each connection point (site) on the infinite chain will be defined by the integer n such that $-\infty < n < +\infty$. On each site n, N' dangling side branches of length d_2 are connected. Here and afterward the cross-sections of all media are considered to be much smaller than the considered wavelength so as to neglect the quantum-size effect (or the subband structure). The respective contributions of media 1 and 2 to the inverse Green's function at the interface space of the composite system are given by Eqs. (2.28)–(2.30).

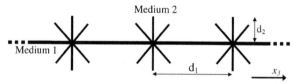

FIG. 2.3 Schematic of the 1D waveguide studied in the present work. The material media are designated by an index i, with i equal 1 for the backbone (*heavy line*) and 2 for the dangling side branches. There are $N'(= 6)$ dangling side branches of length d_2 grafted at equidistant sites separated by a length d_1.

The inverse Green's function of the composite system is then obtained as an infinite banded matrix $[g_\infty(MM)]^{-1}$ defined in the interface domain constituted of all the sites n.

To find the contribution of medium 1 to the diagonal elements of the matrix $[g_\infty(MM)]^{-1}$ one has to take the element $[g_1(0,0)]^{-1} = [g_1(d_i, d_i)]^{-1}$ of Eq. (2.28) and multiply it by 2 (because at each site we have two pasted segments belonging to medium 1). The contribution of medium 2 to the diagonal elements is obtained by calculating the inverse of the matrix given by Eq. (2.28), taking the element $g_2(0,0)[= g_2(d_i, d_i)]$, finding its reciprocal and multiplying it by N'. Therefore, the diagonal elements of the matrix $g_\infty^{-1}(MM)$ are given by $-\{2(F_1 C_1/S_1) + N'(F_2 S_2/C_2)\}$. The off-diagonal elements are given by F_1/S_1 (see Eq. 2.28).

Taking advantage of the translational periodicity of the system in the direction x_3, this matrix can be Fourier transformed as

$$[g_\infty(\mathbf{k}, MM)]^{-1} = 2F_1(-\xi + \cos(kd_1))/S_1, \tag{2.40}$$

where k is the modulus of the 1D reciprocal vector $\mathbf{k}$, d_1 is the period of the system, and $\xi = C_1 + (N'F_2/2F_1)(S_1 S_2/C_2)$.

The dispersion relation of the infinite periodic comb-like waveguide is given by Eq. (2.12), that is, $[g_\infty(\mathbf{k}, MM)]^{-1} = 0$. This relation takes the simple form

$$\cos(kd_1) = C_1 + N'(F_2/2F_1)(S_1 S_2/C_2). \tag{2.41}$$

On the other hand, in the $\mathbf{k}$ space, the Green's function of the infinite homogenous magnonic waveguide is

$$[g_\infty(\mathbf{k}, MM)] = S_1/(2F_1(-\xi + \cos(kd_1))). \tag{2.42}$$

After inverse Fourier transformation, Eq. (2.42) yields [84]

$$g_\infty(n, n') = \frac{S_1}{F_1} \frac{t^{|n-n'|+1}}{t^2 - 1}, \tag{2.43}$$

where the integers n and n' refer to the sites $(-\infty < n, n' < +\infty)$ on the infinite line. The parameter t is defined as

$$t = e^{jkd_1} \tag{2.44}$$

or equivalently

$$t = \begin{cases} \xi - \sqrt{\xi^2 - 1}, & \xi > 1 \\ \xi + \sqrt{\xi^2 - 1}, & \xi < -1 \\ \xi + j\sqrt{1 - \xi^2}, & -1 < \xi < 1 \end{cases} \tag{2.45}$$

with

$$t + \frac{1}{t} = 2\xi \tag{2.46}$$

and

$$|t| < 1. \tag{2.47}$$

Note that the signs in Eqs. (2.44), (2.45) are determined by the facts that inside the passing bands the density of states is positive and outside the passing bands the Green's function has to decay when the distances increase.

2.5.3 Transmission Coefficient of the Finite Comb-Like Structures

Infinite magnonic comb-like structures are not physically realizable but finite comb structures are. Therefore, in this section, we investigate the transmission properties of a finite comb. This comb, as represented in Fig. 2.4, is constructed as follows: a finite piece containing N equidistant groups of dangling side branches is cut out of the infinite periodic system illustrated in Fig. 2.3, and this piece is subsequently connected at its extremities to two semiinfinite leading lines. The finite comb is therefore composed of N' dangling side branches (medium 2) of length d_2 grafted periodically with a spacing period d_1 at N sites on a finite line (medium 1), N and N' being integers. For the sake of simplicity, the semiinfinite leads and medium 1 are assumed to be constituted of the same material. We calculate analytically the transmission coefficient of a bulk spin wave coming from $x_3 = -\infty$.

The system of Fig. 2.4 is constructed from the infinite comb of Fig. 2.3. In a first step, one suppresses the segments linking sites 0 and 1, and sites N and $N + 1$. For this new system composed of a finite comb and two semiinfinite leads, the inverse Green's function at the interface space, $[g_t(MM)]^{-1}$, is an infinite banded matrix defined in the interface domain of all the sites n ($-\infty < n < +\infty$). This matrix is similar to the one associated with the infinite comb. Only a few matrix elements differ, namely, those associated with the sites $n = 0$, $n = 1, n = N$, and $n = N + 1$.

The cleavage operator $\mathbf{V}_{cl}(MM) = [g_t(MM)]^{-1} - [g_\infty(MM)]^{-1}$ [68] is the following 4×4 square matrix defined in the interface domain constituted of sites $0, 1, N, N + 1$

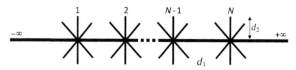

FIG. 2.4 Waveguide with $N' = 6$ dangling side branches of length d_2 grafted at a finite number N of equidistant sites separated by a length d_1 and connected at its extremities to two semiinfinite leading lines.

$$\mathbf{V}_{cl}(MM) = \begin{pmatrix} -A & -B & 0 & 0 \\ -B & -A & 0 & 0 \\ 0 & 0 & -A & -B \\ 0 & 0 & -B & -A \end{pmatrix}, \tag{2.48}$$

where A and B are given by Eqs. (2.29), (2.30). In a second step, two semiinfinite leads constituted of the same material as medium 1 are connected to the extremities $n = 1$ and $n = N$ of the finite comb. With the help of the IRT [67, 68], one deduces that the perturbing operator $\mathbf{V_p}(MM)$ allowing the construction of the system of Fig. 2.4 from the infinite comb is then defined as the 4×4 square matrix (see Eq. 2.27)

$$\mathbf{V_P}(MM) = \begin{pmatrix} -A & -B & 0 & 0 \\ -B & -A - F_1 & 0 & 0 \\ 0 & 0 & -A - F_1 & -B \\ 0 & 0 & -B & -A \end{pmatrix}. \tag{2.49}$$

Using Eqs. (2.43), (2.49), one obtains the matrix operator $\Delta(MM) = \mathbf{I}(MM) + \mathbf{V_p}(MM)\mathbf{g_\infty}(MM)$ in the space M of sites $n, n' = 0, 1, N$, and $N+1$. For the calculation of the transmission coefficient, we need only the matrix elements $\Delta(1, 1)$, $\Delta(1, N)$, $\Delta(N, 1)$, and $\Delta(N, N)$, which can be set in the form of a 2×2 matrix $\Delta_S(MM)$,

$$\Delta_S(MM) = \begin{pmatrix} 1 + Ct & Ct^N \\ Ct^N & 1 + Ct \end{pmatrix} \tag{2.50}$$

with

$$C = -\frac{(t - C_1 + S_1)}{t^2 - 1}. \tag{2.51}$$

The surface Green's function $d_s(MM)$ of the finite comb with two connected semiinfinite leads in the space of sites 1 and N is

$$d_s(MM) = g_s(MM)\Delta_s^{-1}(MM) = \frac{S_1}{F_1}\frac{t}{t^2 - 1}\frac{1}{\det \Delta_s(MM)}$$
$$\times \begin{pmatrix} 1 + Ct(1 - t^{2N-2}) & t^{N-1} \\ t^{N-1} & 1 + Ct(1 - t^{2N-2}) \end{pmatrix} \tag{2.52}$$

with

$$g_s(MM) = \frac{S_1}{F_1}\frac{t}{t^2 - 1}\begin{pmatrix} 1 & t^{N-1} \\ t^{N-1} & 1 \end{pmatrix} \tag{2.53}$$

and

$$\det \Delta_s(MM) = 1 + 2Ct + C^2 t^2 (1 - t^{2N-2}). \tag{2.54}$$

In Eq. (2.53), $g_s(MM)$ is the matrix constituted of elements of $g_\infty(MM)$ associated with sites 1 and N. We now calculate the transmission coefficient with a bulk spin wave coming from $x_3 = -\infty$, $U(x_3) = e^{-\alpha_1 x_3}$. Substituting this incident wave in Eq. (2.13) and considering Eqs. (2.23), (2.52), we obtain the transmitted wave $u(x_3')$, with $x_3' \geq Nd_1$, as

$$u(x_3') = -2S_1 \frac{t^N}{t^2 - 1} \frac{e^{-\alpha_1(x_3' - (N-1)d_1)}}{\det \Delta_s(MM)}. \qquad (2.55)$$

One deduces that the transmission coefficient is

$$T = \left| \frac{2S_1(t^2 - 1)}{t^{-N}(1 - t(C_1 - S_1))^2 - t^N(t - (C_1 - S_1))^2} \right|^2. \qquad (2.56)$$

The latter equation for the case of $N = N' = 1$ can be written as $T = |(2F_1C_2)/(F_2S_2 + 2F_1C_2)|^2$ (see Eq. 2.36).

2.5.4 Magnonic Stop Bands and Transmission Spectrum

We now turn to discussing the numerical results on the band structure and transmission spectrum for the case of identical media ($\alpha_1' = \alpha_2'$) with $d_1 = d_2$ and $N' = 1$. Eq. (2.41) then reduces to a second-order polynomial equation that can be solved for the frequency to give

$$\Omega = \tilde{H} + \{\arccos[(1/3)(\cos(kd_1) \pm (\cos^2(kd_1) + 3)^{1/2})]\}^2, \qquad (2.57)$$

where $\Omega = \omega d_1^2/D_1'$ is the reduced frequency and $\tilde{H} = \gamma_1 H_0 d_1^2/D_1'$. The arccos function appearing on the right-hand side of Eq. (2.57) shows that there are an infinite number of dispersion curves that are repeated periodically. Fig. 2.5A only shows the first seven dispersion curves, in the band structure of the infinite comb composite. There is a complete absolute gap below the lowest band due to the presence of the external field H_0. Other absolute gaps exist between the other six bands. The second and the fifth bands are flat bands, they correspond to localized modes inside each resonator. These modes do not penetrate into the backbone. The tangential points between the third and the fourth bands are degenerate points, they appear at $kd_1 = \pi$ and $-\pi$. Another degenerate point is the tangential point between the sixth and the seventh bands, it appears at $kd_1 = 0$. One can also notice antisymmetry between the third and the fourth bands and between the sixth and seventh bands as well. This antisymmetry is clearly visible when plotting the corresponding transmission factor. Fig. 2.5B shows the frequency dependence of the transmission for $d_2 = d_1$, $N' = 1$, and $N = 10$. The flat bands in Fig. 2.5A, associated with localized modes inside the resonators, do not contribute to the transmission.

The number of oscillations in the transmission factor within the pass bands, which corresponds to the third and the fourth or to the sixth and seventh bands in the band structure, has been noted to be unfailingly $2N - 1$, while this number is $N - 1$ within the pass band which corresponds to the band which has no tangential points with other bands (see also Fig. 2.6).

Fig. 2.6A shows the first five bands for the case of two identical media with $d_2 = 0.4d_1$ and $N' = 1$. The third band is flat, it corresponds to localized modes

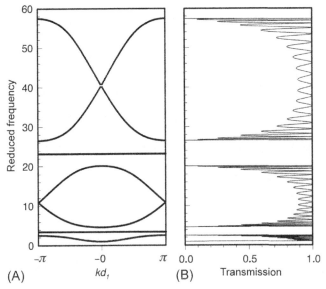

FIG. 2.5 (A) The first seven bands in the magnonic band structure of the infinite periodic comb. We have chosen $d_1 = d_2$, $D'_1 = D'_2$, $\bar{H} = 1$, $N' = 1$, and N infinite. The plot is sketched as the reduced frequency versus the dimensionless quantity kd_1 ($-\pi \leq kd_1 \leq +\pi$), where k is the modulus of the propagation vector. One observes an absolute gap below the first band due to the presence of the external field H_0. (B) Transmission coefficient versus the reduced frequency for a waveguide with $N' = 1$ and $N = 10$. The other parameters are the same as in (A).

inside the resonators. One can notice that decreasing or increasing the length of the resonators removes the degenerate points. The transmission factor has also been influenced by this geometry as it is shown in Fig. 2.6B (N and N' have the same values as in Fig. 2.5B). Interestingly the width of the pass bands (stop bands) decreases (increases) with this other choice of the length d_2. Also, one can also observe on comparing Figs. 2.5B and 2.6B is the number N of sites in the comb necessary to achieve completely formed gaps. This number is of the order of $N = 6$ in Fig. 2.5B where $d_2 = d_1$ (similar results are obtained for $d_2/d_1 = 0.5, 1, 1.5, 2, 2.5, \ldots$), while it becomes of the order of $N > 10$ in Fig. 2.6B with $d_2 = 0.4d_1$.

Fig. 2.7 depicts the effect of variation of the number N of sites on the transmission factor T for two identical media with $d_1 = d_2$. We show the frequency dependence of the transmission for $N' = 1$ and $N = 2, 5$, and 10 in the top, middle, and bottom panels, respectively. One can easily notice that the only important effect of varying N on the transmission is that the pseudogaps turn into full gaps with increasing N. Starting from $N = 5$ (roughly) the gaps are already formed. Moreover, for a given frequency range, there is an optimum value of N above which any additional increment in N leaves the bands practically unaffected.

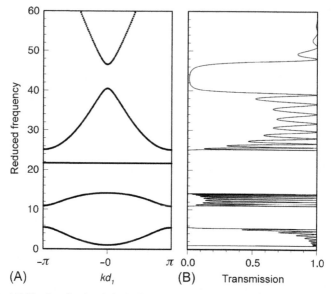

FIG. 2.6 (A) The first five bands in the band structure of the infinite periodic comb with $d_2 = 0.4d_1$, $N' = 1$, and N infinite. The two media are identical. (B) The transmission factor T for $N = 10$, $N' = 1$, and $\tilde{H} = 1$. The other parameters are the same as in Fig. 2.5A.

Next let us discuss the dependence of the transmission factor on the number of dangling side branches. The results are illustrated in Fig. 2.8 for three values of N' in the case of two identical media with $d_1 = d_2$. The top, middle, and bottom panels display the frequency dependence of the transmission for $N = 10$ and $N' = 2$, 4, and 8, respectively. We call attention to the fact that the width of the pass bands (stop bands) decreases (increases) when the number of side branches increases. We can also note that an increase in N' results in an increase in the amplitude of oscillations of the transmission coefficient.

Let us stress that unlike in the usual 2D composite system where the contrast in physical properties between the constituent materials is a critical parameter in determining the existence of the gaps [32], the occurrence of narrow magnonic bands does not require the use of two different materials. In other words, the magnonic structure is tailored within a single homogenous medium, although the boundary conditions impose the restriction that the waves only propagate in the interior of the waveguides.

We end this section with a review of the influence of the geometry on the transmission of the comb systems. We compute the frequency dependence of the transmission factor for $d_2 = 0.7d_1$. The results are displayed in Fig. 2.9, in the top and middle panels. $N'(= 1)$ is kept fixed and N takes the values 4 and 12, respectively. One can notice that varying N results in turning the pseudogaps into full gaps. The convergence to full gaps can be achieved in general for

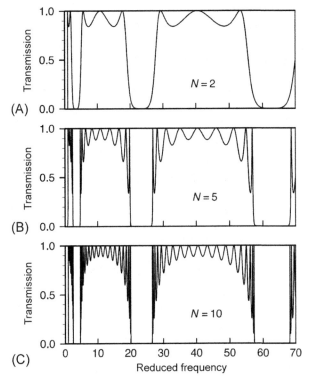

FIG. 2.7 Transmission coefficient versus the reduced frequency for several values of N, $(N' = 1)$. Both media are considered to be identical, $\tilde{H} = 1$ and $d_1 = d_2$. The *top, middle*, and *bottom panels* depict the transmission for $N = 2$, 5, and 10, respectively. Note that increasing N results in turning some of the pseudo-gaps into complete gaps, but leaves their widths virtually intact.

a reasonably small number of sites. In the middle and bottom panels, $N(= 12)$ is kept fixed and N' takes the values 1 and 10, respectively. The most interesting feature is that the width of the pass bands (stop bands) decreases (increases) with increasing N'. On comparing the results of Figs. 2.8 and 2.9 one can notice the same qualitative behavior, but with a more significant widening of the gaps in Fig. 2.9 with increasing N'.

As it is clear from Eq. (2.41), the pass bands described in the previous applications are mainly concentrated around frequencies satisfying either $\sin(\alpha_1 d_1) = 0$ or $\sin(\alpha_2 d_2) = 0$. These conditions, respectively, involve the two characteristic lengths d_1 and d_2, which means that the bands originate either from the periodicity of the system or from the resonance states of each side branch. On the other hand, the narrowness of the bands emphasizes that the condition for constructive interference can only be satisfied in small frequency intervals, in relation with the numbers N and N'. If one first considers the scattering of an incoming wave by the resonators at a single site (i.e., $N = 1$),

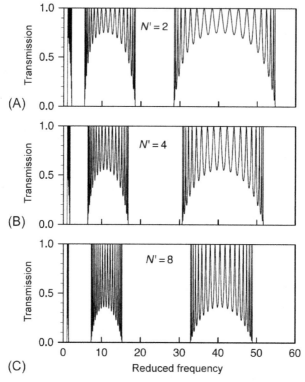

FIG. 2.8 Same as in Fig. 2.7, except that now N' is varied ($N = 10$ is fixed). The *top, middle*, and *bottom panels* depict the transmission for $N' = 2, 4$, and 8, respectively. Noteworthy is the shrinking (widening) of the pass bands (stop bands) with increasing N'. We call attention to the increasing of the oscillation amplitudes with increasing N'.

Eq. (2.56) easily reveals that the transmission will be suppressed if one increases the number N' of side branches (see Eq. 2.56 for the case $N = 1$), except near frequencies where $\sin(\alpha_2 d_2) = 0$. Now grafting the resonators at two or more sites on the backbone opens new channels for transmission at frequencies around the solutions of $\sin(\alpha_1 d_1) = 0$, which means that the condition of constructive interference is allowed in small intervals whose frequencies are related to the new characteristic length d_1. These pass bands slightly increase in width upon increasing the number N of sites. At the same time, for a given N, increasing N' gives rise to tighter conditions for constructive interference, thus contributing to a reduction in the pass bands' widths.

2.5.5 Defect Modes

Here we investigate the existence of localized modes inside the gaps when defect dangling side branches of length d_3 are inserted at one site of the waveguide

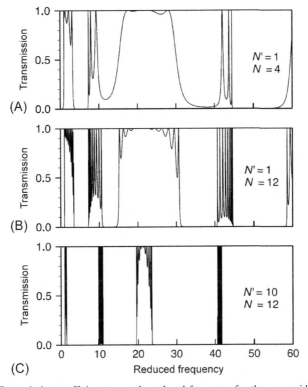

FIG. 2.9 Transmission coefficient versus the reduced frequency for the waveguide with $d_2 = 0.7d_1$. The two media are identical and $\tilde{H} = 1$. In the *top* and *middle panels* $N' = 1$ and N equals 4 and 12 while in the *middle* and *bottom panels* $N = 12$ and N' equals 1 and 10. It is noticeable that the width of the pass bands (stop bands) decreases (increases) with increasing N'.

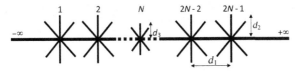

FIG. 2.10 Star-like waveguide with a defect composed of $N'_3 = 6$ side branches of length d_3 introduced in the middle of the finite comb and with $N' = 6$. For simplicity the total number of sites is considered to be odd.

(see Fig. 2.10). With the help of the IRT [67, 68] one can obtain analytically the dispersion relation for the localized states in the case of an infinite comb ($N \longrightarrow \infty$)

$$1 + \chi \frac{t}{t^2 - 1} = 0. \tag{2.58}$$

In the latter equation, the parameter χ characterizes the perturbation introduced in the comb by the defective resonators and is defined as

$$\chi = \frac{S_1}{F_1}(N'(F_2 S_2/C_2) - N_3'(F_3 S_3/C_3)). \tag{2.59}$$

In the preceding relation, N_3' is the number of the defect side branches and the rest of the symbols have their usual meaning.

Next with the help of the previous two equations, we study the effects of variation in the length of the defect branch on the localized states. We assume that our system is composed of identical media and that $N' = N_3' = 1$ and $d_2 = d_1$. Fig. 2.11 gives the frequencies of the localized modes as function of the ratio d_3/d_1. The hatched areas correspond to the bulk bands of the perfect comb. The frequencies of the localized modes are very sensitive to the length d_3. The localized modes emerge from the bulk bands, decrease in frequency with increasing length d_3 and finally merge into a lower bulk band where they become resonant states. At the same time, new localized modes emerge from the bulk bands. Let us note that, for any given reduced frequency in Fig. 2.11, there is a periodic repetition of the modes as function of d_3/d_1.

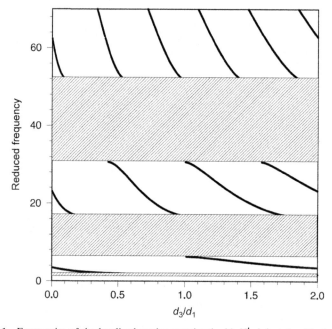

FIG. 2.11 Frequencies of the localized modes associated with N_3' defect dangling branches of length d_3 in an infinite comb. The other parameters are $N' = N_3' = 4$, N infinite, $d_2 = d_1$, and $\tilde{H} = 1$. The system is assumed to be composed of identical media.

Under the assumption that the defect is located in the middle of a finite comb (for simplicity one can take the number of sites to be odd, see Fig. 2.10), one obtains the following result for the transmission factor T through the defective comb

$$T = \left| \frac{2F_1 B_2^2}{(B_1 - F_1)[(B_1 - F_1)(B_3 + 2B_1 + 2N'F_2S_2/C_2) - 2B_2^2]} \right|^2, \quad (2.60)$$

where

$$B_1 = \frac{F_1}{S_1} \frac{[C_1(t - t^{2N-1}) + t^{2N} - 1]}{t(1 - t^{2(N-1)})} \quad (2.61)$$

$$B_2 = \frac{F_1}{S_1} \frac{(t^{N-1} - t^{N+1})}{t(1 - t^{2(N-1)})} \quad (2.62)$$

and

$$B_3 = -N_3' F_3 S_3 / C_3. \quad (2.63)$$

The transmission coefficient is also affected by the presence of a defect inside the comb. In particular, T exhibits narrow peaks associated with the localized modes. In Fig. 2.12, we compare the transmission coefficients for two combs with and without defects. The results are illustrated for $N' = N_3' = 4$, $N = 5$, $d_2 = d_1$, and $d_3 = 1.3\, d_1$. Fig. 2.12B shows one localized mode in the second, the third, and the fourth gaps. The localized mode inside the third forbidden band lies in the middle of the gap while the localized modes in the other two gaps lie closer to the bulk bands. As one can see, the second bulk band is asymmetric due to the proximity of the first localized mode to its left side. The situation is similar for the third pass band but the asymmetry is due to the proximity of the localized mode to its right. In the frequency range displayed in this figure one can see that the peaks corresponding to the localized modes are very narrow. The transmission inside a bulk band can also be significantly affected by the presence of a defect. For instance, in the example shown in Fig. 2.12, T is significantly depressed in the second and third bulk bands.

The influence of N_3' on the transmission factor was also studied. Fig. 2.13 shows a comparison of the transmission coefficients for four combs with N_3' varying from 0 (top panel) to 1, 2, and 4. The results are illustrated for $N' = 2$, $N = 5$, $d_2 = d_1$, and $d_3 = 0.75d_1$. One can notice that the transmission factor in the bands is depressed as N_3' increases. Fig. 2.13B–D shows one localized mode in the third and fourth gaps. One can also observe from Fig. 2.13 that the localized mode is getting closer to the middle of the gap when N_3' increases. One can notice also from Fig. 2.13B and D that the localized modes become more and more confined, that is, the quality factor of the corresponding peaks increases, with increasing N_3'.

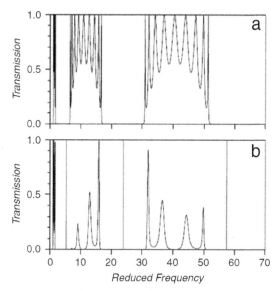

FIG. 2.12 Transmission coefficients versus the reduced frequency for two combs with (B) and without (A) defect. The results are illustrated for $N' = N'_3 = 4$, $N = 5$, $d_2 = d_1$, $d_3 = 1.3d_1$, and $\tilde{H} = 1$. In the frequency range displayed in this figure one can see that the peaks falling in the gap are very narrow. Transmission inside a bulk band is significantly affected by the presence of a defect.

2.5.6   Summary and Concluding Remarks

Large band gaps in a 1D magnonic waveguide with multiple dangling side branches grafted at N equidistant sites were found. These results parallel the calculated spin wave transmission spectra. The existence of the gaps is attributed to the joint effect of periodicity of the system and the resonance states of the grafted dangling side branches (which play the role of resonators). In these systems, the gap width is controlled by the number N and the geometrical parameters including the length of the side branches, the periodicity of the comb as well as the contrast in the physical properties of the side branches and the backbone. Nevertheless, the magnonic band structure exhibits relatively wide gaps for homogeneous systems where the branches and the substrate are constituted of the same material. Devices composed of finite numbers of sites exhibit a behavior similar to that of an infinite periodic comb. Localized states associated with defects in the comb were observed. These defect modes appear as narrow peaks of strong amplitude in the transmission spectra. Since magnetic periodic composites have, in general, wide technical applications, it is anticipated that this new class of materials which can be referred to as "magnonic crystals," may turn out to be of significant value for prospective applications. Especially, one would expect such applications to be feasible in spin wave electronics, since magnon excitations' energies also fall in the

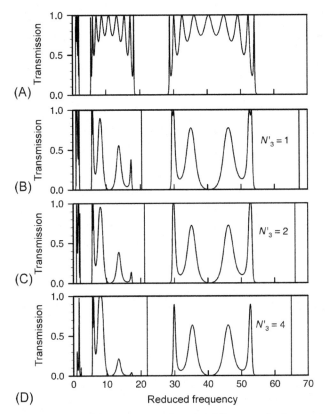

FIG. 2.13 Transmission coefficients, T versus the reduced frequency for four combs with (B), (C), (D) and without (A) defect. In the panels (B), (C), and (D), N'_3 equals 1, 2, and 4, respectively. The results are illustrated for $N' = 2$, $N = 5$, $d_2 = d_1$, $d_3 = 0.75d_1$, and $\tilde{H} = 1$. With increasing N'_3, T is depressed in the bands and the localized mode is approaching the center of the gap.

microwave range. Of special interest is the prospect of achieving a complete band gap. Finally let us report that a novel planar structure of magnonic-crystal waveguides can be found in an excellent paper by Lee et al. [38]. In this structure, which is made of a single magnetic material, the allowed and forbidden bands of propagating dipole-exchange spin waves can be manipulated by the periodic modulation of different widths in thin-film nanostrips. The origin of the presence of several magnonic wide band gaps and the crucial parameters for controlling those band gaps of the order of $\sim$10 GHz are found by micromagnetic numerical and analytical calculations. These results can offer a route to the potential application to the broadband spin wave filters in the gigahertz frequency range.

2.6 EFFECTS OF PINNING FIELDS IN COMB STRUCTURES

2.6.1 Dispersion Relations and Transmission Coefficients

We consider now, in the frame of the long-wavelength Heisenberg model, the effect of a pinning field [69, 70, 85] (acting at the end of the resonators) on the spin wave band gaps and transmission spectra of 1D comb structures studied in the previous section.

Attention was focused on the effects of the pinning fields because they can vary over a wide range of values, depending on the nature of the ends of the resonators and the properties of the magnetic bearing ions. If one has a perfect semiinfinite ferromagnetic wire, in which the surface layer (end) has the same atomic structure and lattice constant as a similar crystallographic plane in the interior of the medium, then the crystal field at a site in the surface layer will have a lower symmetry than the crystal field at an interior site. This fact combined with the presence of spin-orbit coupling can give rise to pinning fields in the surface layer [86]. While one might expect this contribution to the pinning field to be small if the magnetic ion is in an s-state. If the orbital degeneracy is present, one may expect rather strong surface pinning field to be generated even in a crystallographically perfect surface. One can encounter effective pinning field that vary in magnitude over a wide range. Since a pinning field at the ends of the resonators can strongly affect the spin motion near the boundary, therefore the spin wave frequencies can be severely modified.

Closed-form expressions were obtained [70] for the band structure and the transmission coefficients for arbitrary values of the number N of sites and N' of resonators in the comb structures. Stop bands open inside the pass bands due to the effect of the pinning field at the ends of the resonators of the comb.

For a segment constituting a side branch (medium 2) (see Fig. 2.3), the boundary condition at one extremity of the wire is dependent upon the pinning field H_A that gives rise to a local perturbation V_A in the Hamiltonian such that [69] $V_A = -aH_A/M_2$, where a is an interatomic distance and M_2 the magnetization in this medium. With $H_A = 0$, one recovers the case of a free surface. Thus, Eq. (2.28) of a finite wire becomes

$$[g_2(MM)]^{-1} = \begin{pmatrix} A_2 & B_2 \\ B_2 & A_2 + V_A \end{pmatrix}, \tag{2.64}$$

where A_2 and B_2 have the expressions given by Eqs. (2.29), (2.30) and F_i and α_i have the values given by Eqs. (2.24), (2.25).

The inverse surface Green's function of the infinite comb is calculated as in the previous section with the help of the IRT. In the total interface space of the composite, this inverse Green's function is an infinite tridiagonal matrix $[g_\infty(MM)]^{-1}$. In that case, the diagonal and off-diagonal elements of this matrix are, respectively, given by $2A_1 + W$ and B_1 with

$$W = \frac{N'[V_A A_2 + F_2^2]}{(V_A + A_2)}. \tag{2.65}$$

Taking advantage of the translational periodicity of the system in the direction x_3, the previous matrix can be Fourier transformed as

$$[g_\infty(\mathbf{k}, MM)]^{-1} = 2A_1 + W + 2B_1 \cos(kd_1), \qquad (2.66)$$

where k is the modulus of the propagation vector. In this space, the Green's function of the infinite structure is obtained by inverting the previous equation

$$[g_\infty(\mathbf{k}, MM)] = \frac{1}{2A_1 + W + 2B_1 \cos(kd_1)}. \qquad (2.67)$$

The dispersion relation of this waveguide is given by $\cos(kd_1) = \eta(\omega)$ where

$$\eta(\omega) = C_1 + \frac{N'S_1 F_2}{2F_1} \left\{ \frac{F_2 S_2 - C_2 V_A}{F_2 C_2 - S_2 V_A} \right\}. \qquad (2.68)$$

The previous equation for the dispersion relation (after inserting the value of V_A and F_2 inside the bracket) takes the form

$$\eta(\omega) = C_1 + \frac{N'S_1 F_2}{2F_1} \left\{ \frac{\alpha_2(\omega)aS_2 + C_2\varepsilon}{\alpha_2(\omega)aC_2 + S_2\varepsilon} \right\}, \qquad (2.69)$$

where

$$\varepsilon = \gamma H_A a^2 / D_i' \qquad (2.70)$$

is the dimensionless parameter which measures the strength of the pinning field H_A relative to the exchange interaction J. If $\varepsilon = 0$, one recovers the results in Section 2.5. It is straightforward to Fourier analyze back into real space the Green's function $g_\infty(\mathbf{k}, MM)$ and obtain

$$g_\infty(n, n') = \left(\frac{S_1}{F_1} \right) \frac{t^{|n-n'|+1}}{t^2 - 1}, \qquad (2.71)$$

where the integers n and n' refer to the sites $(-\infty < n, n' < +\infty)$ on the infinite waveguide and the parameter t is given by Eq. (2.44). Finally with this definition of the parameter t, the transmission coefficient takes the same form as those given by Eq. (2.56).

2.6.2 Numerical Results and Discussion

The earlier analytical results will be illustrated [70] by a few numerical calculations for some specific examples of *identical media* ($F_1 = F_2$).

Fig. 2.14 displays the projected band structure of an infinite structure for given values of d_1, d_2, a, ε, and D such that $d_1 = d_2 = 1500\,\text{Å}$, $a = 4\,\text{Å}$, $-0.025 \leq \varepsilon \leq 0.025$ and $D = 1.4 \times 10^{-6}$ Hz, respectively [69]. The plot is given as the frequency $\omega - \Omega_0$ (GHz) ($\Omega_0 = \gamma H_0$) versus the dimensionless parameter ε. The shaded areas, corresponding to frequencies for which $|\eta| < 1$, represent bulk bands where spin waves are allowed to propagate in the structure. These areas are separated by mini gaps where the wave propagation is prohibited.

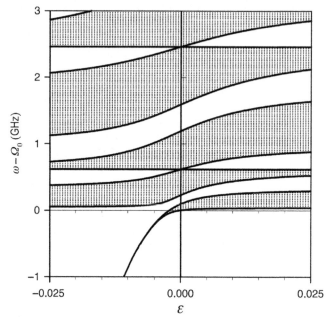

FIG. 2.14 Projected band structure of the comb-like structure as function of ε. The *shaded areas* represent the bulk bands. The gaps created inside the bulk band are related to the strength of the pinning field. With increasing (decreasing) ε all gaps are shifted up (down) to a higher (lower) frequencies. The parameters are $d_1 = d_2 = 1500\,\text{Å}$, $a = 4\,\text{Å}$, $-0.025 \leq \varepsilon \leq 0.025$, $D = 1.4 \times 10^{-6}\,\text{Hz}$, $N' = 1$, and $N \to \infty$. The two media are identical.

In Fig. 2.14, one can distinguish between two types of mini gaps: the gaps created inside the pass bands due to the existence (strength) of the pinning field, and the gaps, occurring for any value of ε, that are related to the periodicity of the structure and the resonance states of the grafted branches (which play the role of resonators). Two interesting points appear in the band structure of Fig. 2.14 with increasing (decreasing) ε: first, shifting the position of the gaps to a higher (lower) frequencies; second, widening the gaps created inside the pass bands due to the existence of the pinning field. Moreover, in the negative range of ε and $\omega - \Omega_0$, there appears a narrow band. This band corresponds to the localized surface states at the ends of the resonators. The increase (decrease) in the length d_2 of the resonators makes this band more narrow (wide). This is due to the weak (strong) interaction between these localized states via the backbone (i.e., more the end of the resonator is far from the backbone, more the surface states are localized). This is also more visible in the negative range of Fig. 2.15 where the length of d_2 is shorter than the one given in Fig. 2.14, namely $d_1 = 1500\,\text{Å}$ and $d_2 = 600\,\text{Å}$, the other parameters being the same as in Fig. 2.14. Let us mention that it is necessary for the existence of the surface spin waves that $\varepsilon < 0$, that is, the pinning field H_A is antiparallel to the magnetization direction [69].

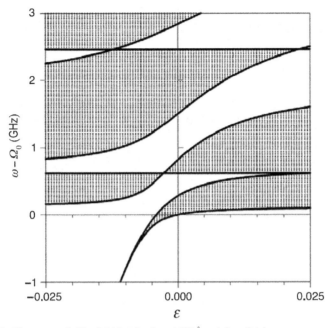

FIG. 2.15   The same as in Fig. 2.14 but for $d_1 = 1500\,\text{Å}$ and $d_2 = 0.4d_1$.

Fig. 2.16 gives the first five dispersion curves in the band structure of the infinite structure for two cases, $\varepsilon = 0$ (solid curve) and $\varepsilon = -0.01$ (dashed curve). The plot is given as $\omega - \Omega_0$ (GHz) versus kd_1, the other parameters being the same as in Fig. 2.14. The first band for $\varepsilon = 0$ disappears with the other choice of ε. A new flat band (which corresponds to localized surface states at the ends of the resonators) appears at $\omega - \Omega_0 \simeq -0.8$ GHz. In the band structure corresponding to $\varepsilon = 0$, the degenerate points between the second and the third bands (which appear at $kd_1 = \pi$ and $-\pi$), and between the fourth and the fifth bands (which appear at $kd_1 = 0$), are removed and a new gap is created in the band structure corresponding to $\varepsilon = -0.01$. The width as well as the frequency position of these gaps depend on the strength of the pinning field.

Let us turn to the study of the transmission power through the structure network. We start with a study of a simple example, namely a waveguide consisting of a unique resonator. The variations of T versus frequency, $\omega - \Omega_0$, are reported in Fig. 2.17A for $\varepsilon = 0$ (solid curve) and $\varepsilon = 0.02$ (dashed curve), the other parameters being as in Fig. 2.14. The pinning field shifts the transmission zeros (which correspond to the eigenmodes of the single resonator) to higher frequencies. These zeros enlarge into gaps when the number of resonators increases.

The transmission rate through a finite-size structure containing $N = 10$ resonators ($N' = 1$) with $\varepsilon = 0, 0.015, -0.005$ is reported in Fig. 2.17B–D,

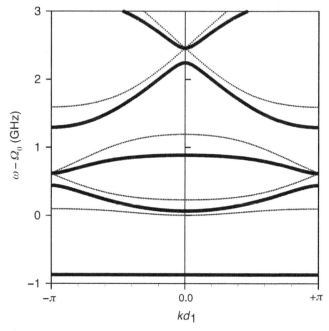

FIG. 2.16 The first five dispersion curves in the spin wave band structure of the infinite comb structure for two cases, $\varepsilon = 0$ (*dotted curve*) and $\varepsilon = 0.01$ (*heavy curve*). The plot is given as $\omega - \Omega_0$ (GHz) versus kd_1, the other parameters being the same as in Fig. 2.14.

respectively. New gaps inside the transmission bands show up in these figures. These new gaps are created by removal of the degenerate points in the band structure (this occurs at small value for ε), see Fig. 2.14. With increasing ε complete gaps are achieved. The gap position is related to the periodicity of the structure. The new gaps shift up to higher frequencies when ε increases. Despite the finite number of resonators in Fig. 2.17, the transmission approaches zero in regions corresponding to the observed gaps in the spin wave band structure of Fig. 2.14. It is worth noticing that the general features discussed in Fig. 2.14 are still valid for any value of N and $d_1 = d_2$. However, the shape of the band structure changes drastically for $d_1 \neq d_2$ (see Fig. 2.15).

2.6.3 Summary

The effect of a pinning field on the spin wave bands and transmission spectra of mono-mode comb structures was reviewed. New gaps are created inside the pass bands due to the strength of the pinning field. Compared to the study presented in Section 2.5, the observed gaps can be made larger with special choice of the pinning field magnitude. The existence of the gaps in the spectra is attributed to the periodicity, the transmission zero associated with a single resonator and the strength of the pinning field. Let us mention that, by associating in

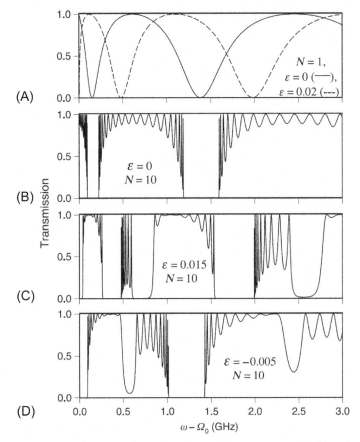

FIG. 2.17 (A) Transmission factor versus frequency (GHz) for a waveguide with one resonator in the case of $\varepsilon = 0$ (*solid curve*) and $\varepsilon = 0.02$ (*dashed curve*). The other parameters being as in Fig. 2.14. (B) Variations of the transmission power through a comb structure versus frequency (GHz) for $N = 10$, $N' = 1$, and $\varepsilon = 0$. The other parameters being as in Fig. 2.14. Figures (C) and (D) are the same as in (B) but for $\varepsilon = 0.015$, -0.005, respectively.

tandem several such structures, one could obtain an ultra-wide gap where the transmission is canceled over a large range of frequencies. In such a structure, the huge gap results from the superposition of the forbidden bands of the individual structures.

2.7   FANO RESONANCES IN A SIMPLE COMB STRUCTURE

We now turn to study, in the frame of the long-wavelength Heisenberg model, Fano resonances in a simple comb structures. This simple device is composed of an infinite 1D mono-mode waveguide (the backbone) along which two or more side resonators are grafted at two sites (see Fig. 2.18B and C). The origin of (and other related studies to) Fano resonances was discussed in the introduction.

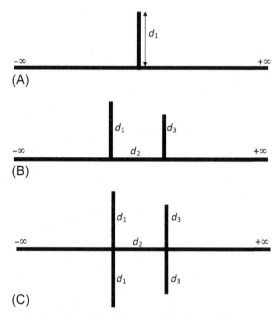

FIG. 2.18 Schematic illustration of the 1D waveguide with dangling resonators studied in the present section. The 1D media constituting the infinite mono-mode guide and the finite segments are assumed to be of the same material. (A) An infinite line with one grafted segment of length d_1. (B) The same as in (A) but for an infinite line with two segments of lengths d_1 and d_3 grafted at two sites separated by a segment of length d_2. (C) The same as in (B) but for two side resonators grafted at two sites.

2.7.1 Theoretical Discussion

Before addressing the problem of the simple structure presented in this section (see Fig. 2.18), it is helpful to know the surface elements of its elementary constituents, namely, the Green's function of a finite segment of length d_i, $i = 1$, 2, 3, and of a semiinfinite medium (lead). The finite segment of length d_2 is bounded by two free surfaces located at $x_3 = 0$ and $x_3 = d_2$. These surface elements can be written in the form of a (2×2) matrix $g_2(MM)$, within the interface space $M = \{0, +d_2\}$. The inverse of this matrix takes the following form [68] (see Eqs. 2.28–2.32)

$$[g_2(MM)]^{-1} = \begin{pmatrix} \dfrac{-F_2 C_2}{S_2} & \dfrac{F_2}{S_2} \\ \dfrac{F_2}{S_2} & \dfrac{-F_2 C_2}{S_2} \end{pmatrix}. \tag{2.72}$$

The inverse of the surface Green's functions of the dangling resonators grafted at the sites $\{0\}$ and $\{d_2\}$ is given by $[g_1(0,0)]^{-1} = -NF_1 C_1/S_1$ and $[g_3(d_2, d_2)]^{-1} = -N' F_3 C_3 / S_3$, where $C_i = \cos(\alpha'_i d_i)$, $S_i = j\sin(\alpha'_i d_i) = j\, S'_i$,

$i = 1, 3$ (see Eqs. 2.28, 2.29). Here N and N' are the number of side branches on both sides of the finite segment of length d_2. The inverse of the surface Green's functions of the two semiinfinite ferromagnetic leads surrounding the whole structure is given by $[g_s(0,0)]^{-1} = [g_s(d_2, d_2)]^{-1} = -F_s$ (see Eq. 2.27). In what follows, we assume that the semiinfinite leads and mediums (segments) 1, 2, and 3 are constituted of the same material (i.e., $F_1 = F_2 = F_3 = F_s = F = D'\alpha/\gamma M_0$, $D' = (2Ja^2 M_0)/(\gamma \hbar^2)$, $\alpha = j\sqrt{(\omega - \gamma H_0)/D'} = j\alpha'$). We report on results of calculated transmission coefficients and phase or phase time as a function of frequency. Using the Green's function method [68], the expression giving the inverse of the Green's function of the whole system depicted in Fig. 2.18 can be obtained from a linear superposition of the above inverse Green's functions of the constituent, namely

$$[g(MM)]^{-1} = F \begin{pmatrix} \dfrac{-C_2}{S_2} - \dfrac{NS_1}{C_1} - 1 & \dfrac{1}{S_2} \\[2ex] \dfrac{1}{S_2} & \dfrac{-C_2}{S_2} - \dfrac{N'S_3}{C_3} - 1 \end{pmatrix}. \tag{2.73}$$

The transmission function is given by [68] $t = -2Fg(0, d_2)$, or equivalently

$$t = \frac{2C_1 C_3}{\beta_1 + j\beta_2}, \tag{2.74}$$

where

$$\beta_1 = 2C_1 C_2 C_3 - S_2'(NS_1' C_3 + N' C_1 S_3') \tag{2.75}$$

and

$$\beta_2 = 2C_1 S_2' C_3 + N' C_1 C_2 S_3' + NS_1'(C_2 C_3 - N' S_2' S_3'). \tag{2.76}$$

From the expression of t (Eq. 2.74), one can deduce the transmission amplitude

$$T = \frac{4C_1^2 C_3^2}{\beta_1^2 + \beta_2^2} \tag{2.77}$$

as well as the phase

$$\varphi = \arctan(\beta_2/\beta_1) + \pi\Theta[C_1 C_3], \tag{2.78}$$

where Θ means the Heaviside function. From Eqs. (2.74), (2.77), one can notice that the transmission zeros are induced by the side branches (i.e., $C_1 = 0$ or $C_3 = 0$). When the expression $C_1 C_3$ changes sign at some frequencies denoted by ω_n, then the phase (Eq. 2.78) exhibits a jump of π.

Another interesting quantity is the first derivative of φ with respect to the frequency which is related to the delay time taken by the magnons to traverse the structure. This quantity, called phase time, is defined by [87–91]

$$\tau_\varphi = \frac{d\varphi}{d\omega} \tag{2.79}$$

and can be written as

$$\tau_\varphi = \frac{d}{d\omega} \arctan(\beta_2/\beta_1) + \pi \sum_n sgn\left[\frac{d}{d\omega}(C_1 C_3)_{\omega=\omega_n}\right]\delta(\omega - \omega_n), \quad (2.80)$$

where *sgn* means the sign function. Furthermore, the density of states of the present composite system from which we have subtracted the density of states of the semiinfinite leads is given by [91]

$$\Delta n(\omega) = \frac{1}{\pi}\frac{d}{d\omega}\arctan(\beta_2/\beta_1). \quad (2.81)$$

Because of the second term in the right-hand side of Eq. (2.80), one can deduce that $\tau_\varphi \neq \pi\Delta n(\omega)$ as τ_φ (Eq. 2.80) may exhibit δ functions at the transmission zeros that do not exist in the variation of the density of states (Eq. 2.81). However, if the system does not exhibit transmission zeros or the expression $C_1 C_3$ does not change sign at some frequencies, then $\Theta[C_1 C_3] = 0$ and $\tau_\varphi = \pi\Delta n(\omega)$. All these cases will be illustrated following.

It should be pointed out that the validity of our results is subject to the requirement that the cross-section of the waveguide being negligible compared to their length and to the propagation wavelength. The assumption of mono-mode propagation is then satisfied.

2.7.2 Transmission Gaps and Fano Resonances

Before addressing the problem of the whole structure described earlier, let us first recall briefly the results of a particular case necessary for the understanding of the spin wave propagation in the structures shown in Fig. 2.18, namely if $d_2 = 0$, $N = 1$, and $N' = 0$, we obtain the transmission function of a simple structure consisting of one resonator grafted on an infinite guide (see Fig. 2.18A): $t = C_1/(C_1 + jS'_1/2)$. This expression enables us to deduce the transmission coefficient $T = |t|^2 = 4C_1^2/(4C_1^2 + S_1'^2)$ and the phase $\varphi = \pi\Theta(C_1) - \arctan(S'_1/2C_1)$. We can see that the transmission coefficient equals zero when $C_1 = 0$ (i.e., $\alpha'd_1 = (l'+0.5)\pi$), where l' is a positive integer. The corresponding frequencies will be $\omega_g = \gamma H_0 + D'((l' + 0.5)\pi/d_1)^2$, or symbolically $\Omega_g = \tilde{H}' + ((l' + 0.5)\pi)^2$, where $\Omega_g = \omega_g d_1^2/D'$ is a reduced frequency, and $\tilde{H}' = \gamma H_0 d_1^2/D'$. From these results one can notice that for this composite system there exist an infinite set of forbidden frequencies Ω_g corresponding to the eigenmodes of the grafted finite branch. This grafted branch behaves as a resonator and this simple composite system filters out the modes Ω_g. This phenomenon is related to the resonances associated with the finite additional path offered to the spin wave propagation. The variation of T versus the reduced frequency $\Omega_1 = \omega d_1^2/D'$ is reported in Fig. 2.19A (which is exactly the same as Fig. 2.2). T is equal to zero when $\alpha'd_1$ is an odd multiple of $\pi/2$ and reaches its maximum value of 1 when $\alpha'd_1$ is a multiple of π. The variation of the phase versus the reduced frequency (Fig. 2.19B) shows an abrupt

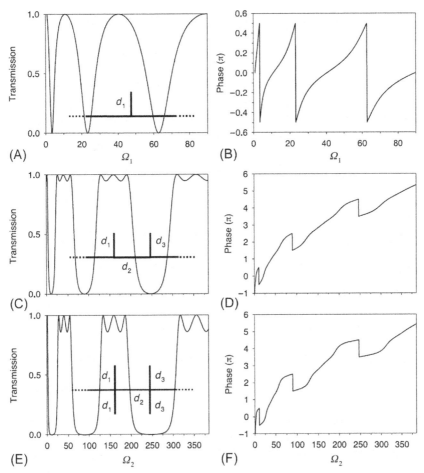

FIG. 2.19 (A) Transmission coefficient versus the reduced frequency Ω_1 for the structure depicted in Fig. 2.18A. For convenience $\tilde{H}'$ is considered to be 1. (B) The same as in (A) but for the variation of the phase. (C) Transmission coefficient versus the reduced frequency Ω_2 for the structure depicted in Fig. 2.18B with $d_2 = 2d_3 = 2d_1$ and $N = N' = 1$. For convenience $\tilde{H}$ is considered to be 1. (D) The same as in (C) but for the variation of the phase. (E) The same as in (C) but for the structure depicted in Fig. 2.18C ($N = N' = 2$). (F) The same as in (E) but for the variation of the phase.

change of π at the transmission zeros and therefore the corresponding phase time is different from the density of states as mentioned earlier.

For the structures shown in Fig. 2.18B and C, Eq. (2.74) clearly shows that the transmission zeros are due only to the dangling resonators (i.e., when $C_1 = 0$ or $C_3 = 0$). Fig. 2.19C gives the transmission coefficient in presence of two identical dangling resonators (i.e., $N = N' = 1$ and $d_1 = d_3 = 0.5d_2$). One can notice that the transmission coefficient presents well-defined dips induced by the grafted branches. These dips transform into large transmission gaps when the number of branches increases as it is illustrated in Fig. 2.19E for $N = N' = 2$. It is worth to mention that because of the existence of two resonators, one can expect two phase drops of π (i.e., 2π) at the transmission zeros given by $C_1 = C_3 = 0$ (i.e., $\Omega_2 = \omega d_2^2/D' = \tilde{H} + ((l' + 0.5)\pi)^2$, $\tilde{H} = \gamma H_0 d_2^2/D'$, $l' = 0, 1, 2, \ldots$). However, one can see in Fig. 2.19D and F that the phase presents only a phase drop of π. This is due to existence of a resonant state with zero width at these values of Ω_2 which induce a phase jump of π, these resonances collapse when $d_1 = d_3$ is taken exactly equal to $0.5d_2$. To enlarge these resonances, we have to take d_1 and d_3 slightly different from $0.5d_2$. Indeed, at $\Omega_2 = \tilde{H} + (l\pi)^2$, $l = 1, 2, \ldots$ and for $N = N' = 1$, the expression of the transmission function (Eq. 2.74) becomes

$$t = \pm \frac{2C_1 C_3}{2C_1 C_3 + jN \sin[\alpha'(d_1 + d_3)]}. \tag{2.82}$$

In particular, if $\alpha'(d_1 + d_3) = m\pi$, $\alpha' d_1 \neq (m_1 + 0.5)\pi$, and $\alpha' d_3 \neq (m_2 + 0.5)\pi$ (m, m_1, and m_2 are integers), one obtains a resonance that reaches unity (i.e., $T = 1$). An example corresponding to this situation is given in Fig. 2.20A where $d_1 = 0.46d_2$ and $d_3 = 0.54d_2$ (with $d_1 + d_3 = d_2$). One can notice that the resonance at $\Omega_2 = \tilde{H} + \pi^2$ is squeezed between two zeros (indicated by solid circles on the abscissa of Fig. 2.20A) induced by the dangling resonators as it is also illustrated in the plot describing the variation of the phase (Fig. 2.20C). The width of this resonance increases as far as d_1 and d_3 deviate from $0.5d_2$. In the particular case where $\alpha' d_1 = (m_1 + 0.5)\pi$ and $\alpha' d_3 = (m_2 + 0.5)\pi$, the numerator and denominator of t (Eq. 2.82) vanishes altogether. In this case, the resonance as well as the two zeros induced by the resonators fall at the same position, then the resonance collapses, the transmission coefficient vanishes and the phase drops by π as it is shown in Fig. 2.19D and F.

The resonance in Fig. 2.20A shows the same characteristics as a Fano resonance but with two zeros of transmission around the resonance instead of one as it is usually the case [45–48]. Indeed, one can obtain an approximate analytical expression for the transmission function (Eq. 2.74) in the vicinity of the resonance. A Taylor expansion around $\alpha' d_2 = \pi$ (i.e., $\alpha' d_2 = \pi + \varepsilon$ with $\varepsilon/\pi \ll 1$) enables us to obtain

$$t = \frac{-\zeta \zeta'}{2N\varepsilon^2 + \zeta \zeta' - j\varepsilon[2(N + N^2) + \zeta \zeta'(N^2 - 2)/2]}, \tag{2.83}$$

where $\zeta = 2\Delta + \varepsilon(1 + 2\Delta/\pi)$, $\zeta' = -2\Delta + \varepsilon(1 - 2\Delta/\pi)$, and Δ is the detuning of d_1 and d_3 from $0.5d_2$ (i.e., $\Delta = \pi(0.5 - d_1/d_2) = \pi(-0.5 + d_3/d_2)$). Using Eq. (2.83), one can show that the transmission coefficient T can be written (following the Fano line shape [45–48]) in the form

$$T = A \frac{(\varepsilon + q_1 \Gamma)^2 (\varepsilon - q_2 \Gamma)^2}{\varepsilon^2 + \Gamma^2} \tag{2.84}$$

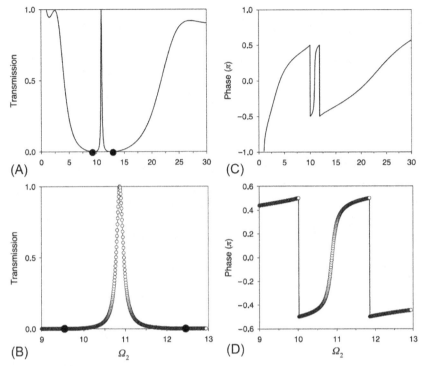

FIG. 2.20 (A) The same as in Fig. 2.19C but the lengths of the resonators are taken such that $d_1 = 0.46d_2$ and $d_3 = 0.54d_2$ and $N = N' = 1$. *Solid circles* on the abscissa indicate the positions of the transmission zeros induced by the dangling resonators on both sides of the resonance. (C) The same as in (A) but for the variation of the phase. (B) and (D) give the approximate results (*open circles*) around the resonance.

where $A = (1 - 4\Delta^2/\pi^2)^2/4[N(N+1) + \Delta^2(2 - N^2)]^2$. $\Gamma = 2\Delta^2/[N(N+1) + \Delta^2(2 - N^2)]$ characterizes the width of the resonance falling at $\varepsilon = 0$. $q_1 = [N(N+1) + \Delta^2(2 - N^2)]/\Delta(1 + 2\Delta/\pi)$ and $q_2 = [N(N+1) + \Delta^2(2 - N^2)]/\Delta(1 - 2\Delta/\pi)$ are the coupling parameters, they give qualitatively the interference between the bound states and the propagating continuum states [45–51].

One can notice that when increasing Δ, Γ increases and $q_1(q_2)$ decreases. The results of the approximate expression (Eq. 2.84) are shown in Fig. 2.20B by open circles. These results are in accordance with the exact ones (solid lines) and clearly show that the resonance is of Fano type with $q_1 \simeq 14.85$, $q_2 \simeq 17.43$ and width $2\Gamma \simeq 0.03$. The commonly studied Fano resonances are asymmetric because of the presence of only one transmission zero near the resonance. In addition, in the electronic counterparts studies, a perturbation is often introduced to the system in order to create the resonance state [45–52]. However, the previous calculation shows that *without introducing any perturbation in the structure, one can find a well-defined symmetric Fano resonance* with width 2Γ and coupling parameters q_1 and q_2 that can be adjusted by tailoring the lengths of the resonators (i.e., Δ). Eq. (2.83) enables us also to deduce an approximate expression for the phase as

$$\varphi = -\arctan\left\{\frac{\varepsilon[2(N+N^2)+\zeta\zeta'(N^2-2)/2]}{2N\varepsilon^2+\zeta\zeta'}\right\} + \pi\Theta(\zeta) + \pi\Theta(\zeta'). \quad (2.85)$$

This function is plotted by open circles in Fig. 2.20D and clearly shows two abrupt phase changes of π at $\zeta = 0$ and $\zeta' = 0$ (i.e., $\varepsilon_1 = -q_1\Gamma$ and $\varepsilon_2 = q_2\Gamma$) in accordance with the exact results (solid line).

One can also create an asymmetric Fano resonance by adjusting the transmission zeros on only one side of the resonance, this can be obtained by considering a structure where the resonators are supposed to be identical with lengths slightly different from $0.5d_2$. This is shown in Fig. 2.21A for $d_1 = d_3 = 0.46d_2$ and $N = N' = 1$. Indeed, an analytical Taylor expansion around $\alpha'd_2 = \pi$ enables us to write the transmission function (Eq. 2.74) as

$$t = \frac{-2\xi^2}{(\xi - jN)(N\varepsilon + 2\xi + j\varepsilon\xi)}, \quad (2.86)$$

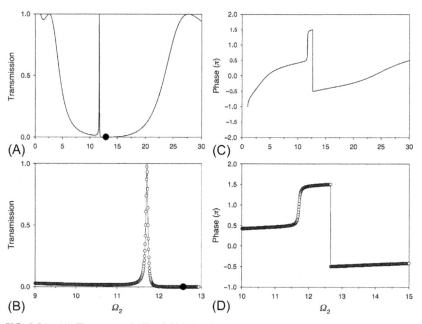

FIG. 2.21 (A) The same as in Fig. 2.20A but the resonators are taken to be of identical lengths $d_1 = d_3 = 0.46d_2$. (C) The same as in (A) but for the variation of the phase. (B) and (D) give the approximate results (*open circles*) around the resonance.

where $\xi = \Delta + 0.5\,\varepsilon(1 + 2\Delta/\pi)$ and Δ is the detuning of the lengths of the two resonators from $0.5d_2$ (i.e., $\Delta = \pi(d_1/d_2 - 0.5)$). From the expression of t (Eq. 2.86), one can deduce the following Fano line-shape transmission coefficient

$$T = \frac{B}{N^2 + \xi^2}\frac{(\varepsilon - \varepsilon_R + q\Gamma)^4}{(\varepsilon - \varepsilon_R)^2 + \Gamma^2} \simeq \frac{B}{N^2}\frac{(\varepsilon - \varepsilon_R + q\Gamma)^4}{(\varepsilon - \varepsilon_R)^2 + \Gamma^2}, \qquad (2.87)$$

where $B = (1 + 2\Delta/\pi)^4/4(N + 1 + 2\Delta/\pi)^2$.

$$q = (N + 1 + 2\Delta/\pi)^2/\Delta(1 + 2\Delta/\pi) \qquad (2.88)$$

is the Fano parameter.

$$\Gamma = 2\Delta^2/N^2\left[1 + \frac{1}{N}(1 + 2\Delta/\pi)\right]^3 \qquad (2.89)$$

and

$$\varepsilon_R = -2\Delta/(N + 1 + 2\Delta/\pi) \qquad (2.90)$$

characterize the width and the shift of the resonance, respectively.

One can notice that the resonance shifts slightly from $\alpha'd_2 = \pi$ and its width is small as compared to the preceding case, this is in accordance with the numerical results of Figs. 2.20A and 2.21A. Also q increases when Δ decreases and tends to infinity when Δ vanishes. In this case the resonance falls at $\varepsilon_R = 0$ and, as expected, its width 2Γ reduces to zero (see Fig. 2.19C). The results of the approximate expression (Eq. 2.87) are sketched (open circles) in Fig. 2.21B for $\Delta = \pi(d_1/d_2 - 0.5) = -0.04\pi$ (i.e., $d_1/d_2 = 0.46$) and $N = N' = 1$. These results are in accordance with the exact ones (solid lines) and clearly show that the resonance is of Fano type with $|q| \simeq 32$ and width $2\Gamma \simeq 0.0089$. Concerning the evolution of the phase of the spin waves in this structure, one can notice from Eq. (2.74) that the numerator of the transmission function t vanishes when $C_1 = C_3 = 0$ (or equivalently $\xi = 0$ in the approximate result (Eq. 2.86) at $\alpha'd_2 = (\pi/2)d_2/d_1$ ($\Omega_2 = 12.66$) indicated by a filled circle on the abscissa of Fig. 2.21A). The transmission zeros induced by the two identical resonators fall at the same frequency, therefore the phase (Fig. 2.21C and D) shows a phase drop of 2π at these frequencies. Indeed, as the phase is defined modulo 2π, the 2π phase drop can be observed if we take into account the absorption in the system [92, 93].

In order to show the profile of the Fano resonances as function of the parameter Δ (or equivalently d_1/d_2), Fig. 2.22A gives the characteristic features of the resonances as function of the reduced frequency Ω_2 for d_1/d_2 around 0.5 and for an asymmetric resonance. One can notice that the position of the resonance decreases as function of d_1/d_2, its asymmetric Fano profile becomes symmetric and changes sign for $d_1/d_2 \simeq 0.5$. In other words, the parameter q responsible for the asymmetric Fano profile of the resonance diverges and changes sign around $d_1/d_2 = 0.5$. The width of the resonance decreases when

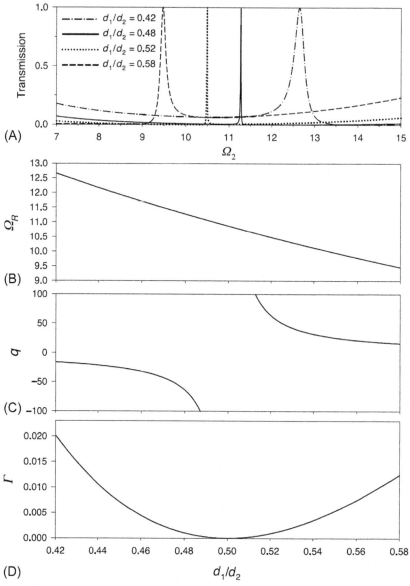

FIG. 2.22 (A) The same as in Fig. 2.21A but for different values of d_1/d_2. (B–D) Variations of the quantities $(\Omega_R = \tilde{H} + (\pi + \varepsilon_R)^2, \tilde{H} = 1)$ (Eq. 2.90), q (Eq. 2.88), and Γ (Eq. 2.89) as a function of d_1/d_2 around $d_1/d_2 = 0.5$.

d_1/d_2 tends to 0.5 and vanishes when d_1/d_2 is exactly equal 0.5 giving rise to the collapse of the resonance (see Fig. 2.22A). These results are well illustrated by the plots of the approximate expressions of $\Omega_R = \tilde{H} + (\pi + \varepsilon_R)^2$ (Fig. 2.22B), q (Fig. 2.22C), and Γ (Fig. 2.22D) around $d_1/d_2 = 0.5$.

Until now, we have concentrated our analysis on the waveguide structure with only one dangling resonator (i.e., $N = N' = 1$). The advantage of such a structure lies in the facility to be designed experimentally. However, the analytical approximate expressions (Eqs. 2.84, 2.87–2.90) clearly show that the resonances remain of Fano type even for $N = N'$ different from one. Indeed, Fig. 2.23 gives the dependence of the transmission rate of both symmetric (Fig. 2.23A) and asymmetric (Fig. 2.23B) Fano resonances for different values of $N = N'$. Fig. 2.23A displays the transmission amplitude T (Eq. 2.77) as a function of Ω_2 for $N = N' = 1, 2$, and 5, the other parameters are $d_1 = 0.46d_2$, $d_3 = 0.54d_2$, and $\tilde{H} = 1$. The results of the approximate expression (Eq. 2.84) are sketched by open circles. Even though the resonances fall at the same frequency $\Omega_g \simeq 10.88$ (i.e., $\alpha'_2 d_2 = \pi$), their widths decrease as function of N and their q parameter increases giving rise to a symmetric resonance of

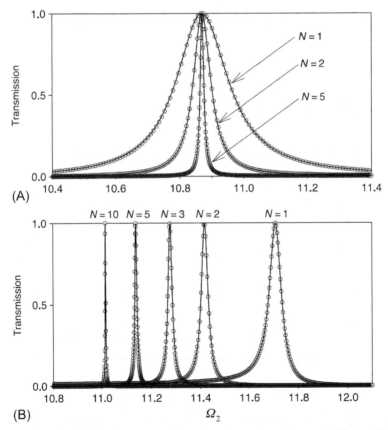

(A)

(B)

Ω_2

FIG. 2.23 (A) The Fano profile of the symmetric resonance depicted in Fig. 2.20A for different values of $N = N'$. (B) The same as in (A) but for the asymmetric Fano resonance depicted in Fig. 2.21A.

Breight-Wigner type. These results are in accordance with the approximate results of Eq. (2.84).

In the same way, Fig. 2.23B depicts the effect of variation in the number N of dangling side branches on the transmission rate T for a structure constituted of two identical dangling side branches with $d_1 = d_3 = 0.46d_2$. One can notice that, contrary to the results of Fig. 2.23A, the position of the resonance decreases as function of N and tends to $\Omega_2 \simeq 10.88$ (i.e., $\alpha'_2 d_2 = \pi$) when N goes to infinity (see Eq. 2.90). However, the width Γ of the resonance and its q parameter exhibit similar behavior as in Fig. 2.23A when N increases (see Eqs. 2.88, 2.89).

In Fig. 2.24A, the transmission rate through the structures shown in Fig. 2.18B ($N = N' = 1$, solid line) and C ($N = N' = 2$, dashed line) is redrawn for the sake of comparison with Fig. 2.24B giving the variation of the density of states. The lengths of the resonators are chosen such that $d_1 = d_3 = 0.5d_2$. A well-defined gap is obtained when the number of resonators is increased. Such a stop band could be useful in constructing a rejecting signal device. In Fig. 2.24B, one can notice that the density of states is strongly reduced in the transmission gap regions, in particular, when the number of dangling resonators increases. An analysis of the phase time is given in Fig. 2.24C. This quantity gives information on the time spent by the magnon inside the structure before its transmission. Because of the existence of the transmission zeros, the phase time gives rise to delta functions around the transmission zeros at $\Omega_2 = 10.87$, 89.3, and 247.74, according to Eq. (2.80). These delta functions have been enlarged by adding a small imaginary part to the pulsation ω, which plays the role of absorption in the system. Such negative delta peaks have been shown experimentally in a simple photonic [92] and phononic [93] waveguide, giving rise to the so-called superluminal velocity. Fig. 2.24B and C clearly shows, in accordance with Eqs. (2.80), (2.81), that except the frequencies lying around the transmission zeros, the density of states and the phase time exhibit almost the same behavior.

2.7.3 Summary

In summary, we have clearly demonstrated that a simple geometry of a 1D mono-mode waveguide with dangling side resonators on both sides can pave the way to the derivation of gaps in the spin wave propagation. The existence of the stop bands in the spectrum is attributed to the zeros of transmission associated with the dangling resonators. The width of the transmission gaps depends on the number of the side resonators grafted on both sides of the backbone. Besides the transmission gaps, we have shown the existence of asymmetric and symmetric Fano resonances that may lie near the vicinity of a transmission zero or be squeezed between two transmission zeros. These resonances are obtained by tailoring the lengths of the different branches constituting the structure and for different values $N = N'$ of the dangling side branches. A study of the phase

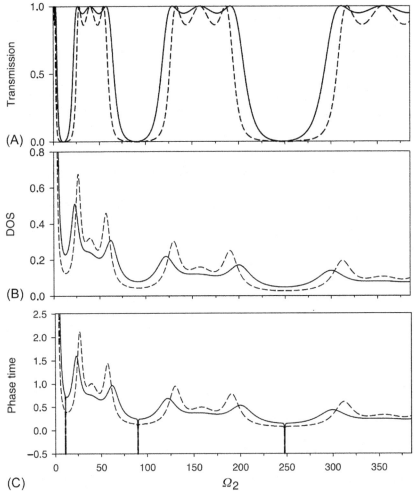

FIG. 2.24 (A) Transmission coefficient versus the reduced frequency Ω_2 for the structures depicted in Fig. 2.19C (*solid line*) and E (*dashed line*). (B) The same as in (A) but for the variation of the density of states (in units of $d_2^2/2D'$). (C) The same as in (B) but for the variation of the phase time (in units of $d_2^2/2D'$). The parameters are $d_1 = d_3 = 0.5d_2$ and $\bar{H} = 1$.

of the transmission function enables us to deduce several properties on the spin wave propagation, through such structures, such as the phase times and therefore the density of states. The phase time calculation are, in general, the same as the density of states, except for the frequencies lying around the transmission zeros where the phase time may exhibit additional negative delta peaks.

The advantage of the simple magnonic waveguide model presented in this section consists in finding simple analytical expressions. These expressions enable us to discuss the existence of Fano resonances as well as the effect of the different segments lengths and the number of dangling side branches in tailoring these resonances without incorporating a defect in the structure as it is usually the case in the electronic counterparts studies [45, 46, 49, 52].

2.8 MAGNONIC ANALOG OF ELECTROMAGNETIC INDUCED TRANSPARENCY IN DETUNED MAGNETIC CIRCUIT

In this section, we are interested in showing the analog of electromagnetic induced transparency (EIT) resonance in a simple magnetic device made of two dangling side stubs grafted at the same site along a waveguide. By detuning the lengths of the two stubs, a resonance can be squeezed between two transmission zeros induced by the two resonators. We give detailed analytical expressions for the transmission and reflection coefficients of the whole structure as a function of those of the individual resonators taken alone. The destructive interference between the waves in the two resonators may give rise to a sharp resonance peak with high Q factor in the transmission. We give an explicit expression of the transmission coefficient around the MIT (analog of EIT) resonance following a Fano-like form. In particular, we give an explicit expression of the position, width, and Fano parameter of the resonances as a function of the detuning between the stubs. The analytical results are obtained by means of the Green's function method.

2.8.1 Motivations

Recently magnonic crystals, which have a regular (periodic) distribution of magnetic scattering centers, have been shown to possess a distinct and interesting array of magnetic properties. Indeed, 1D, 2D, and 3D structures in which propagation of spin waves (magnons) are forbidden have attracted much attention [33, 94, 95]. This is related to a number of advantages exhibited by magnonic crystals in comparison with their photonic crystals' counterparts [96]. The wavelength of a spin wave and, hence, the properties of such crystals depend on the external magnetic field and can be controlled by this field. The wavelength of propagating spin waves for a wide class of ferromagnetic materials in the microwave range is on the order of tens or even hundreds of micrometers. The phase and group velocities of spin waves are also functions of the structure size and the applied external field and may vary over a wide range [32, 97–102]. The interest in these band gaps is related to the potential applications of magnon-transport devices and is supported by the advanced progress in nanofabrication technology [94]. These magnonic band-gap materials can have many practical applications such as spin injection into devices [29, 30]. On experimental grounds, arrays of very long ferromagnetic nanowires made of Ni, permalloy (or Fe), and Co with diameters in the range of 30–500 nm have been created [102–104]. The realization of such nanowires can reasonably be considered, to an excellent approximation, as mono-mode waveguides.

The 1D structures are attractive since their production are more feasible and they require only simple analytical and numerical calculations.

In the previous sections (see also [64, 70, 105, 106]), we have reported band structures and transmission spectra of periodically dangling stubs on an infinite 1D mono-mode waveguide. It was shown that these systems may exhibit two types of forbidden bands: (i) gaps originating from the periodicity of the structure (Bragg gaps) and (ii) gaps originating from the transmission zeros associated with local resonances in the stubs. These gaps appear even if the stubs and the waveguide are composed of the same material [33].

In addition to the band gaps, a great interest has been paid to the so-called EIT resonance [107, 108]. EIT is the phenomenon where a sharp transparent window associated with steep dispersion is induced into opaque atomic media. These systems have shown potential applications for slow light, sensing, and optical data storage [108]. However, it was demonstrated that EIT resonances are not restricted to quantum systems and can be extended to classical systems such as plasmonic materials and planar metamaterials [109, 110], photonic crystal waveguides coupled to cavities [111, 112], coupled microresonators [113, 114], acoustic [115–117], and photonic circuits [118, 119]. However, little work has been devoted to magnonic counterpart systems [106]. To our knowledge, the first paper [106] dealing with a theoretical demonstration of MIT resonance in magnonics has been published by some of us in a simple 1D structure made by two side detuned resonators grafted at two sites on a waveguide.

These resonances were first theoretically described by Fano [45] when he studied the inelastic auto ionizing resonances in atoms. The symmetric and asymmetric line shapes have been also reported in the electronic transport in mesoscopic systems using the Aharonov-Bohm system [120, 121]. Mainly, the subject of these studies was to use these interferometric systems to show the conditions for the existence and the collapse of Fano resonances as a function of the applied current-voltage and magnetic flux. The analogy between scattering properties of electrons, phonons, photons, and magnons suggests that this type of feature can also appear in magnonic systems [62].

Here, we present the possibility of existence of MIT resonance in a simple structure composed by two resonators grafted at the same point on 1D mono-mode waveguide (see Fig. 2.25). We show analytically and numerically that this simple structure can exhibit transmission gaps and MIT-like resonance. In particular, we show that the transmission and reflection amplitudes of the spin waves in the two stubs can be written as a function of the transmission and reflection coefficients of each resonator taken alone. This enables to deduce that the MIT resonance results from a destructive interference between the waves in the two interacting stubs. In addition, we demonstrate that the transmission amplitude through such a system can be written following the Fano-like shape around these resonances. In particular, we give an explicit expression of the Fano parameters as well as the width of the MIT resonance as a function of the geometrical parameters of the system.

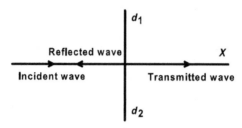

FIG. 2.25 Schematic illustration of the 1D magnonic waveguide with two dangling resonators of lengths d_1 and d_2 grafted at the same site on the waveguide. The waveguide and resonators are made by the same material.

2.8.2 Transmission and Reflection Coefficients

Our theoretical calculations are performed within the frame of long-wavelength Heisenberg model and the IRT of continuous media. The IRT allows the calculation of the Green's function of any composite material and then, total and local densities of states as well as transmission and reflection coefficients. In what follows, we shall avoid the details of these calculations which are similar to those given in the previous sections and give only the expressions of the transmission and reflection coefficients.

Let us consider an incident wave $m_{inc}(x) = e^{-jk_R x}$ launched in the left semiinfinite waveguide (Fig. 2.25), where $x \equiv x_3$, $k_R = \sqrt{(\omega - \gamma H_0)/D}$ is the incident wavevector, and $D = (2Ja^2 M_0)/(\gamma \hbar^2)$ [33].

When the boundary conditions at the stub ends are free (see Fig. 2.25), the transmitted wave in the right semiinfinite waveguide is given by [33]

$$m_{tra}(x) = t\, e^{-jk_R x}, \tag{2.91}$$

where

$$t = \frac{2k_R C_1 C_2}{2k_R C_1 C_2 - jkS} \tag{2.92}$$

is the transmission amplitude with $C_i = \cos(kd_i)$, $S_i = \sin(kd_i)$, $S = \sin(kd)$ $(i = 1, 2)$ and $d = d_1 + d_2$ is the total length of the two resonators. $k = k_R + jk_I$ where $k_I = \acute{\varGamma}/2\sqrt{D(\omega - \gamma H_0)}$ is the imaginary part of the wavevector in the stubs.

By the same way, the reflected wave in the left semiinfinite waveguide (Fig. 2.25) is given by

$$m_{ref}(x) = r\, e^{jkx}, \tag{2.93}$$

where

$$r = \frac{jkS}{2k_R C_1 C_2 - jkS} \tag{2.94}$$

is the reflection amplitude.

From the expressions of t (Eq. 2.92) and r (Eq. 2.94), one can, respectively, calculate the transmission and reflection coefficients $T = |t|^2$ and $R = |r|^2$. Also in the case of lossless waveguides (i.e., $\Gamma = 0$ or $k_I = 0$), one can deduce easily the conservation energy, namely $R + T = 1$.

In order to understand the MIT phenomenon in the double stub structure (Fig. 2.25), it is important to relate the transmission and reflection coefficients of the whole structure to those of its elementary constituents, namely, each stub taken separately. From Eqs. (2.92), (2.94), one can deduce the transmission and reflection coefficients of one attached resonator of length d_1 as follows

$$t_1 = \frac{k_R C_1}{k_R C_1 - jk S_1/2} \quad \text{and} \quad r_1 = \frac{ik S_1/2}{k_R C_1 - jk S_1/2}. \tag{2.95}$$

One can notice from Eqs. (2.92), (2.94), (2.95) that the reflection and transmission coefficients of the whole structure can be written as a function of those of the individual resonators (Eq. 2.95)

$$t = \frac{t_1 t_2}{1 - r_1 r_2} \tag{2.96}$$

and

$$r = \frac{r_1 + r_2 + 2r_1 r_2}{1 - r_1 r_2}. \tag{2.97}$$

Eq. (2.96) clearly shows that the transmission zeros of the double stub structure are exactly the same transmission zeros as those of each stub taken individually. However, the reflection (Eq. 2.97) appears as the sum of the reflection of each stub alone plus a third term representing the interference between the two stubs. Indeed, if we call Ω_1 and Ω_2 two closed frequencies for which the transmission vanishes (Ω_1 and Ω_2 are induced by the stubs 1 and 2, respectively), then one can show easily that the reflection coefficient always vanishes (total transmission) for a given frequency Ω_3 lying between Ω_1 and Ω_2. Here, it is worth mentioning that blocked boundary conditions can also be achieved by taking into account the effect of the pinning field at the ends of the side resonators [111]. Since the presence of the pinning field can strongly affect the spin motion near the boundary, therefore the spin wave frequencies can be severely modified. Such an effect may lead to new features, in comparison with the free boundary condition at the stub ends.

In what follows, we shall show theoretically the existence of MIT resonances in lossless materials (i.e., $k_I = 0$). In this case, one can deduce easily from Eqs. (2.92), (2.94) the conservation of energy, namely $R + T = 1$. Also we will show below how the lengths of the two stubs should be chosen in order to realize the MIT resonances.

Eq. (2.92) clearly shows that the transmission zeros are due only to the dangling resonators, these occur when $C_1 = 0$ or $C_2 = 0$ (i.e., $kd_1 = (m_1 + 0.5)\pi$ or $kd_2 = (m_2 + 0.5)\pi$). These frequencies correspond, respectively, to the eigenmodes of the stubs 1 and 2 with blocked boundary condition on one side and free boundary condition on the other side [33]. Now, if $k(d_1+d_2) = m\pi$ (i.e., $S = 0$), but $kd_1 \neq (m_1 + 0.5)\pi$ and $kd_2 \neq (m_2 + 0.5)\pi$ (i.e., $C_1 \neq 0$ and $C_2 \neq 0$), then one obtains a resonance where the transmission reaches unity (i.e., $T = 1$).

In the particular case where $kd_1 = (m_1 + 0.5)\pi$ and $kd_2 = (m_2 + 0.5)\pi$, then $k(d_1 + d_2) = (m_1 + m_2 + 1)\pi$ and the numerator and denominator of t (Eq. 2.92) vanish altogether. In this case, the resonance as well as the two zeros induced by the resonators fall at the same position, then the resonance collapses and the transmission coefficient vanishes. These modes coincide with those of the long stub of length $d = d_1 + d_2$ with free boundary conditions on both sides as $S = 0$. The previous conditions are satisfied if d_1 and d_2 are chosen such that

$$d_1^0 = \frac{(2m_1 + 1)d}{2(m_1 + m_2) + 2} \tag{2.98}$$

and

$$d_2^0 = \frac{(2m_2 + 1)d}{2(m_1 + m_2) + 2}. \tag{2.99}$$

The previous expressions show how the lengths of the two stubs should be chosen in order to realize the MIT resonance. In what follows we consider the first modes of the two stubs given by $m_1 = m_2 = 0$ and keep the total length d constant in order to fix the resonance at the same frequency. In this case, the earlier lengths of the two stubs become $d_1^0 = d_2^0 = d/2$. Also, we shall use the dimensionless frequency $\Omega = \tilde{H} + (kd)^2$ where $\tilde{H} = \gamma H_0 d^2 / D$.

In order to show the possibility of existence of MIT resonance (i.e., a resonance squeezed between two transmission zeros), we have to take d_1 and d_2 slightly different from $d/2$ such that $d_1 = d/2 - \delta/2$ and $d_2 = d/2 + \delta/2$ (where $\delta = d_2 - d_1$ and $d = d_1 + d_2$). An example corresponding to this situation is given in Fig. 2.26A for $\delta = 0.16d$. One can notice that the resonance

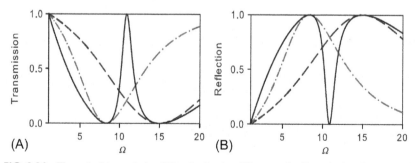

FIG. 2.26 Theoretical transmission (A) and reflection (B) versus the dimensionless frequency Ω for $\delta = d_2 - d_1 = 0.16d$ where d_1 and d_2 are the lengths of the two stubs and $d = d_1 + d_2$. The *blue (dashed) and red (dashed-dotted) curves* in (A) and (B) correspond, respectively, to each stub alone with $d_2 = 0.58d$ and $d_1 = 0.42d$, whereas the *black solid curves* correspond to the two stubs present together.

at $\Omega = \Omega_r = \pi^2 + 1$ is squeezed between two transmission zeros induced by the two stubs. As predicted, in the absence of loss, the intensity of this resonance reaches unity (Fig. 2.26A) while the reflection vanishes (Fig. 2.26B). As a matter of comparison, we have also plotted the transmission amplitudes for each stub attached alone using Eq. (2.95) (see the blue (dashed) and red (dashed-dotted) curves in Fig. 2.26A and B). As mentioned previously (Eq. 2.97), one can notice that the reflection coefficient of the whole structure is the same as the reflection coefficient of each resonator around the stub resonances falling to zero at $\Omega = 8.24$ and $\Omega = 14.82$ (i.e., $C_1 = 0$ and $C_2 = 0$), respectively. Whereas, at $\Omega = \Omega_r = 10.86$ (i.e., $S = 0$) the stubs interfere destructively giving rise to a zero reflection (or total transmission). These results are in accordance with Eq. (2.96) and the discussion given earlier where total transmission induced by the two stubs taken together always occurs between two transmission zeros induced by the two stubs. In the absence of loss, this phenomenon can be qualified as MIT resonance.

2.8.3 Analytical Results Around the MIT Resonances

In order to give a quantitative study of the shape, width, and intensity of the MIT resonance, we shall give in this section approximate expressions of the different quantities studied in the previous paragraph around the MIT resonance. Indeed, a Taylor expansion around Ω_r (i.e., $\Omega = \Omega_r + \varepsilon$) enables us to write the transmission amplitude T (Eq. 2.92) following the Fano line-shape or MIT-like resonance [31].

$$T = B \frac{(\varepsilon + q_1 \Gamma)^2 (\varepsilon - q_2 \Gamma)^2}{\varepsilon(\varepsilon + \beta) + \Gamma^2}, \tag{2.100}$$

where $B = \dfrac{\left(\frac{1}{4} - \frac{\Delta^2}{\pi^2}\right)^2}{\frac{1}{4} + \frac{4\Delta^4}{\pi^2}}$, $\beta = \dfrac{\frac{4\Delta^4}{\pi}}{\frac{1}{4} + \frac{4\Delta^4}{\pi^2}}$, and $\Delta = \pi\delta/2d$.

The full width at half maximum of the MIT resonance falling at $\varepsilon = 0$ (i.e., $\Omega = \Omega_r$) is given by

$$\Gamma = \sqrt{\frac{\Delta^4}{\frac{1}{4} + \frac{4\Delta^4}{\pi^2}}}. \tag{2.101}$$

$q_1 = \dfrac{\Delta}{\Gamma\left(\frac{1}{2} + \frac{\Delta}{\pi}\right)}$ and $q_2 = \dfrac{\Delta}{\Gamma\left(\frac{1}{2} - \frac{\Delta}{\pi}\right)}$ are the Fano parameters; they give qualitatively the strength of the interference between the bound state and the propagating continuum states. One can notice that when increasing Δ (or δ), Γ increases and q decreases. As predicted, T_{max} reaches unity when $\varepsilon = 0$ (i.e., at the resonance).

An example of the results of the approximate expression (Eq. 2.100) is shown in Fig. 2.27A by open circles for $\Delta = 0.16d$ and clearly shows that the resonance is Fano-like with $q_1 = 1.872$, $q_2 = 2.585$ and width $2\Gamma = 0.462$.

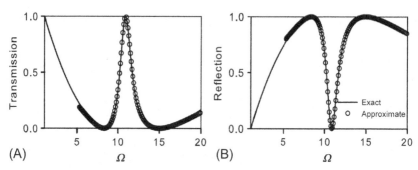

FIG. 2.27 Theoretical transmission (A) and reflection (B) versus the dimensionless frequency Ω for $\delta = d_2 - d_1 = 0.16d$ where d_1 and d_2 are the lengths of the two stubs and $d = d_1 + d_2$. The *solid lines* correspond to the exact results whereas the *circles* correspond to the approximate results.

It is well known that the introduction of a resonance within a band gap of a periodic crystal requires the presence of a defect within the perfect system [38]. However, the earlier calculation shows that without introducing any perturbation in the structure, one can find a well-defined MIT resonance with a width 2Γ and coupling parameters q_1 and q_2 that can be adjusted by tailoring the lengths of the resonators (i.e., Δ). Similarly a Taylor expansion of the reflection amplitude (Eq. 2.94) around Ω_r enables us to write R as follows

$$R = F \frac{\frac{\varepsilon^2}{4}}{\varepsilon(\varepsilon + \beta) + \Gamma^2}, \tag{2.102}$$

where $F = \frac{1}{\frac{1}{4} + \frac{4\Delta^4}{\pi^2}}$. Eq. (2.102) shows that the reflection coefficient take the minimum value at Ω_r such that $R = 0$.

An example of the approximate curves of the reflection rate (Eq. 2.102) is given in Fig. 2.27B by open circles for $\Delta = 0.16d$. The exact results are well fitted by the approximate expression (Eq. 2.102).

In Fig. 2.28 we have displayed the variation of the quality factor Q of the MIT resonance defined by $Q = \Omega_r/\Gamma$ as a function of Δ. The width Γ of the resonance is given by Eq. (2.101). When Γ increases, Q decreases and goes to a limit around $\simeq 4.8$ for large values of δ. For $\delta = 0$ (i.e., $d_1 = d_2 = 0.5d$) the MIT resonance collapses in the transmission spectra and the system behaves like an opaque medium around $\Omega = \Omega_r$. This kind of hidden resonance characterized by an infinite quality factor (or infinite lifetime) (Fig. 2.28), gives rise to the so-called bound in continuum state [122]. This state is confined in the vertical stubs and do not interact with the horizontal waveguide.

2.8.4 Summary

We have proposed a simple magnetic circuit that enables to obtain a magnonic analog of EIT resonance called MIT. The system is composed of two magnetic

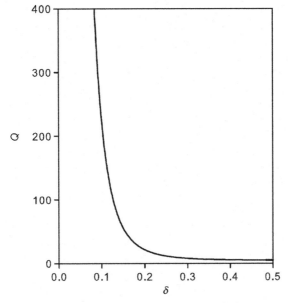

FIG. 2.28 Theoretical variation of the quality factor of the transmission resonance around Ω_r as a function of δ.

stubs inserted at the same site along a waveguide. We have shown that such resonance can be obtained as a result of a destructive interference between the waves in the two stubs when the latter are detuned appropriately. We have performed simple analytical expressions of the transmission and reflection coefficients which enable us to discuss deeply the conditions for the existence of MIT resonance as well as the effect of the different stub lengths in tailoring these resonances. Also, we have shown that the transmission coefficient can be written following a Fano-like form which enables us to deduce the width as well as the Fano parameters of the resonance as a function of the detuning Δ. We believe that these result bring a new piece of work in the field of spin wave transport in 1D waveguide structures. These magnonic circuits may have useful applications for the design of integrated devices such as narrow-frequency optical or microwave filters, high-speed switches, multiplexers, storage devices, and so on.

2.9 CONCLUSION AND PERSPECTIVES

In this chapter, we reviewed the state of the art of a new fast developing field called "magnonic circuits." Indeed all particles, bosons (magnon, photon, phonon, plasmon, etc.) and fermions (electrons, etc.) have a wave associated with them. When coherent and mono-mode propagation of such waves is made possible in linear waveguides, many new circuits and devices can be made out of such guides. We reviewed only a few circuits actually under investigation. A simple model of the tunneling between two wires and of a magnon multiplexer was also proposed [105]. No doubt that many others devices will be conceived in the future decades, as well with micro as with nano dimensions. One can start and imagine even atomic structures for any specific application.

So this chapter looks like a starting point for many applications. Its backbone is a simple analytical theory for the propagation of coherent spin waves in mono-mode circuits. This simple approach starts to be complemented by more realistic numerical simulations and by experiments for some long-wavelength spin waves. A lot remains to be done for microwaves, especially those used in modern telecommunications. Let us just mention that investigations on nonlinear effects in such structures are also realized; see, for example, Zielinski et al. [123].

REFERENCES

[1] Z.Q. Zheng, C.M. Falco, J.B. Ketterson, I.K. Schuller, Magnetization of compositionally modulated CuNi films, Appl. Phys. Lett. 38 (1981) 424.

[2] T. Jarlborg, A.J. Freeman, Magnetism of metallic superlattices, J. Appl. Phys. 53 (1982) 8041.

[3] K. Flevaris, J.B. Ketterson, J.E. Hilliard, Magnetic properties of compositionally modulated thin films, J. Appl. Phys. 53 (1982) 8046.

[4] T. Shinjo, K. Kawaguchi, R. Yamamoto, N. Hosaito, T. Takada, J. Phys. (Paris) Colloq. 45 (1984) C5-367.

[5] A.J. Freeman, J.-H. Xu, S. Ohnishi, T. Jarlborg, J. Phys. (Paris) Colloq. 45 (1984) C5-C369.

[6] R. Krishnan, W. Jantz, Ferromagnetic resonance studies in compositionally modulated cobalt-niobium films, Solid State Commun. 50 (1984) 533.

[7] R.E. Camley, T.S. Rahman, D.L. Mills, Magnetic excitations in layered media: Spin waves and the light-scattering spectrum, Phys. Rev. B 27 (1983) 261.

[8] F. Herman, P. Lambin, O. Jepsen, Electronic and magnetic structure of ultrathin cobalt-chromium superlattices, Phys. Rev. B 31 (1985) 4394.

[9] K. Mika, P. Grunberg, Dipolar spin-wave modes of a ferromagnetic multilayer with alternating directions of magnetization, Phys. Rev. B 31 (1985) 4465.

[10] L. Dobrzynski, B. Djafari-Rouhani, H. Puszkarski, Theory of bulk and surface magnons in Heisenberg ferromagnetic superlattices, Phys. Rev. B 33 (1986) 3251.

[11] A. Kueny, M.R. Khan, I.K. Schuller, M. Grimsditch, Magnons in superlattices: A light scattering study, Phys. Rev. B 29 (1984) 2879.

[12] M.N. Baibich, J.M. Broto, A. Fert, F. Nguyen Van Dau, F. Petroff, P. Etienne, G. Creuzet, A. Friederich, J. Chazelas, Giant magnetoresistance of (001) Fe/(001) Cr magnetic superlattices, Phys. Rev. Lett. 61 (1988) 2472.

[13] G. Binasch, P. Grunberg, F. Saurenbach, W. Zinn, Enhanced magnetoresistance in layered magnetic structures with antiferromagnetic interlayer exchange, Phys. Rev. B 39 (1989) 4828.

[14] A.V. Chumak, A.A. Serga, B. Hillebrands, M.P. Kostylev, Scattering of backward spin waves in a one-dimensional magnonic crystal, Appl. Phys. Lett. 93 (2008) 022508.

[15] A.V. Chumak, P. Pirro, A.A. Serga, M.P. Kostylev, R.L. Stamps, H. Schultheiss, K. Vogt, S.J. Hermsdoerfer, B. Laegel, P.A. Beck, B. Hillebrands, Spin-wave propagation in a microstructured magnonic crystal, Appl. Phys. Lett. 95 (2009) 262508.

[16] A.V. Chumak, A.A. Serga, S. Wolff, B. Hillebrands, M.P. Kostylev, Design and optimization of one-dimensional ferrite-film based magnonic crystals, J. Appl. Phys. 105 (2009) 083906.

[17] A.V. Chumak, A.A. Serga, S. Wolff, B. Hillebrands, M.P. Kostylev, Scattering of surface and volume spin waves in a magnonic crystal, Appl. Phys. Lett. 94 (2009) 172511.

[18] A.D. Karenowska, A.V. Chumak, A.A. Serga, J.F. Gregg, B. Hillebrands, Magnonic crystal based forced dominant wavenumber selection in a spin-wave active ring, Appl. Phys. Lett. 96 (2010) 082505.

[19] Z.K. Wang, V.L. Zhang, H.S. Lim, S.C. Ng, M.H. Kuok, S. Jain, A.O. Adeyeye, Nanostructured magnonic crystals with size-tunable bandgaps, ACS Nano 4 (2010) 643.

[20] M. Greven, R.J. Birgeneau, U.J. Wiese, Monte Carlo study of correlations in quantum spin ladders, Phys. Rev. Lett. 77 (1996) 1865.

[21] S. Chakravarty, Dimensional crossover in quantum antiferromagnets, Phys. Rev. Lett. 77 (1996) 4446.

[22] D.G. Shelton, A.A. Nersesyan, A.M. Tsvelik, Phys. Rev. B 53 (1996) 8521.

[23] J. Piekarewicz, J.R. Shepard, Dynamic spin response for Heisenberg ladders, Phys. Rev. B 57 (1998) 10260.

[24] P. Rai-Chaudhry (Ed.), The handbook of microlithography, micromachining and microfabrication, in: SPIE, 1996.

[25] X.F. Wang, P. Vasilopoulos, F.M. Peeters, Spin-current modulation and square-wave transmission through periodically stubbed electron waveguides, Phys. Rev. B 65 (2002) 165217.

[26] R. Arias, D.L. Mills, Theory of spin excitations and the microwave response of cylindrical ferromagnetic nanowires, Phys. Rev. B 63 (2001) 134439.

[27] A. Encinas-Oropesa, M. Demand, L. Piraux, I. Huynen, U. Ebels, Dipolar interactions in arrays of nickel nanowires studied by ferromagnetic resonance, Phys. Rev. B 63 (2001) 104415.

[28] J. Jorzick, S.O. Demokritov, C. Mathieu, B. Hillebrands, B. Bartenlian, C. Chappert, F. Rousseaux, A.N. Slavin, Brillouin light scattering from quantized spin waves in micron–size magnetic wires, Phys. Rev. B 60 (1999) 15194.

[29] D.P. DiVincenzo, D. Loss, Quantum information is physical, Superlattice Microstruct. 23 (1998) 419.

[30] B.E. Kane, A silicon-based nuclear spin quantum computer, Nature (London) 393 (1998) 133.

[31] A.G. Mal'shukov, K.A. Chao, Waveguide diffusion modes and slowdown of D'yakonov-Perel'spin relaxation in narrow two-dimensional semiconductor channels, Phys. Rev. B 61 (2000) R2413.

[32] J.O. Vasseur, L. Dobrzynski, B. Djafari-Rouhani, H. Puszkarski, Magnon band structure of periodic composites, Phys. Rev. B. 54 (1996) 1043.

[33] H. Al-Wahsh, A. Akjouj, B. Djafari-Rouhani, L. Dobrzynski, Magnonic circuits and crystals, Surf. Sci. Rep. 66 (2011) 29.

[34] M. Krawczyk, J.C. Levy, D. Mercier, H. Puszkarski, Forbidden frequency gaps in magnonic spectra of ferromagnetic layered composites, Phys. Lett. A. 282 (2001) 186.

[35] S.L. Vysotskiĭ, S.A. Nikitov, Y.A. Filimonov, Magnetostatic spin waves in two-dimensional periodic structures (magnetophoton crystals), J. Exp. Theor. Phys. 101 (2005) 547.

[36] S.A. Nikitov, P. Tailhades, C.S. Tsai, Spin waves in periodic magnetic structures—magnonic crystals, J. Magn. Magn. Mater. 236 (2001) 320.

[37] Y.V. Gulyaev, S.A. Nikitov, Magnonic crystals and spin waves in periodic structures, Dokl. Akad. Nauk. 380 (2001) 469.

[38] K.-S. Lee, D.-S. Han, S.-K. Kim, Physical origin and generic control of magnonic band gaps of dipole-exchange spin waves in width-modulated nanostrip waveguides, Phys. Rev. Lett. 102 (2009) 127202.

[39] A. Figotin, I. Vitebsky, Nonreciprocal magnetic photonic crystals, Phys. Rev. E 63 (2001) 066609.

[40] D.S. Deng, X.F. Jin, R. Tao, Magnon energy gap in a periodic anisotropic magnetic superlattice, Phys. Rev. B 66 (2002) 104435.

[41] C. Goffaux, J. Sánchez-Dehesa, Two-dimensional phononic crystals studied using a variational method: Application to lattices of locally resonant materials, Phys. Rev. B 67 (2003) 144301.

[42] C. Goffaux, J. Sánchez-Dehesa, Evidence of Fano-like interference phenomena in locally resonant materials, Phys. Rev. Lett. 88 (2002) 225502.

[43] Z. Liu, C.T. Chan, P. Sheng, Analytic model of phononic crystals with local resonances, Phys. Rev. B 71 (2005) 014103.

[44] Y.A. Kosevich, C. Goffaux, J. Sánchez-Dehesa, Fano-like resonance phenomena by flexural shell modes in sound transmission through two-dimensional periodic arrays of thin-walled hollow cylinders, Phys. Rev. B 74 (2006) 012301.

[45] U. Fano, Effects of configuration interaction on intensities and phase shifts, Phys. Rev. 124 (1961) 1866.

[46] D. Goldhaber-Gordon, H. Shtrikman, D. Mahalu, D. Abush-Magder, U. Meirav, M.A. Kastner, Kondo effect in a single-electron transistor, Nature (London) 391 (1988) 156.

[47] J. Gres, D. Goldhaber-Gordon, S. Heemeyer, M.A. Kastner, H. Shtrikman, D. Mahalu, U. Meirav, Fano resonances in electronic transport through a single-electron transistor, Phys. Rev. B 62 (2000) 2188.

[48] Y. Ji, M. Heiblum, H. Shtrikman, Transmission phase of a quantum dot with Kondo correlation near the unitary limit, Phys. Rev. Lett. 88 (2002) 076601.

[49] K. Kobayashi, H. Aikawa, S. Katsumoto, Y. Iye, Tuning of the Fano effect through a quantum dot in an Aharonov-Bohm interferometer, Phys. Rev. Lett. 88 (2002) 256806.

[50] J. Kim, J.-R. Kim, J.-O. Lee, J.W. Park, H.M. So, N. Kim, K. Kang, K.-H. Yoo, J.-J. Kim, Fano resonance in crossed carbon nanotubes, Phys. Rev. Lett. 90 (2003) 166403.

[51] B. Babic, C. Schönenberger, Observation of Fano resonances in single-wall carbon nanotubes, Phys. Rev. B 70 (2004) 195408.

[52] Y.S. Joe, A.M. Satanin, G. Klimeck, Interactions of Fano resonances in the transmission of an Aharonov-Bohm ring with two embedded quantum dots in the presence of a magnetic field, Phys. Rev. B 72 (2005) 115310.

[53] M.L. Ladron de Guevara, F. Claro, P.A. Orellana, Ghost Fano resonance in a double quantum dot molecule attached to leads, Phys. Rev. B 67 (2003) 195335.

[54] K.K. Voo, C.S. Chu, Fano resonance in transport through a mesoscopic two-lead ring, Phys. Rev. B 72 (2005) 165307.

[55] G. Hackenbroich, H.A. Weidenmller, Transmission through a quantum dot in an Aharonov-Bohm ring, Phys. Rev. Lett. 76 (1996) 110.

[56] J. Wu, B.L. Gu, H. Chen, W. Duan, Y. Kawazoe, Resonant tunneling in an Aharonov-Bohm ring with a quantum dot, Phys. Rev. Lett. 80 (1998) 1952.

[57] T. Taniguchi, M. Buttiker, Friedel phases and phases of transmission amplitudes in quantum scattering systems, Phys. Rev. B 60 (1999) 13814.

[58] H.W. Lee, Generic transmission zeros and in-phase resonances in time-reversal symmetric single channel transport, Phys. Rev. Lett. 82 (1999) 2358.

[59] H. Al-Wahsh, E.H. El Boudouti, B. Djafari-Rouhani, A. Akjouj, L. Dobrzynski, Transmission gaps and sharp resonant states in the electronic transport through a simple mesoscopic device, Phys. Rev. B 75 (2007) 125313.

[60] A. Yacoby, M. Heiblum, D. Mahalu, H. Shtrikman, Coherence and phase sensitive measurements in a quantum dot, Phys. Rev. Lett. 74 (1995) 4047.

[61] R. Schuster, E. Buks, M. Heiblum, D. Mahalu, V. Umansky, H. Shtrikman, Phase measurement in a quantum dot via a double-slit interference experiment, Nature (London) 385 (1997) 417.

[62] J.O. Vasseur, A. Akjouj, L. Dobrzynski, B. Djafari-Rouhani, E.H. El Boudouti, Photon, electron, magnon, phonon and plasmon mono-mode circuits, Surf. Sci. Rep. 54 (2004) 1.

[63] H. Al-Wahsh, A. Akjouj, B. Djafari-Rouhani, L. Dobrzynski, in: J.S. Levy (Ed.), Nanostructures and Their Magnetic Properties, Research Signpost, Kerala, India 2009, p. 173.

[64] H. Al-Wahsh, A. Akjouj, B. Djafari-Rouhani, J.O. Vasseur, L. Dobrzynski, P.A. Deymier, Large magnonic band gaps and defect modes in one-dimensional comblike structures, Phys. Rev. B 59 (1999) 8709.

[65] L. Dobrzynski, Interface response theory of continuous composite materials, Surf. Sci. 180 (1987) 489.

[66] L. Dobrzynski, Interface response theory of electromagnetism in composite dielectric materials, Surf. Sci. 180 (1987) 505.

[67] L. Dobrzynski, Interface response theory of discrete composite systems, Surf. Sci. Rep. 6 (1986) 119.

[68] L. Dobrzynski, Interface response theory of continuous composite systems, Surf. Sci. Rep. 11 (1990) 139.

[69] R.C. Moul, M.G. Cottam, A macroscopic theory of the response functions for a semi-infinite Heisenberg ferromagnet, and the application to light scattering, J. Phys. C 12 (1979) 5191.

[70] H. Al-Wahsh, A. Akjouj, B. Djafari-Rouhani, A. Mir, L. Dobrzynski, Effect of pinning fields on the spin wave band gaps in comblike structures, Eur. Phys. J. B 37 (2004) 499.

[71] L. Dobrzynski, Interface-response theory of electromagnetism in dielectric superlattices, Phys. Rev. B 37 (1988) 8027.

[72] L. Dobrzynski, H. Puszkarski, Eigenvectors of composite systems. I. General theory, J. Phys. Condens. Matter 1 (1989) 1239.

[73] F. Garcia-Moliner, J. Rubio, A new method in the quantum theory of surface states, J. Phys. C 2 (1969) 1789.

[74] T.G. Phillips, H.M. Rosenberg, Spin waves in ferromagnets, Rep. Prog. Phys. 29 (1966) 285.

[75] L. Dobrzynski, J. Mendialdua, A. Rodriguez, S. Bolibo, M. More, Interface response theory of composite elastic media, J. Phys. (France) 50 (1989) 2563.

[76] A. Akjouj, B. Sylla, L. Dobrzynski, Introductiona la théorie des systemes composites: exemples simples de matériaux lamellaires, Ann. Phys. (Paris) 18 (1993) 363.

[77] M.L. Bah, A. Akjouj, L. Dobrzynski, Response functions in layered dielectric media, Surf. Sci. Rep. 16 (1992) 97–131.

[78] W. Porod, Z. Shao, C.S. Lent, Transmission resonances and zeros in quantum waveguides with resonantly coupled cavities, Appl. Phys. Lett. 61 (1992) 1350.

[79] L. Dobrzynski, B. Djafari-Rouhani, P. Zielinski, A. Akjouj, B. Sylla, E. Oumghar, Resonant Phonons in Adsorbates, Acta Phys. Pol. A 89 (1996) 139.

[80] P.A. Deymier, E. Oumghar, J.O. Vasseur, B. Djafari-Rouhani, L. Dobrzynski, Simple models of adsorbed polymers: Vibrational properties, Prog. Surf. Sci. 53 (1997) 179.

[81] J.O. Vasseur, P.A. Deymier, B. Djafari-Rouhani, L. Dobrzynski, A. Akjouj, Absolute band gaps and electromagnetic transmission in quasi-one-dimensional comb structures, Phys. Rev. B 55 (1997) 10434.

[82] L. Dobrzynski, A. Akjouj, B. Djafari-Rouhani, J.O. Vasseur, J. Zemmouri, Giant gaps in photonic band structures, Phys. Rev. B 57 (1998) R9388.

[83] M.S. Kushwaha, A. Akjouj, B. Djafari-Rouhani, L. Dobrzynski, J.O. Vasseur, Acoustic spectral gaps and discrete transmisson in slender tubes, Solid State Commun. 106 (1998) 659.

[84] E.N. Economou, Green's Functions in Quantum Physics, Springer-Verlag, Berlin, 1983, p. 79.

[85] M.G. Cottam, A magnetostatic theory of the response functions for a finite-thickness ferromagnetic slab and the application to Brillouin scattering, J. Phys. C 12 (1979) 1709.

[86] C. Kittel, Excitation of spin waves in a ferromagnet by a uniform rf field, Phys. Rev. 110 (1958) 1295.

[87] M. Büttiker, R. Landauer, Traversal time for tunneling, Phys. Rev. Lett. 49 (1982) 1739.

[88] M. Büttiker, Larmor precession and the traversal time for tunneling, Phys. Rev. B 27 (1983) 6178.

[89] E.H. Hauge, J.A. Stovneng, Tunneling times: a critical review, Rev. Mod. Phys. 61 (1989) 917.

[90] V. Laude, P. Tournois, Superluminal asymptotic tunneling times through one-dimensional photonic bandgaps in quarter-wave-stack dielectric mirrors, J. Opt. Soc. Am. B 16 (1999) 194.

[91] M.L.H. Lahlaouti, A. Akjouj, B. Djafari-Rouhani, L. Dobrzynski, M. Hammouchi, E.H. El Boudouti, A. Nougaoui, B. Kharbouch, Theoretical analysis of the density of states and phase times: Application to resonant electromagnetic modes in finite superlattices, Phys. Rev. B 63 (2001) 35312.

[92] E.H. El Boudouti, N. Fettouhi, A. Akjouj, B. Djafari-Rouhani, A. Mir, J. Vasseur, L. Dobrzynski, J. Zemmouri, Experimental and theoretical evidence for the existence of photonic bandgaps and selective transmissions in serial loop structures, J. Appl. Phys. 95 (2004) 1102.

[93] W.M. Robertson, J. Pappafotis, P. Flannigan, Sound beyond the speed of light: Measurement of negative group velocity in an acoustic loop filter, Appl. Phys. Lett. 90 (2007) 014102.

[94] S.O. Demokritov, A.N. Slavin (Eds.), Magnonics: From Fundamentals to Applications, vol. 125, Springer, Berlin, 2013.

[95] M. Krawczyk, D. Grundler, Review and prospects of magnonic crystals and devices with reprogrammable band structure, J. Phys. Condens. Matter 26 (2014) 123202.

[96] J.D. Joannopoulos, S.G. Johnson, J.N. Winn, R.D. Meade, Photonic Crystals: Molding the Flow of Light, second ed., Princeton University Press, Princeton, NJ, 2008.

[97] S.L. Vysotskii, S.A. Nikitov, Y.A. Filimonov, Magnetostatic spin waves in two-dimensional periodic structures, JETP 101 (2005) 547.

[98] D.S. Deng, X.F. Jin, R. Tao, Magnon energy gap in a periodic anisotropic magnetic superlattice, Phys. Rev. B 66 (2002) 104435.

[99] J.C. Slonczewski, Current-driven excitation of magnetic multilayers, J. Magn. Magn. Mater. 159 (1996) L1.

[100] V.L. Zhang, F.S. Ma, H.H. Pan, C.S. Lin, H.S. Lim, S.C. Ng, M.H. Kuok, S. Jain, A.O. Adeyeye, Observation of dual magnonic and phononic bandgaps in bi-component nanostructured crystals, Appl. Phys. Lett. 100 (2012) 163118.

[101] M.A. Morozova, A.Y. Sharaevskaya, A.V. Sadovnikov, S.V. Grishin, D.V. Romanenko, E.N. Beginin, Y.P. Sharaevskii, S.A. Nikitov, Band gap formation and control in coupled periodic ferromagnetic structures, J. Appl. Phys. 120 (2016) 223901.

[102] V.L. Zhang, H.S. Lim, S.C. Ng, M.H. Kuok, X. Zhou, A.O. Adeyeye, Spin-wave dispersion of nanostructured magnonic crystals with periodic defects, AIP Adv. 6 (2016) 115106.

[103] A. Encinas-Oropesa, M. Demand, L. Piraux, I. Huynen, U. Ebels, Dipolar interactions in arrays of nickel nanowires studied by ferromagnetic resonance, Phys. Rev. B 63 (2001) 104415; R. Arias, D.L. Mills, Theory of spin excitations and the microwave response of cylindrical ferromagnetic nanowires, Phys. Rev. B 63 (2001) 134439.

[104] V.L. Zhang, H.S. Lim, C.S. Lin, Z.K. Wang, S.C. Ng, M.H. Kuok, S. Jain, A.O. Adeyeye, M.G. Cottam, Ferromagnetic and antiferromagnetic spin-wave dispersions in a dipole-exchange coupled bi-component magnonic crystal, Appl. Phys. Lett. 99 (2011) 143118.

[105] A. Akjouj, A. Mir, B. Djafari-Rouhani, J.O. Vasseur, L. Dobrzynski, H. Al-Wahsh, P.A. Deymier, Giant magnonic band gaps and defect modes in serial stub structures: application to the tunneling between two wires, Surf. Sci. 482–485 (2001) 1062.

[106] H. Al-Wahsh, E.H. El Boudouti, B. Djafari-Rouhani, A. Akjouj, T. Mrabti, L. Dobrzynski, Evidence of Fano-like resonances in mono-mode magnetic circuits, Phys. Rev. B 78 (2008) 075401.

[107] K.J. Boller, A. Imamoglu, S.E. Harris, Observation of electromagnetically induced transparency, Phys. Rev. Lett. 66 (1991) 2593.

[108] M. Fleischhauer1, A. Imamoglu, J.P. Marangos, Electromagnetically induced transparency: Optics in coherent media, Rev. Mod. Phys. 77 (2005) 633.

[109] V.A. Fedotov, M. Rose, S.L. Prosvirnin, N. Papasimakis, N.I. Zheludev, Sharp trapped-mode resonances in planar metamaterials with a broken structural symmetry, Phys. Rev. Lett. 99 (2007) 147401.

[110] S. Zhang, D.A. Genov, Y. Wang, M. Liu, X. Zhang, Plasmon-induced transparency in metamaterials, Phys. Rev. Lett. 101 (2008) 047401.

[111] S. Fan, J.D. Joannopoulos, Analysis of guided resonances in photonic crystal slabs, Phys. Rev. B 65 (2002) 235112.

[112] M.F. Yanik, W. Suh, Z. Wang, S. Fan, Stopping light in a waveguide with an all-optical analog of electromagnetically induced transparency, Phys. Rev. Lett. 93 (2004) 233903.

[113] L. Maleki, A.B. Matsko, A.A. Savchenkov, V.S. Ilchenko, Tunable delay line with interacting whispering-gallery-mode resonators, Opt. Lett. 29 (2004) 626.

[114] D.D. Smith, H. Chang, K.A. Fuller, A.T. Rosenberger, R.W. Boyd, Coupled-resonator-induced transparency, Phys. Rev. A 69 (2004) 063804.

[115] E.H. El Boudouti, T. Mrabti, H. Al-Wahsh, B. Djafari-Rouhani, A. Akjouj, L. Dobrzynski, Transmission gaps and Fano resonances in an acoustic waveguide: analytical model, J. Phys. Condens. Matter 20 (2008) 255212.

[116] A. Santillan, S.I. Bozhevolnyi, Acoustic transparency and slow sound using detuned acoustic resonators, Phys. Rev. B 84 (2011) 064304.

[117] A. Merkel, G. Theocharis, O. Richoux, V. Romero-García, V. Pagneux, Control of acoustic absorption in one-dimensional scattering by resonant scatterers, Appl. Phys. Lett. 107 (2015) 244102.

[118] A. Mouadili, E.H. El Boudouti, A. Soltani, A. Talbi, A. Akjouj, B. Djafari-Rouhani, Theoretical and experimental evidence of Fano-like resonances in simple monomode photonic circuits, J. Appl. Phys. 113 (2013) 164101.

[119] A. Mouadili, E.H. El Boudouti, A. Soltani, A. Talbi, B. DjafariRouhani, A. Akjouj, K. Haddadi, Electromagnetically induced absorption in detuned stub waveguides: a simple analytical and experimental model, J. Phys. Condens. Matter 26 (2014) 505901.

[120] Y.S. Joe, A.M. Satanin, G. Klimeck, Interactions of Fano resonances in the transmission of an Aharonov-Bohm ring with two embedded quantum dots in the presence of a magnetic field, Phys. Rev. B 72 (2005) 115310.

[121] M.L. Ladron de Guevara, F. Claro, P.A. Orellana, Ghost Fano resonance in a double quantum dot molecule attached to leads, Phys. Rev. B 67 (2003) 195335.

[122] C.W. Hsu, B. Zhen, A.D. Stone, J.D. Joannopoulos, M. Soljačić, Bound states in the continuum, Nat. Rev. Mater. 1 (2016) 16048.

[123] P. Zielinski, A. Kulak, L. Dobrzynski, B. Djafari-Rouhani, Propagation of waves and chaos in transmission line with strongly anharmonic dangling resonator, Eur. Phys. J. B 32 (2003) 73.

FURTHER READING

[124] C.H. Raymond Ooi, C.H. Kam, Controlling quantum resonances in photonic crystals and thin films with electromagnetically induced transparency, Phys. Rev. B 81 (2010) 195119.

[125] B.-B. Li, Y.-F. Xiao, C.-L. Zou, X.-F. Jiang, Y.-C. Liu, F.-W. Sun, Y. Li, Q. Gong, Experimental controlling of Fano resonance in indirectly coupled whispering-gallery microresonators, Appl. Phys. Lett. 100 (2012) 021108.

Chapter 3

Magnon Mono-Mode Circuits: Serial Loop Structures

Housni Al-Wahsh*,†, Abdellatif Akjouj*, Leonard Dobrzyński*
and Bahram Djafari-Rouhani*

*Department of Physics, Faculty of Sciences and Technologies, Institute of Electronics, Microelectronics and Nanotechnology, UMR CNRS 8520, Lille University, Villeneuve d'Ascq Cedex, France †Faculty of Engineering, Benha University, Cairo, Egypt

Chapter Outline

3.1 Introduction	111	
3.2 Symmetric Serial Loop Structures	114	
3.2.1 Theoretical Model	114	
3.2.2 Numerical Results and Discussion	119	
3.2.3 Summary	127	
3.3 Asymmetric Serial Loop Structures	128	

3.4 Stop Bands and Defect Modes in a Magnonic Chain of Cells Showing Single-Cell Spectral Gaps   131
3.5 General Conclusions and Prospectives   138
Acknowledgments   138
References   138

3.1 INTRODUCTION

An exciting development in materials science is the appearance of new samples of alternating thin layers of two different materials, with the thickness and composition of each element subject to precise control. The resulting entity may possess new physical properties. Most particularly, by means of sputtering technique, specimens were prepared from two metals, each of which is present as a layer with thickness from a few angstroms to several hundred angstroms [1–6]. The nature of the spin waves (and their particle-like analog, magnons) spectrum of such systems started to be studied theoretically [2, 3, 7–10] and experimentally by light scattering [11]. Since these early works on magnetic superlattices, very exciting developments followed. Our aim here is not to review them here. Let us mostly cite those who lead to giant magneto

Magnonics. https://doi.org/10.1016/B978-0-12-813366-8.00003-0

resistance and to Albert Fert [12] and Peter Grunberg [13] Nobel prize. Recent experimental developments [14–19], enabling to master the scattering and transmission of surface spin waves, start paving the ways for the use of magnons for circuits.

Low-dimensional spin systems (i.e., magnetic configuration with a dimensionality less than three) have attracted enthusiastic attention in the past few years [20–23]. This was related to both the fundamental interest and the potential applications of spin-transport devices, and is supported by the advanced progress in nanofabrication technology [24, 25]. Arrays of very long ferromagnetic nanowires of (e.g., Ni, permalloy, Co) with diameters in the range of 30–500 nm have been created [26, 27]. These are very uniform in cross-section, with lengths in the range of 20 μm. Thus they are realizations of nanowires which one can reasonably view as infinite in length, to excellent approximation. In addition to the static and spin-transport properties of magnetic nanowire arrays, the dynamical properties of magnetic nanostructures are also of considerable interest in both fundamental research as well as applied research [27]. The study of spin waves is a powerful tool for probing the dynamic properties of magnetic media in general and those of laterally patterned magnetic structures in particular [28]. Spin waves in magnetic thin films have attracted great interest in the 1960s and 1970s. In the 1980s, the research focus shifted to magnetic superlattices. Along the growth direction, these superlattices can be viewed as one-dimensional (1D) magnonic crystals. They consist of a sequence of layers with alternating magnetic properties and are probably the best-studied systems in which the spectrum of magnons has a band structure and contains band gaps [29, 30].

Spin systems, which have a regular distribution of scattering centers, have been seen to possess a distinct and interesting array of magnonic properties, perhaps most strikingly frequency band gaps within which magnons cannot propagate through the structure, called magnonic band gaps [31–40]. The interest in these band gaps is related to a number of advantages the magnonic crystals have in comparison with photonic crystals [41, 42]. The wavelength of a spin wave and, hence, the properties of such crystals depend on the external magnetic field and can be controlled by this field [43, 44]. The wavelength of propagating spin waves for a wide class of ferromagnetic materials in the microwave frequency range is on the order of tens or even hundreds of micrometers and can be decreased to the nanoscale [45, 46]. The phase and group velocities of spin waves are also functions of the structure size and the applied external field and may vary over a wide range [33]. As a rule, the velocity of spin waves is several orders of magnitude smaller than the velocity of electromagnetic waves in a given medium. Such crystals may have a planar geometry, which can be extremely important for designing integrated devices such as narrow-frequency optical or microwave information processing devices and high-speed switches.

In this chapter, we study three network structures, namely a serial loop structure (SLS) where the dangling side branches of the comb-like structure (CLS), presented in Chapter 2, are replaced by symmetric loops, asymmetric serial loop structure (ASLS) where the side branches are replaced by asymmetric loops, and a combination of the CLS and the SLS. We present results on the magnonic band structures and transmission coefficients. The geometry of the structures presented in this chapter has the peculiar property of giving rise to transmission gaps. These transmission gaps occur at particular frequencies that are related to the length and to the physical characteristics of the constituents. These frequencies broaden into absolute gaps as the loop number N increases. The results reported here demonstrate that the widths of the pass bands (and hence of the stop bands) in the serial loop magnonic band structure can be controlled by modifying appropriately the geometry and the chemical nature of the network's constituents. In addition to the excitation spectra of SLSs, we have also calculated the transmission spectra of finite ones. Finally, we address the issue of the existence of localized states in the forbidden bands of the magnonic band structure. Such localized states result from the presence of a defect segment inside the waveguide. Let us stress that in a comb structure [47], an important difficulty lies in the technical realization of the boundary condition at the free ends of the resonators, while this problem is avoided in SLSs. Let us finally mention that the study presented in this chapter is conducted within the frame of the Dobrzynski's interface response theory [48–51]. The basic concepts and the fundamental equations of this theory, and its application to deduce the necessary response operators, to study the dynamical properties of magnetic composites with interfaces, are presented in Chapter 1 of the phononic book of this series.

In Section 3.2, we deal with infinite and finite SLSs. First, we use the semiclassical torque equation for the magnetization, given in Chapter 2, and the Green function method to write down the magnetic GF for an infinite Heisenberg ferromagnetic medium. We then calculate the dispersion relation for SLSs and the transmission coefficients with and without defects. We then illustrate these analytical results by numerical examples with emphasis on the effect of the geometry on the band gap and the transmission spectrum of the networks. Finally we turn to the existence of localized modes within the gaps. In Section 3.3, we present similar results to those reviewed before for the ASLS structure where the symmetric loops are replaced by asymmetric loops. A combination of the two structures (CLS and SLS) will be discussed in Section 3.4. The main conclusions and prospectives are summarized in Section 3.5.

It is worthy to mention that the aim of this chapter is to give a comprehensive review of a simple theoretical approach to the propagation properties of magnons in 1D mono-mode circuits (SLSs); a short review of this can be found in Al-Wahsh et al. [52]. We will give here all the details necessary for graduate students to understand this fast-growing field.

3.2 SYMMETRIC SERIAL LOOP STRUCTURES

We review in this section the magnonic band structures and transmission coefficients for symmetric SLSs [53, 54].

3.2.1 Theoretical Model

3.2.1.1 One-Dimensional Infinite SLSs

The 1D infinite serial loop waveguide can be modeled as an infinite number of unit cells pasted together (see Fig. 3.1A). Each cell is composed of a finite wire (medium 1) of length d_1 in the direction x_3, connected to a symmetric loop "ring" (medium 2) of length $2d_2$ (each loop is constructed of two wires with the same length d_2). The period of the structure is $D = d_1 + d_2$. The interface domain is constituted of all the connection points between finite segments and loops. A space position along the x_3 axis in medium i belonging to the unit cell n is indicated by (n, i, x_3), where n "cell number" is an integer such that $-\infty < n < +\infty$, i the medium index, and $-d_i/2 \leq x_3 \leq +d_i/2$. Due to the translational periodicity of the system in the direction x_3 one can define a wave vector k along the axis of the waveguide associated with the period D. With these ingredients, one can derive analytically the dispersion relation of the loop structure, as well as the transmission coefficient through a waveguide containing a finite number of loops.

First, we recall that the surface elements of the inverse of the Green's functions of a finite wire of length d_i, bounded by two free surfaces located at $x_3 = -d_i/2$ and $x_3 = +d_i/2$ are given by (see Chapter 2):

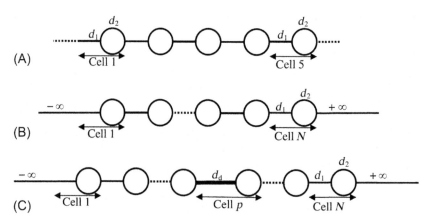

FIG. 3.1 (A) Schematic of the 1D SLS. The media are designated by an index i, with i equal 1 for the finite branch and 2 for the loop. Each loop has a length $2d_2$ and is distant by d_1 from neighboring loops. Each cell is composed of a finite branch and the loop connected to its right extreme. (B) Waveguide with finite N loops separated by a length d_1 and connected at its extremities to two semiinfinite leading lines (medium 3). (C) Same as in (B) except that a defect branch of length d_d (heavy line) is introduced in the cell p of the waveguide.

$$[g_i(M_iM_i)]^{-1} = \begin{pmatrix} A_i & B_i \\ B_i & A_i \end{pmatrix}, \tag{3.1}$$

where

$$A_i = -F_iC_i/S_i \tag{3.2}$$

$$B_i = F_i/S_i \tag{3.3}$$

$$C_i = \cosh(\alpha_i(\omega)d_i) \tag{3.4}$$

$$S_i = \sinh(\alpha_i(\omega)d_i) \tag{3.5}$$

$$F_i = D_i'\alpha_i(\omega)/\gamma_iM_i \tag{3.6}$$

$$\alpha_i(\omega) = \sqrt{-(\omega - \gamma_iH_0)/D_i'} \tag{3.7}$$

and

$$D_i' = 2Ja^2M_i/\gamma_i\hbar^2. \tag{3.8}$$

In the previous equations, M_i, H_0, ω, J, and γ_i represent, respectively, the spontaneous magnetization, the static external field in the x_1-direction, the frequency of the spin wave, the exchange interaction between neighboring magnetic sites in the simple cubic lattice of lattice parameter a constituting the ferromagnetic medium and the gyromagnetic ratio.

Within the total interface space of the infinite loop structure, the inverse of the matrix giving all the interface elements of its Green's function g is an infinite tridiagonal matrix formed by linear superposition of the elements $[g_i(MM)]^{-1}$. Taking into account the respective contributions of media 1 and 2 in the interface domain constituted of all the sites $(n, i, \pm d_1/2)$, this matrix takes the following form

$$[g(MM)]^{-1} = \begin{pmatrix} B_1 & A_1 + 2A_2 & 2B_2 & \cdots & \cdots & \cdots \\ \cdots & 2B_2 & A_1 + 2A_2 & B_1 & \cdots & \cdots \\ \cdots & \cdots & B_1 & A_1 + 2A_2 & 2B_2 & \cdots \\ \cdots & \cdots & \cdots & 2B_2 & A_1 + 2A_2 & B_1 \end{pmatrix}. \tag{3.9}$$

Taking advantage of the translational periodicity of the system in the direction x_3, the previous matrix can be Fourier transformed as

$$[g(k, MM)]^{-1} = \begin{pmatrix} A_1 + 2A_2 & B_1 + 2B_2e^{-jkD} \\ B_1 + 2B_2e^{jkD} & A_1 + 2A_2 \end{pmatrix}, \tag{3.10}$$

where k is the modulus of the 1D propagation vector. In this space, the Green's function of the infinite loop structure is obtained by inverting the previous matrix, that is,

$$g(k, MM) = \frac{1/2}{\cos(kD) - \xi} \begin{pmatrix} Y_1 & \frac{S_2}{2F_2} + \frac{S_1}{F_1}e^{-jkD} \\ \frac{S_2}{2F_2} + \frac{S_1}{F_1}e^{jkD} & Y_1 \end{pmatrix}, \tag{3.11}$$

where

$$\xi = C_1 C_2 - \frac{1}{2} \left(\frac{F_1}{2F_2} + \frac{2F_2}{F_1} \right) S_1 S_2 \qquad (3.12)$$

and

$$Y_1 = \frac{C_2 S_1}{F_1} + \frac{C_1 S_2}{2F_2}. \qquad (3.13)$$

The dispersion relation of the infinite loop waveguide is given by

$$\cos(kD) = \xi. \qquad (3.14)$$

It is also straightforward to Fourier analysis back into real space all the elements of $g(k, MM)$ and obtain all the interface elements of g in the following form:

$$g(n, 1, +d_1/2; n', 1, +d_1/2) = g(n, 1, -d_1/2; n', 1, -d_1/2) = Y_1 \frac{t^{|n-n'|+1}}{t^2 - 1} \qquad (3.15)$$

$$g(n, 1, +d_1/2; n', 1, -d_1/2) = \frac{S_2}{2F_2} \frac{t^{|n-n'|+1}}{t^2 - 1} + \frac{S_1}{F_1} \frac{t^{|n-n'+1|+1}}{t^2 - 1} \qquad (3.16)$$

$$g(n, 1, -d_1/2; n', 1, +d_1/2) = \frac{S_2}{2F_2} \frac{t^{|n-n'|+1}}{t^2 - 1} + \frac{S_1}{F_1} \frac{t^{|n-n'-1|+1}}{t^2 - 1}, \qquad (3.17)$$

where the integers n and n' refer to the cell number $(-\infty < n, n' < +\infty)$ on the infinite waveguide and the parameter t is given by

$$t = e^{jkD}. \qquad (3.18)$$

3.2.1.2 Transmission Coefficient of the Finite Loop Structures

We review now the transmission properties of a finite loop structure. This structure is constructed as follows: a finite piece containing N equidistant loops is cut out of the infinite periodic system illustrated in Fig. 3.1, and this piece is subsequently connected at its extremities to two semiinfinite leading lines (medium 3). The finite loop structure is therefore composed of N loops (medium 2, $2d_2$ is the length of each loop) pasted periodically with a finite segment (medium 1) of length d_1. We calculate analytically the transmission coefficient of a bulk spin-wave coming from $x_3 = -\infty$. The system of Fig. 3.1B is built of the infinite loop structure illustrated in Fig. 3.1A.

In a first step, one suppresses the segment linking the loops laying in the cell 0 and in the cell 1 as well as the segment linking the loops laying in cell N and in cell $N + 1$. For this new system composed of a finite loop structure and two semiinfinite leads, the inverse Green's function in the interface space, $[g_t(MM)]^{-1}$, is an infinite banded matrix defined in the interface domain of all sites $(n, i, \pm d_1/2)$, $-\infty < n < +\infty$. This matrix is similar to the one associated with the infinite loop structure. Only a few matrix elements differ, namely, those associated with the interface M sites $(1, 1, -d_1/2)$, $(1, 1, +d_1/2)$, $(N + 1, 1, -d_1/2)$, and $(N + 1, 1, +d_1/2)$. The cleavage operator:

$$V_{cl}(MM) = [g_t(MM)]^{-1} - [g(MM)]^{-1} \tag{3.19}$$

is the following 4×4 square matrix defined in the corresponding interface domain:

$$V_{cl}(MM) = \begin{pmatrix} -A_1 & -B_1 & 0 & 0 \\ -B_1 & -A_1 & 0 & 0 \\ 0 & 0 & -A_1 & -B_1 \\ 0 & 0 & -B_1 & -A_1 \end{pmatrix}. \tag{3.20}$$

Using Eqs. (3.15)–(3.17), (3.20), one obtains the matrix operator:

$$\Delta (MM) = I(MM) + V_{cl}(MM)g(MM) \tag{3.21}$$

in the space M. For the calculation of the transmission coefficient, we need only the matrix elements within the space M' formed by the sites $(1, 1, +d_1/2)$ and $(N + 1, 1, -d_1/2)$, which can be set in the form of the 2×2 matrix:

$$\Delta_s (M'M') = \begin{pmatrix} 1 + \frac{-t(A_1 Y_1 + B_1 Y_2')}{t^2 - 1} & \frac{-t^N(B_1 Y_1 t + A_1 Y_2)}{t^2 - 1} \\ \frac{-t^N(B_1 Y_1 t + A_1 Y_2)}{t^2 - 1} & 1 + \frac{-t(A_1 Y_1 + B_1 Y_2')}{t^2 - 1} \end{pmatrix}, \tag{3.22}$$

where

$$Y_2 = \frac{S_2}{2F_2}t + \frac{S_1}{F_1} \tag{3.23}$$

and

$$Y_2' = \frac{S_2}{2F_2} + \frac{S_1}{F_1}t. \tag{3.24}$$

The inverse surface Green's function $[d_s(M'M')]^{-1}$ of the finite loop structure in the M' space is given by

$$[d_s(M'M')]^{-1} = \Delta_s(M'M')[g_s(M'M')]^{-1}, \tag{3.25}$$

where

$$g_s(M'M') = \frac{t}{t^2 - 1} \begin{pmatrix} Y_1 & t^{N-1}Y_2 \\ t^{N-1}Y_2 & Y_1 \end{pmatrix} \tag{3.26}$$

is the matrix constituted of elements of $g(MM)$ associated with sites $(1, 1, +d_1/2)$ and $(N + 1, 1, -d_1/2)$. From Eq. (3.25), simple algebra leads to

$$[d_s(M'M')]^{-1} = \begin{pmatrix} A(N) & B(N) \\ B(N) & A(N) \end{pmatrix}, \tag{3.27}$$

where

$$A(N) = \frac{Y_1 Y_4 - t^{2N-2} Y_3 Y_2}{Y_1^2 - t^{2N-2} Y_2^2} \tag{3.28}$$

$$B(N) = \frac{-Y_2 Y_4 + Y_3 Y_1}{Y_1^2 - t^{2N-2} Y_2^2} t^{N-1} \tag{3.29}$$

$$Y_3 = C_1 - C_2 t \tag{3.30}$$

and

$$Y_4 = -\frac{1}{t} + C_1 C_2 + \frac{F_1}{2F_2} S_1 S_2. \tag{3.31}$$

In a second step, two semiinfinite leads (medium 3) are connected to the extremities of the finite loop structure. The inverse surface Green's function $[d(MM)]^{-1}$ of the finite serial loops with two connected semiinfinite leads is given by

$$[d(MM)]^{-1} = \begin{pmatrix} A(N) - F_3 & B(N) \\ B(N) & A(N) - F_3 \end{pmatrix} \tag{3.32}$$

and consequently

$$d(MM) = \frac{1}{(A(N) - F_3)^2 - B(N)^2} \begin{pmatrix} A(N) - F_3 & -B(N) \\ -B(N) & A(N) - F_3 \end{pmatrix}, \tag{3.33}$$

where F_3 is the inverse surface Green's function of the semiinfinite lead.

We now calculate the transmission coefficient with a bulk spin-wave coming from, $x_3 = -\infty$, $U(x) = e^{-\alpha_3 x_3}$. The transmission coefficient takes the form:

$$T = \left| \frac{2F_3 B(N)}{(A(N) - F_3)^2 - B^2(N)} \right|^2. \tag{3.34}$$

3.2.1.3 Transmission Coefficient of a Structure With a Defect

The following section focuses on the existence of localized modes present within the gaps when a finite wire of length d_1 is replaced by a segment of length $d_d \neq d_1$ in one cell of the waveguide (see Fig. 3.1C). The localized states in the case of an infinite loop structure ($N \to \infty$) were obtained to be given by

$$\left\{ 1 + \frac{t}{t^2 - 1} (Y_1 - Y_2')(A_d - B_d + B_1 - A_1) \right\}$$
$$\times \left\{ 1 + \frac{t}{t^2 - 1} (Y_1 + Y_2')(A_d + B_d - B_1 - A_1) \right\} = 0, \tag{3.35}$$

where A_d and B_d have the same definitions as A_i and B_i given by Eqs. (3.2), (3.3).

Assuming that the defect is located in any cell of a finite loop structure (see Fig. 3.1C), an analytical expression for the transmission factor T through the defective waveguide was found to be

$$T = \left| \frac{2F_3 B(N) B(p) B_d}{Y_5 Y_5' - B_d^2 (A(N) - F_3)(A(p) - F_3)} \right|^2, \tag{3.36}$$

where

$$Y_5 = B^2(N) - (A(N) - F_3)(A(N) + A_d) \tag{3.37}$$

and

$$Y_5' = B^2(p) - (A(p) - F_3)(A(p) + A_d). \tag{3.38}$$

The integer p, such as $1 < p \leq N$, refers to the position of the defect cell in the waveguide and the rest of the symbols have their usual meaning.

3.2.2 Numerical Results and Discussion

These analytical results will be now illustrated by a few specific examples of dispersion relations and transmission factors with and without defect for structures made out of *identical media* $(F_1 = F_2 = F_3 = F_d)$.

3.2.2.1 Magnonic Band Gaps and Transmission Spectra

Consider first a simple example, namely a waveguide consisting of a single loop. Eq. (3.34) for the transmission factor T in the case of $N = 1$ can be written as

$$T = \frac{16}{25 - 9 \cos^2(\alpha_2' d_2)}, \tag{3.39}$$

where α' is given by Eq. (3.7). This equation is identical to those given by Xia [55]. The transmission coefficient *reaches its minimum value* of 16/25 when $\cos(\alpha_2' d_2) = 0$, that is,

$$\alpha_2' d_2 = \left(m + \frac{1}{2} \right) \pi. \tag{3.40}$$

The corresponding frequency is then

$$\omega_g = \gamma_2 H_0 + D_2' \left[(m + 1/2) \frac{\pi}{d_2} \right]^2, \tag{3.41}$$

where m is a positive integer. The previous equation can be rewritten in the dimensionless form:

$$\Omega_g = H_g + [(m + 1/2)\pi]^2 \tag{3.42}$$

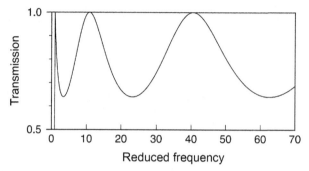

FIG. 3.2 Transmission coefficient versus reduced frequency for a waveguide with one loop in the case of identical media 2 and 3. For convenience H_g is considered to be 1.

with $\Omega_g = \omega_g d_2^2 / D_2'$ and $H_g = \gamma_2 H_0 d_2^2 / D_2'$. It is noticeable from Eq. (3.39) that T reaches its maximum value of 1 when $\alpha_2' d_2$ is a multiple of π. The variations of T versus the reduced frequency, $\Omega = \omega d_2^2 / D_2'$, are reported in Fig. 3.2.

We now turn to the numerical results of the band structure and transmission coefficient for $N > 1$. In Fig. 3.3A the first five dispersion curves are shown in the band structure of the infinite serial loop composite with $d_1 = d_2$ and $N \longrightarrow \infty$. The plots are given for the reduced frequency $\Omega = \omega d_1^2 / D_1' = \tilde{H} - \alpha_1^2 d_1^2$, with $\tilde{H} = \gamma_1 H_0 d_1^2 / D_1'$, versus the dimensionless wave vector kD ($-\pi \le kD \le +\pi$). There is a complete absolute gap below the lowest band due to the presence of the external field H_0. There exist other absolute gaps, between the first and the second bands, and the third and the fourth bands. The tangential points between the second and the third bands, on one hand, and the fourth and fifth bands, on the other hand, are degenerate points and they appear at $kD = 0$. Asymmetry is revealed between the second and the third bands as well as between the fourth and the fifth bands. This is reflected in the plot of the transmission factor. Fig. 3.3B shows the frequency dependence of the transmission for $d_2 = d_1$ and $N = 10$. The number of oscillations in the transmission factor within the pass bands, which corresponds to the second and the third bands or to the fourth and fifth bands, has been noted to be unfailingly $2N - 1$. This number is $N - 1$ within the first pass band, which corresponds to the band that has no tangential points with any other bands (see also Fig. 3.3D).

In order to study the influence of the geometry of the loop structure on its magnonic band structure, the band structure for $d_2/d_1 \neq 1$ is presented. For instance, Fig. 3.3C shows the first three bands for $d_2 = 0.3 d_1$. It is noticeable that in this case the degenerate (tangential) points are removed and the band gaps get wider. A comparison between Fig. 3.3A and C reveals that the number of dispersion curves in the reduced frequency range, going from 0 to 60, decreases when the ratio d_2/d_1 decreases. In other words, introducing such a deficiency in the geometry of the waveguide leads to a widening of the stop bands.

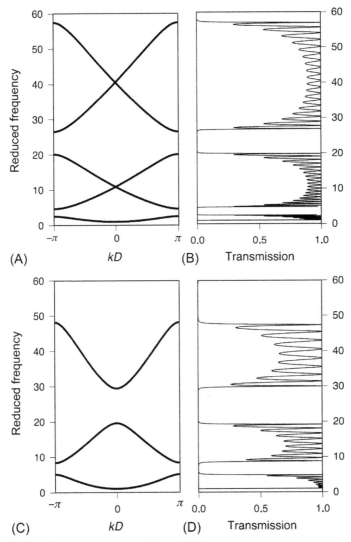

FIG. 3.3 (A) The magnonic band structure of the infinite SLS. We have chosen $d_1 = d_2$, $D'_1 = D'_2$, $\tilde{H} = 1$, and $N \to \infty$. One observes an absolute gap below the first band due to the presence of the external field H_0. (B) Transmission spectrum versus the reduced frequency for a waveguide with $N = 10$. The other parameters are the same as in (A). (C) Same as in (A) but for $d_2 = 0.3d_1$. (D) Same as in (B) but for $d_2 = 0.3d_1$.

The transmission factor is also influenced by this change of geometry. This phenomenon is illustrated in Fig. 3.3D for $N = 10$. Interestingly, the width of the pass bands (stop bands) decreases (increases) with this length d_2. Let us also underline the fact that the number N of loops in the waveguide is important

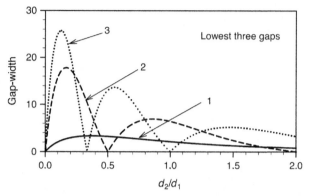

FIG. 3.4 The variation of the gap width of the first three magnonic band gaps appearing in the band structure as a function of d_2/d_1. The numbers 1 to 3 refer to the first, second, and third gaps, with increasing frequencies, in Fig. 3.3.

in achieving completely formed gaps. This number is of the order of $N \approx 6$ where $d_2 = d_1$ (similar results are obtained for $d_2/d_1 = 0.5, 1, 1.5, 2, 2.5, \ldots$), while it is of the order of $N \approx 8$ when $d_2 = 0.3\,d_1$.

Fig. 3.4 gives the evolution of the gap width of the three lowest band gaps in the loop structure of Fig. 3.3. The gap width represents the difference between the top and bottom frequencies of the absolute band gaps in function of d_2/d_1 in the range $0 < d_2/d_1 < 2$. As shown in Fig. 3.4, the second and the third gaps, respectively, close up for $d_2/d_1 = 0.5$ and 2, and $d_2/d_1 = 1/3$ and 1. All the three gaps present a maximum width for $d_2/d_1 < 1$.

Fig. 3.5 depicts the effect of the number N of loops on the transmission spectrum T for a finite loop structure with $d_2/d_1 = 1$. We show the frequency dependence of the transmission for $N = 2$, 6, and 9 in the top, middle, and bottom panels, respectively. One can see in the top panel for $N = 2$ that the transmission factor still does not reach the zero value, that is, there exists a partial transmission within the pseudo-gaps shown in this picture. It is apparent also that as N increases these pseudo-gaps in the transmission factor turn into full gaps. However, one does not need exceedingly large values of N. Indeed, at the relatively small value of $N = 6$, the gaps exist (note that the gap edge is, however, not yet sharp for $N = 6$). Moreover, for a given frequency range, there is an optimum value of the loop number above which any additional increase leaves the bands practically unaffected.

We end this section with a study of the influence of the number of loops on the behavior of the transmission factor when the ratio $d_2/d_1 \neq 1$. The results are displayed in Fig. 3.6 where $d_2/d_1 = 0.3$ is kept fixed, while N takes the values 2, 6, and 9 in the top, middle, and bottom panels, respectively. Again we notice that the shrinking (widening) of the pass bands (stop bands) reveals the same behavior as described in Fig. 3.5. For two different lengths d_1 and d_2, the

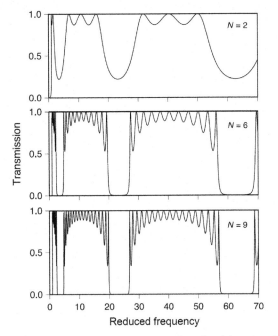

FIG. 3.5 Evolution of the transmission spectrum versus the reduced frequency for several values of N. We have chosen $\tilde{H} = 1$ and $d_1 = d_2$. The *top, middle,* and *bottom panels* depict the transmission for $N = 2, 6,$ and 9, respectively. Note that increasing N results in turning some of the pseudo-gaps into complete gaps.

oscillations amplitude decreases while the oscillation number is $N - 1$. Thus, the convergence to full gaps can be achieved in most cases for a reasonably small number of loops.

3.2.2.2 Defect Modes and Selective Transmission

The existence of localized states within the gaps when a defect is inserted in one cell of the waveguide was discussed. Using Eq. (3.35), their existence was studied as a function of the defect length. As mentioned before, the system was composed of identical media and $d_2/d_1 = 0.3$.

Fig. 3.7 gives the frequencies of the localized modes as function of the ratio d_d/d_1, where d_d is the length of the defect branch. The hatched areas correspond to the bulk bands of the perfect infinite loop structure. The frequencies of the localized modes are very sensitive to the length d_d. The localized modes emerge from the bulk bands, decrease in frequency with increasing length d_d, and finally, merge into a lower bulk band where they become resonant states. At the same time, new localized modes emerge from the bulk bands. One can notice that for any given reduced frequency in Fig. 3.7 there is a periodic repetition of the modes as functions of d_d/d_1. The transmission spectrum is also affected by

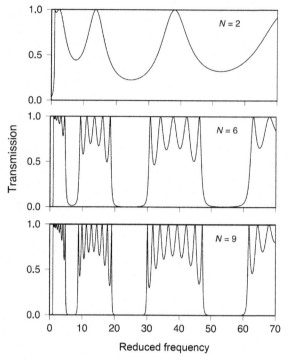

FIG. 3.6 Same as in Fig. 3.5 but for $d_2 = 0.3d_1$. Attention is drawn to the increasing of the oscillation amplitudes with this choice of the length $d_2 \neq d_1$.

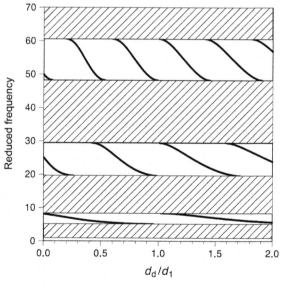

FIG. 3.7 Reduced frequencies of the localized modes associated with a defect branch of length d_d, inserted in an infinite SLS. The other parameters are $N \longrightarrow \infty$, $d_2 = 0.3d_1$, and $\tilde{H} = 1$. The system is assumed to be composed of identical media.

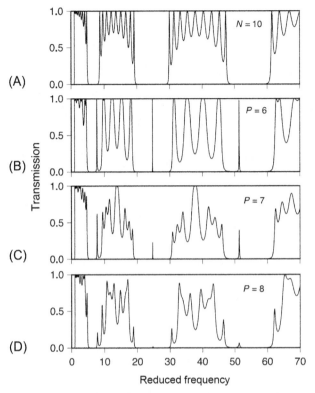

FIG. 3.8 (A) Transmission spectrum versus the reduced frequency for a 10-loop structure without defect. The other parameters are $\tilde{H} = 1$ and $d_2 = 0.3d_1$. (B) Same as in (A) but with one defect of length $d_d = 1.3d_1$ located in the middle of the waveguide, that is, in the sixth cell. (C) and (D) Same as in (B) but the defect is located in the seventh and eighth cells, respectively. In the frequency range displayed in (B), one can see that the peaks falling in the gaps are very narrow and present a strong amplitude. Transmission inside a bulk band is nontrivially affected by the presence of a defect. One notes the increase of the oscillation amplitudes. The panels (C) and (D) show that shifting the defect from the center leads to a depression in the transmission factor inside the pass bands and reduces the amplitude of the peaks associated with the localized modes.

the presence of a defect branch inside the finite loop structure. In particular, T (see Eq. 3.36) exhibits narrow peaks associated with the localized modes.

In Fig. 3.8, we compare the transmission coefficient for finite loop structures with and without defect. The results are illustrated for $N = 10$ and $d_2/d_1 = 0.3$. Fig. 3.8A presents the transmission through a nondefective waveguide in the reduced frequency range going from 0 to 70 and is similar to Fig. 3.6 (bottom panel). One observes in Fig. 3.8B that the presence of a defect branch of length $d_d = 1.3d_1$ in the middle of the waveguide (i.e., $p = 6$) gives rise to localized modes in the second, third, and fourth gaps. The localized mode inside the third forbidden band lies in the middle of the gap while the localized modes in the other two gaps lie closer to the bulk bands. The second bulk band is asymmetric due to the proximity of the first localized mode to its left side. The situation is similar for the third pass band, but the asymmetry is due to the proximity of the localized mode to its right. In the frequency range displayed in this figure, one can see that the peaks corresponding to the localized modes are very narrow. The localized mode situated in the middle of the gap is more confined, that is, the quality factor of the corresponding peak is greater. The transmission inside the pass bands is also affected nontrivially by the presence of a defect. The amplitude of the oscillations is much higher in the perturbed waveguide than in the perfect one. This behavior is due to the presence of the defect branch in the middle of the waveguide. This forces the system to behave as two linked identical loop structures waveguides with five loops. Each of these five-loop structures contributes in the same manner to the transmission depicted in Fig. 3.8B. In particular, the transmission occurring inside the pass bands through each five loops presents the same number of oscillations with identical amplitude. This can have a constructive effect on the transmission of the defective waveguide, and can explain why the oscillations are of stronger amplitude in Fig. 3.8B than those observed in Fig. 3.8A. Finally, the last two panels of Fig. 3.8 demonstrate the influence of the position of the defect unit cell p on the transmission factor. In Fig. 3.8C and D, the defect has been displaced from the center of the structure (see Fig. 3.8B) to the seventh and eighth cells, respectively. In Fig. 3.8C, there are six loops to the left of the defect and four to its right, while in Fig. 3.8D there are seven to the left of the defect and only three to its right. Unlike what is shown in Fig. 3.8B, the 10 SLS behaves here as two linked waveguides with different numbers of loops. These two linked waveguides contribute in a different manner to the transmission of the defective structure, plotted in Fig. 3.8C and D. In particular, the maxima and minima of transmission associated with each linked waveguide do not overlap. This may have a destructive effect on the transmission of the defective structure. In summary, further is the location of the defect branch from the middle of the waveguide, stronger is the depression in the transmission spectrum. In parallel, closer is the defect from the center, more confined is the peak associated with the localized state.

We end this section with a study of the evolution of the intensity of the gap states as function of the defect length d_d/d_1. The results are illustrated for three position of the defect in Fig. 3.9A. The plots are given for $N = 10$ and $d_2/d_1 = 0.3$. The hatched areas correspond to the pass bands of the structure. The dot, dash, and full lines display the defect length dependence of the gap mode intensities for $p = 6$, 7, and 8, respectively. The intensity of gap modes increases, with the centralization of the defect inside the waveguide, and vice versa. One can also notice that for $p = 6$, that is, for the defect branch placed in the middle of the structure, the intensity of the gap modes is equal to 1. This property is commonly verified when the composite system is symmetric while it may only happen under special conditions if the composite system is asymmetric. Fig. 3.9B gives the frequencies of the localized modes as function of the ratio d_d/d_1. The hatched areas correspond to the bulk bands of the perfect infinite loop structure.

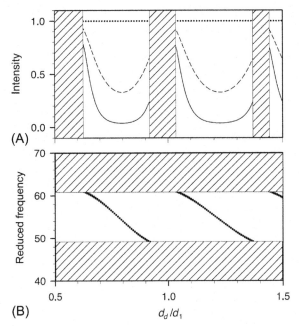

FIG. 3.9   (A) Variation of the intensity of the transmitted gap modes as function of the defect length d_d, inserted in an SLS. The defect is situated on the sixth (*dotted lines*), seventh (*dashed lines*), and eighth (*full lines*) unit cells, respectively. The results are illustrated for $N = 10$, $d_2 = 0.3d_1$, and $\tilde{H} = 1$. (B) Frequencies of the gap modes associated with a defect branch of length d_d.

3.2.3   Summary

In this section we reviewed a theoretical investigation of the magnonic band structure of 1D loop structures. Closed-form expressions were obtained for the band structure as well as for the transmission coefficients for arbitrary values of the number N of loops in the structure. Absolute band gaps exist in the spin wave band structure of an infinite loop structure. The calculated transmission coefficient of magnons in finite loop structures parallels the band spectrum of the infinite periodic loop structures. The existence of gaps in the spectra is attributed to the periodicity. In these systems, the gap width is controlled by the geometrical parameters as well as by the contrast in the physical properties of the constituents of the waveguide. Nevertheless, the magnonic band structure exhibits relatively wide gaps for homogeneous systems where the loops and the segments are constituted of the same material. Waveguides composed of finite numbers of loops exhibit a behavior similar to that of infinite periodic loop structures.

Analytical and numerical results on localized modes in perturbed waveguides were also reported. These localized modes result from the presence of a defect branch in the structure. These modes appear as narrow peaks of strong amplitude in the transmission spectrum when the defect is located in the middle of the waveguide. By changing the location of the defect in the waveguide, one affects significantly the transmission factor.

3.3 ASYMMETRIC SERIAL LOOP STRUCTURES

In this section, we propose a different geometry, called ASLS, of a mono-mode networked waveguide [56]. The structure is composed of asymmetric loops pasted together with segments of finite length (see Fig. 3.10A). Such structure may exhibit new features, in comparison with the CLSs (Chapter 2) and the symmetric SLS waveguides; for example, the existence of larger gaps, the avoidance of the constraint on the boundary condition at the end of the side branches (in the case of CLSs), appearance of quasiquantized bands with inserting a defect, achieving complete gaps for a small number of loops. These new features (which could be of potential interest in waveguide structures) are essentially due the asymmetry of the loop structure, which is quite different from the case of CLSs or SLSs. We report on results of calculated band structures and transmission coefficients. We also show that the width of the band gaps may be enlarged by coupling several ASLSs of different physical characteristics.

The 1D infinite ASLS can be modeled as an infinite number of unit cells pasted together. In each unit cell, the two arms of the ring have different lengths d_2 (of medium 2) and d_3 (of medium 3). This results in asymmetric loop of length $d_2 + d_3$ (see Fig. 3.10A), which is pasted to a segment of length d_1 (of medium 1). We focus on homogeneous ASLS where the media 1, 2, and 3 are made of the same material. The dispersion relation of the infinite ASLS, which relates the pulsation of the spin wave ω to the Bloch wave vector k, can be derived using the Green's function method [51]. It can be written as $\cos(kd) = \eta(\omega)$ where d is the period of the structure and

$$\eta(\omega) = \frac{\sin(\alpha' d_1)((5/4)\cos(\alpha' L) - (1/4)\cos(\alpha' \Delta L) - 1) + \cos(\alpha' d_1)\sin(\alpha' L)}{2\sin(\alpha' L/2)\cos(\alpha' \Delta L/2)}.$$

(3.43)

Here $L = d_2 + d_3$, $\Delta L = d_2 - d_3$, and α' is given by Eq. (3.7).

Fig. 3.10B displays the projected band structure (the plot is given as the reduced frequency $\Omega = \omega/D' = \tilde{H} + \alpha'^2$, with $\tilde{H} = \gamma H_0/D'$, versus ΔL) of an infinite ASLS for given values of L, d_1, and $\tilde{H}$ such that $L = 2$, $d_1 = 1$, and $\tilde{H} = 1$, respectively. There is a complete absolute gap below the lowest band due to the presence of the external field H_0. The shaded areas, corresponding to frequencies for which $|\eta| < 1$, represent bulk bands where spin waves are allowed to propagate in the structure. These areas are separated by minigaps where the wave propagation is prohibited. Inside these gaps, the dashed lines show the frequencies for which the denominator of Eq. (3.43) vanishes: the dashed horizontal and curved lines, which correspond to the vanishing of $\sin(\alpha' L/2)$ and $\cos(\alpha' \Delta L/2)$, respectively, define the frequencies at which the transmission through a single asymmetric loop becomes exactly equal to 0. In

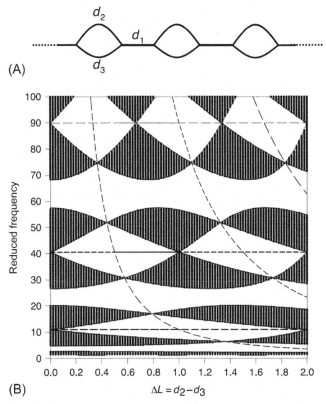

(A)

(B)

FIG. 3.10 (A) Schematic of the 1D ASLS. The 1D media constituting the loop and the finite segments are assumed to be of the same material. The lengths of the three wires are denoted d_1, d_2, and d_3, respectively. (B) Projected band structure of the ASLS as function of $\Delta L = d_2 - d_3$ for $d_1 = 1$ and $L = d_2 + d_3 = 2$. The *shaded areas* represent the bulk bands. The *dashed curves* indicate the frequencies for which the denominator of η (Eq. 3.43) vanishes.

Fig. 3.10B, one can distinguish between two types of minigaps: those of lozenge pattern that originate from the crossings of the zero transmission lines, and the gaps around $\Omega = 25$ or 60 (occurring for any value of ΔL) that are related to the periodicity of the structure. One interesting point to notice in the band structure of Fig. 3.10B, namely, at certain values of ΔL (e.g., $\Delta L \approx 0.44$), one can obtain a series of narrow minibands separated by large gaps; this is because the points at which the minibands close, align more or less vertically in such a way that a few successive bands may become very narrow.

We now turn to study the transmission coefficient. We start with a study of a simple example, namely a waveguide consisting of a unique asymmetric loop. The transmission factor T can be written as:

$$T = \left| \frac{2(S_2 + S_3)S_2S_3}{(C_2S_3 + C_3S_2 + S_2S_3)^2 - (S_2 + S_3)^2} \right|^2, \tag{3.44}$$

where C_i and S_i, $i = 2, 3$, are given by Eqs. (3.4), (3.5). The transmission is equal to 0 only when $S_2 + S_3 = 2\sin(\alpha'L/2)\cos(\alpha'\Delta L/2) = 0$. The zero frequencies, corresponding to the eigenmodes of a single loop, are given by:

$$\Omega_g = \tilde{H} + \left[\frac{(2m+1)\pi}{\Delta L}\right]^2 \tag{3.45}$$

$$\Omega_g = \tilde{H} + \left[\frac{4n\pi}{L}\right]^2, \tag{3.46}$$

where m and n are integers, and $\Omega_g = \omega_g/D'$. The variations of T versus the reduced frequency, $\Omega = \omega/D'$, are reported in Fig. 3.11B for $d_2 = 1$, $d_3 = 0.5$

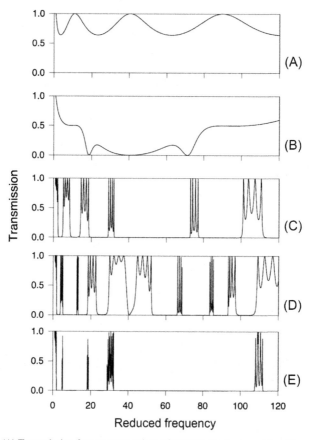

FIG. 3.11 (A) Transmission factor versus reduced frequency for a waveguide with one loop in the case of SLS ($d_2 = d_3$). (B) The same as in (A) but for ASLS with $\Delta L = 0.5$ and $L = 1.5$. For convenience $\tilde{H}$ is considered to be 1. (C) Variations of the transmission power through ASLS for $N = 5$ loops, $d_1 = 1$, $L = 2$, and $\Delta L = 0.44$. (D) The same as in (C), but for $d_1 = 1.5$, $L = 2$, and $\Delta L = 1.1$. (E) Transmission power through a tandem structure build of the above ASLSs (C) and (D) (the plot is given for $N = 20$). The superposition of the forbidden bands in (C) and (D) is well seen in (E).

(i.e., $\Delta L = 0.5$). In the case where the number of the asymmetric loops is greater than 1, the zeros of the transmission coefficient enlarge into gaps.

In the particular case of a symmetric loop ($d_2 = d_3$, i.e., $\Delta L = 0$), the transmission coefficient is given by Eq. (3.39). In contrast to the transmission coefficient of an asymmetric single loop, the transmission of a symmetric one never reaches zero values (see Fig. 3.11A). That is why, in symmetric SLS, the gaps originate only from the periodicity. On the contrary, in ASLS, the gaps are due to the conjugate effect of the periodicity and the zero transmission associated with a single asymmetric loop, which plays the role of a resonator.

The transmission rate through a finite size of ASLS containing $N = 5$ loops with $\Delta L = 0.44$, $L = 2$, and $d_1 = 1$ is reported in the middle panel of Fig. 3.11C. Clearly, the existence of wide gaps separated by narrow bands show up. Despite the finite number of loops in Fig. 3.11C, the transmission approaches zero in regions corresponding to the observed gaps in the magnonic band structure of Fig. 3.10B. It is worth noticing that the general features discussed in Fig. 3.10B are still valid for any values of d_1 and L and various ΔL. However, the shape of the band structure changes drastically for fixed values of d_1 and ΔL and various L. Fig. 3.11D shows the transmission power for another different ASLS with $d_1 = 1.5$, $L = 2$, and $\Delta L = 1.1$.

Now, by associating in tandem the above ASLSs, one obtains (Fig. 3.11E) an ultrawide gap where the transmission is canceled over a large range of frequencies going from $\Omega \simeq 35$ to $\Omega \simeq 105$. In this structure, the huge gap results from the superposition of the forbidden bands of the individual ASLS (Fig. 3.11C and D).

If a defect is included in the structure, a state can be created in the gap. A defect in ASLSs can be realized by replacing a finite wire of length d_1 by a segment of length $d_f \neq d_1$ in one cell of the waveguide. The transmission spectrum versus the reduced frequency for a structure with eight asymmetric loops and a defect segment of length $d_f = 0.1d_1$ located in the middle of the structure, is depicted in Fig. 3.12. The frequency of the defect mode inside the gap depends on the length of the defect segment whereas the intensity of the peak in the transmission spectrum depends on the number N of loops in the ASLS.

3.4 STOP BANDS AND DEFECT MODES IN A MAGNONIC CHAIN OF CELLS SHOWING SINGLE-CELL SPECTRAL GAPS

In this section, we introduce another different geometry, called *SLSs with dangling resonators*, of a mono-mode networked waveguide [57]. The structure is composed of symmetric loops (rings) pasted together with segments of finite length and connected with dangling resonators; see Fig. 3.13A. Such a structure may exhibit new features, in comparison with the CLS and the SLS waveguides; for example, the existence of larger gaps, appearance of quasiquantized bands

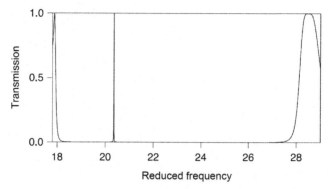

FIG. 3.12 Transmission spectrum versus the reduced frequency for a eight-loop ASLS with one defect segment of length $d_f = 0.1d_1$ located in the middle of the waveguide. The other parameters are considered to be $\tilde{H} = 1$, $d_1 = 1$, $L = 2$, and $\Delta L = 1$. In the frequency range displayed one can see that the peaks falling in the gaps are very narrow and present a strong amplitude.

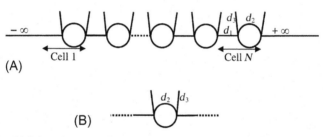

FIG. 3.13 (A) Schematic of the 1D SLS with dangling resonators studied in the present section. The media are designated by an index i, with i equal to 1 for the finite branch, 2 for the loop, and 3 for the resonators (the side branches). Each loop has a length $2d_2$ and is distant by d_1 from neighboring loops. Each cell is composed of a finite branch and the loop connected to its right extreme. The resonators are connected at the points of connection of the loop with the finite segments. (B) Waveguide with one loop, with dangling side resonators, connected at its extremities to two semiinfinite leading lines.

without inserting a defect, *achieving a complete stop band for a single loop.* These features (which could be of potential interest in waveguide structures) are essentially due to the existence of the resonators connected with the loops (which is quite different from the case of CLSs or SLSs). We report on results of calculated band structures and transmission coefficients. We also show that the width of the stop bands may be enlarged by coupling several loop structures with dangling resonators of different physical parameters.

The 1D infinite SLS with dangling resonators can be modeled as an infinite number of unit cells pasted together (see Fig. 3.13A). Each unit cell is composed of a finite wire (medium 1) of length d_1 in the direction of propagation, connected to a loop "ring" (medium 2) of length $2d_2$ (each loop is constructed of two wires with the same length d_2), and the loop is connected (at the points of connection to the finite wire of medium 1) to a finite number of dangling side segments (medium 3) of length d_3. The period of the SLS with dangling resonators structure is $D = d_1 + d_2$.

We focus on *homogeneous SLS with dangling resonators where the media 1, 2, and 3 are made of the same material.* The dispersion relation of the infinite SLS with dangling resonators, which relates the pulsation of the spin wave ω to the Bloch wave vector k, can be derived using the Green's function method [51]. It can be written as $\cos(kD) = \xi(\omega)$ where

$$\xi(\omega) = C_1C_2 + \frac{5}{4}S_1S_2 + N'\left(\frac{S_1S_2}{4}\frac{S_3}{C_3}\right)\left(N'\frac{S_3}{C_3} + 4\left[\frac{C_1}{2S_1} + \frac{C_2}{S_2}\right]\right). \quad (3.47)$$

Here C_i and S_i, $i = 1, 2, 3$, are given by Eqs. (3.4), (3.5). *Interestingly, if $N' = 0$ or $d_3 = 0$ we recover the results obtained in Section 2.2 for SLS.*

Fig. 3.14A and B displays the projected band structure of an infinite SLS with dangling resonators for given values of d_1, d_2, d_3, $\tilde{H}$, and N' such that

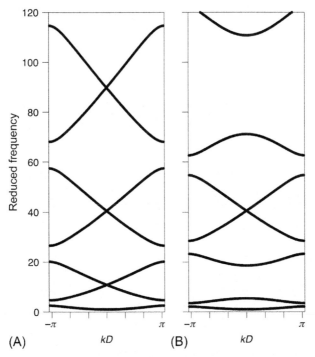

(A) kD (B) kD

FIG. 3.14 (A) The spin wave band structure of the infinite SLS without resonators ($N' = 0$). We have chosen $d_1 = 1$, $d_2 = d_1$, $\tilde{H} = 1$, and $N \to \infty$. One observes an absolute stop band below the first band due to the presence of the external field H_0. (B) The same as in (A) but for SLS with dangling resonators, with $N' = 1$, $d_3 = 0.5d_1$. Note the wide stop band observed in (B) (in comparison with (A)) due to the existence of the resonators connected to the loops.

$d_1 = 1$, $d_2 = d_1$, $d_3 = 0.5d_1$, $\tilde{H} = 1$, and $N' = 0, 1$, respectively. The plot is given as the reduced frequency $\Omega = \omega/D' = \tilde{H} + \alpha'^2$, with $\tilde{H} = \gamma H_0/D'$ versus the dimensionless wave vector kD ($-\pi \leq kD \leq +\pi$). There is a complete absolute stop band (where the propagation of spin waves is prohibited) below the lowest band due to the presence of the external field H_0. There are also other stop bands with different widths (besides the asymmetry in the pass bands) in the spectrum shown in the figure. Introducing the resonator into the waveguide (Fig. 3.14B) the degenerate points between the second and third bands and as well between the sixth and seventh bands which appear at $kD = 0$ (in Fig. 3.14A) are removed and wide stop bands are observed. *The width of these stop bands is increased with an increase the number N' of the resonators.* The above observations are also reflected in the plot of the transmission factor.

Now we turn to a study of the transmission rate through the 1D SLS with dangling resonators waveguides. We begin with a study of a simple example, namely, a structure consisting of a single loop with a finite number of resonators (see Fig. 3.14B). The transmission factor T can be written as:

$$T = \left| \frac{S_2}{1 - (C_2 + 0.5S_2 + 0.5N'S_2T_3)^2} \right|^2, \tag{3.48}$$

where $T_3 = S_3/C_3$. Again if $N' = 0$ or $d_3 = 0$ we recover the results obtained in Section 2.2 for SLS. The transmission is *equal to 0* when $S_2 = 0$ or $C_3 = 0$. The zero frequencies, corresponding to the eigenmodes of a single loop, are given by

$$\Omega_g = \tilde{H} + [m\pi/d_2]^2 \tag{3.49}$$

and

$$\Omega_g = \tilde{H} + [(2m' + 1)\pi/2d_3]^2, \tag{3.50}$$

where m and m' are integers, and $\Omega_g = \omega_g/D'$. The variations of T versus the reduced frequency, $\Omega = \omega/D'$, are reported in Fig. 3.15A for $d_2 = 1$, $d_3 = 0.5d_2$, and $N' = 1$. In the case where the number of the resonators N' is greater than 1, and the number of loops is still 1, the zeros of the transmission coefficient enlarge into stop bands. This point is shown in Fig. 3.15B for $d_2 = 1$, $d_3 = 0.5d_2$, and $N' = 2$.

In the particular case of a single loop with zero number of resonators (i.e., $N' = 0$), the inverse of the transmission coefficient becomes $T^{-1} = (25 - 9\cos^2(\alpha'd_2))/16$ (see Eq. 3.39). In contrast to the transmission coefficient of a single loop with dangling resonators, the transmission power of a single loop without resonators never reaches zero values. That's why, in SLSs, the stop bands originate only from the periodicity. On the contrary, in SLS with dangling resonators, the stop bands are due to the conjugate effect of the periodicity and the zero transmission associated with the side branches which play the role of a resonator. Table 3.1 presents the width of the region where $T \ll 0.001$ (i.e., of practically stop bands) in the case of systems studied in the pervious sections

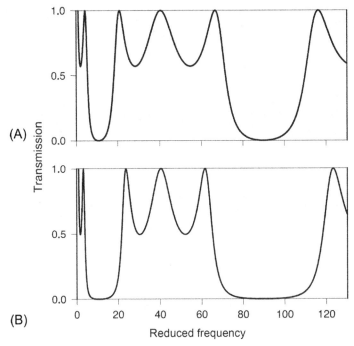

FIG. 3.15 (A) Transmission rate versus reduced frequency for a waveguide composed of one loop with $N' = 1$. (B) The same as in (A) but for $N' = 2$. For convenience $\tilde{H}$ is considered to be 1, $d_2 = 1$, and $d_3 = 0.5d_2$.

TABLE 3.1 Width of the Stop Band $\Delta\Omega$ (Which Corresponds to the Second Lowest Zero of Transmission) of the Individual Cell for the Systems Studied in Refs. [47, 53] (CLSs; SLS) and in the Present Section

System	$\Delta\Omega$
Comb-like structures with one resonator [47]	$\simeq 0.5$
Serial loop structure—one loop [53]	0
Serial loop structure with dangling resonators—one loop with one resonator	$\simeq 5$
Serial loop structure with dangling resonators—one loop with two resonators	$\simeq 12$

together with the present system. The crucial influence of the number N' of the resonators on the stop band width can be easily appreciated.

The transmission rate through a finite size of SLS with dangling resonators containing $N = 4$ loops with $N' = 3$, $d_1 = 1$, $d_2 = d_1$, and $d_3 = 0.5d_1$ is

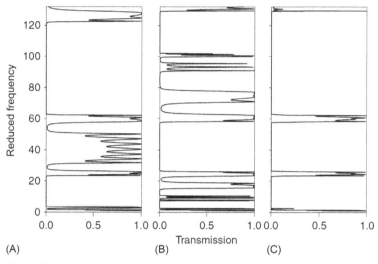

FIG. 3.16 Variations of the transmission power through an SLS with dangling resonators for $N = 4$ loops, $N' = 3$ resonators, $d_1 = 1$, $d_2 = d_1$, and $d_3 = 0.5d_1$. (B) The same as in (A), but for $d_3 = 0.75d_1$. (C) Transmission power through a tandem structure build of the above SLS with dangling resonators (A) and (B) (the plot is given for $N = 8$). The superposition of the forbidden bands in (A) and (B) is well seen in (C).

reported in Fig. 3.16A. The existence of wide stop bands separated by narrow pass bands shown clearly in this figure. Despite the finite number of loops in Fig. 3.16A, the transmission factor approaches zero in regions corresponding to the observed gaps in the magnonic band structure. Fig. 3.16B shows the transmission power for another different SLS with dangling resonators with $N = 4$, $N' = 3$, $d_1 = 1$, $d_2 = d_1$, and $d_3 = 0.75d_1$. By associating in tandem the above SLS with dangling resonators, one obtains (Fig. 3.16C) an ultra wide stop band where the transmission is canceled over a large range of frequencies going from $\Omega \simeq 63-130$. In this structure, the huge stop band results from the superposition of the forbidden bands of the individual SLS with dangling resonators (Fig. 3.16A and B).

Now, if a defect is included in the structure, a localized state can be observed in the stop band. A defect in SLS with dangling resonators can be realized by replacing a finite wire of length d_1 by a segment of length $d_f \neq d_1$ in one cell of the structure. The transmission power versus the reduced frequency for a structure with 10 loops, 3 resonators, and a defect segment of length $d_f = 0.5d_1$ located in the middle of the structure is reported in Fig. 3.17. The frequency of the defect mode inside the stop band depends on the length of the defect segment whereas the intensity of the peak in the transmission spectrum depends on the numbers N and N' in the SLS with dangling resonators. It is worth pointing out again that in all of our calculations, the cross-section of all media is considered

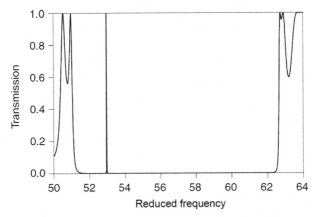

FIG. 3.17 Transmission spectrum versus the reduced frequency for a 10-loop SLS with dangling resonators, with one defect segment of length $d_f = 0.5d_1$ located in the middle of the waveguide. The other parameters considered are $\tilde{H} = 1$, $d_1 = 1$, $d_2 = 0.5d_1$, $d_3 = 0.5d_1$, and $N' = 3$. In the frequency range displayed in this figure one can see that the peaks falling in the gaps are very narrow and present a strong amplitude.

to be much smaller than the considered wavelength of the magnons, so as to neglect the quantum-size effect (or the subband structure).

In Sections 2.3 and 2.4, we have considered a different 1D mono-mode structures exhibiting very large magnonic band gaps. A theoretical investigation of the magnonic band structure of 1D ASLS and 1D SLS with dangling resonators using a Green's function method is presented. Absolute band gaps exist in the spin wave band structure of an infinite ASLS (or SLS with dangling resonators). Compared with other 1D networks such as CLS waveguides, the observed gaps in ASLS (SLS with dangling resonators) are significantly larger. The calculated transmission rate of magnons in finite loop structures parallels the band spectrum of the infinite periodic ASLS (SLS with dangling resonators). The existence of the gaps in the spectrum of ASLS (SLS with dangling resonators) is attributed to the conjugate effect of the periodicity and the zero transmission associated with a single asymmetric loop (dangling side branches) which plays the role of a resonator. In these systems, the gap width is controlled by the geometrical parameters. Numerical results on localized modes in perturbed waveguides were also reported. Since it is generally the case that magnetic periodic networks have wide technical applications, it is anticipated that this new class of materials, which can be referred to as "magnonic crystals," will turn out to be of significant value for prospective applications. One would expect such applications to be feasible in spintronic devices, since magnon excitation energies also fall in the microwave range.

Finally, an important incentive to study magnonic crystals now is their prospective application to the construction of logic systems using the wave nature of spin excitations. Many papers on this subject have been published, the discussed effects of possible practical application including wave front reversal (phase conjugation) of surface magnetostatic waves [58], spin wave interference [59–61], or the possibilities of controlling spin wave phase (an analog of the Aharonov-Bohm effect) [62]. Other prospective applications are based on negative magnetic refraction coefficient (left-handed spin waves) [63] or magnetostatic wave focusing [64, 65].

3.5 GENERAL CONCLUSIONS AND PROSPECTIVES

We presented from a tutorial point of view the state of art of a new fast-developing field, one could call magnon circuits. Indeed, all particles, bosons (magnon, photon, phonon, plasmon, etc.) and fermions (electrons, etc.) have a wave associated with them. When coherent and mono-mode propagations of such waves is made possible in linear guides, many new circuits and devices can be made out of such guides. We discussed only a few circuits actually under investigation. A simple model of the tunneling between two wires and of a magnon multiplexer was also proposed [66]. No doubt that many other devices will be conceived in the future decades, as well with micro as with nano dimensions. One can start and imagine even atomic structures for any specific application.

Such circuits can by designed with the help of all magnetic crystals. One kind of magnetic materials of great interest is the man-made 2D magnonic crystals also reviewed in the present book. No doubt, however, that this will be in the next decades a fast-growing field of investigation.

So this chapter looks like a starting point for many applications. Its backbone is a simple analytical theory for the propagation of coherent spin waves in mono-mode circuits. This simple approach starts to be complemented by more realistic numerical simulations and by experiments for some long-wavelength spin waves. A lot remains to be done for microwaves, especially those used in modern telecommunications. Let us finally mention that investigations on nonlinear effects in such structures are also reported; see, for example, Zielinski et al. [67].

ACKNOWLEDGMENTS

H. Al-Wahsh gratefully acknowledges the hospitality of the Université de Lille Nord de France.

REFERENCES

[1] Z.Q. Zheng, C.M. Falco, J.B. Ketterson, I.K. Schuller, Magnetization of compositionally modulated CuNi films, Appl. Phys. Lett. 38 (1981) 424.
[2] T. Jarlborg, A.J. Freeman, Magnetism of metallic superlattices, J. Appl. Phys. 53 (1982) 8041.

[3] K. Flevaris, J.B. Ketterson, J.E. Hilliard, Magnetic properties of compositionally modulated thin films, J. Appl. Phys. 53 (1982) 8046.

[4] T. Shinjo, K. Kawaguchi, R. Yamamoto, N. Hosaito, T. Takada, J. Phys. (Paris) Colloq. 45 (1984) C5-367.

[5] A.J. Freeman, J.-H. Xu, S. Ohnishi, T. Jarlborg, J. Phys. (Paris) Colloq. 45 (1984) C5-369.

[6] R. Krishnan, W. Jantz, Ferromagnetic resonance studies in compositionally modulated cobalt-niobium films, Solid State Commun. 50 (1984) 533.

[7] R.E. Camley, T.S. Rahman, D.L. Mills, Magnetic excitations in layered media: spin waves and the light-scattering spectrum, Phys. Rev. B 27 (1983) 261.

[8] F. Herman, P. Lambin, O. Jepsen, Electronic and magnetic structure of ultrathin cobalt-chromium superlattices, Phys. Rev. B 31 (1985) 4394.

[9] K. Mika, P. Grunberg, Dipolar spin-wave modes of a ferromagnetic multilayer with alternating directions of magnetization, Phys. Rev. B 31 (1985) 4465.

[10] L. Dobrzynski, B. Djafari-Rouhani, H. Puszkarski, Theory of bulk and surface magnons in Heisenberg ferromagnetic superlattices, Phys. Rev. B 33 (1986) 3251.

[11] A. Kueny, M.R. Khan, I.K. Schuller, M. Grimsditch, Magnons in superlattices: a light scattering study, Phys. Rev. B 29 (1984) 2879.

[12] M.N. Baibich, J.M. Broto, A. Fert, F. Nguyen Van Dau, F. Petroff, P. Etienne, G. Creuzet, A. Friederich, J. Chazelas, Giant magnetoresistance of (001)Fe/(001)Cr magnetic superlattices, Phys. Rev. Lett. 61 (1988) 2472.

[13] G. Binasch, P. Grunberg, F. Saurenbach, W. Zinn, Enhanced magnetoresistance in layered magnetic structures with antiferromagnetic interlayer exchange, Phys. Rev. B 39 (1989) 4828.

[14] A.V. Chumak, A.A. Serga, B. Hillebrands, M.P. Kostylev, Scattering of backward spin waves in a one-dimensional magnonic crystal, Appl. Phys. Lett. 93 (2008) 022508.

[15] A.V. Chumak, P. Pirro, A.A. Serga, M.P. Kostylev, R.L. Stamps, H. Schultheiss, K. Vogt, S.J. Hermsdoerfer, B. Laegel, P.A. Beck, B. Hillebrands, Spin-wave propagation in a microstructured magnonic crystal, Appl. Phys. Lett. 95 (2009) 262508.

[16] A.V. Chumak, A.A. Serga, S. Wolff, B. Hillebrands, M.P. Kostylev, Design and optimization of one-dimensional ferrite-film based magnonic crystals, J. Appl. Phys. 105 (2009) 083906.

[17] A.V. Chumak, A.A. Serga, S. Wolff, B. Hillebrands, M.P. Kostylev, Scattering of surface and volume spin waves in a magnonic crystal, Appl. Phys. Lett. 94 (2009) 172511.

[18] A.D. Karenowska, A.V. Chumak, A.A. Serga, J.F. Gregg, B. Hillebrands, Magnonic crystal based forced dominant wavenumber selection in a spin-wave active ring, Appl. Phys. Lett. 96 (2010) 082505.

[19] Z.K. Wang, V.L. Zhang, H.S. Lim, S.C. Ng, M.H. Kuok, S. Jain, A.O. Adeyeye, ACSNANO 4 (2010) 643.

[20] M. Greven, R.J. Birgeneau, U.J. Wiese, Monte Carlo study of correlations in quantum spin ladders, Phys. Rev. Lett. 77 (1996) 1865.

[21] S. Chakravarty, Dimensional crossover in quantum antiferromagnets, Phys. Rev. Lett. 77 (1996) 4446.

[22] D.G. Shelton, A.A. Nersesyan, A.M. Tsvelik, Antiferromagnetic spin ladders: crossover between spin S=1/2 and S=1 chains, Phys. Rev. B 53 (1996) 8521.

[23] J. Piekarewicz, J.R. Shepard, Dynamic spin response for Heisenberg ladders, Phys. Rev. B 57 (1998) 10260.

[24] P. Rai-Chaudhry (Ed.), The handbook of microlithography, micromachining and microfabrication, in: SPIE, 1996.

[25] X.F. Wang, P. Vasilopoulos, F.M. Peeters, Spin-current modulation and square-wave transmission through periodically stubbed electron waveguides, Phys. Rev. B 65 (2002) 165217.

[26] R. Arias, D.L. Mills, Theory of spin excitations and the microwave response of cylindrical ferromagnetic nanowires, Phys. Rev. B 63 (2001) 134439.

[27] A. Encinas-Oropesa, M. Demand, L. Piraux, I. Huynen, U. Ebels, Dipolar interactions in arrays of nickel nanowires studied by ferromagnetic resonance, Phys. Rev. B 63 (2001) 104415.

[28] J. Jorzick, S.O. Demokritov, C. Mathieu, B. Hillebrands, B. Bartenlian, C. Chappert, F. Rousseaux, A.N. Slavin, Brillouin light scattering from quantized spin waves in micron-size magnetic wires, Phys. Rev. B 60 (1999) 15194.

[29] S.O. Demokritov, A.N. Slavin (Eds.), Magnonics: From Fundamentals to Applications Topics in Applied Physics, vol. 125, Springer, Berlin, 2013.

[30] S.O. Demokritov (Ed.), Spin Wave Confinement, Pan Stanford Publishing (Pte. Ltd.), Singapore, 2009.

[31] J.O. Vasseur, L. Dobrzynski, B. Djafari-Rouhani, H. Puszkarski, Magnon band structure of periodic composites, Phys. Rev. B. 54 (1996) 1043.

[32] M. Krawczyk, J.C. Levy, D. Mercier, H. Puszkarski, Forbidden frequency gaps in magnonic spectra of ferromagnetic layered composites, Phys. Lett. A 282 (2001) 186.

[33] S.L. Vysotskiĭ, S.A. Nikitov, Y.A. Filimonov, Magnetostatic spin waves in two-dimensional periodic structures (magnetophoton crystals), J. Exp. Theor. Phys. 101 (2005) 547.

[34] S.A. Nikitov, P. Tailhades, C.S. Tsai, Spin waves in periodic magnetic structures—magnonic crystals, J. Magn. Magn. Mater. 236 (2001) 320.

[35] Y.V. Gulyaev, S.A. Nikitov, Magnonic crystals and spin waves in periodic structures, Dokl. Akad. Nauk. 380 (2001) 469.

[36] K.-S. Lee, D.-S. Han, S.-K. Kim, Physical origin and generic control of magnonic band gaps of dipole-exchange spin waves in width-modulated nanostrip waveguides, Phys. Rev. Lett. 102 (2009) 127202.

[37] A. Figotin, I. Vitebsky, Nonreciprocal magnetic photonic crystals, Phys. Rev. E 63 (2001) 066609.

[38] D.S. Deng, X.F. Jin, R. Tao, Magnon energy gap in a periodic anisotropic magnetic superlattice, Phys. Rev. B 66 (2002) 104435.

[39] G. Gubbiotti, R. Silvani, S. Tacchi, M. Madami, M. Carlotti, Z. Yang, A.O. Adeyeye, M. Kostylev, Tailoring the spin waves band structure of 1D magnonic crystals consisting of L-shaped iron/permalloy nanowires, J. Phys. D Appl. Phys. 50 (2017) 105002.

[40] M. Krawczyk, D. Grundler, Review and prospects of magnonic crystals and devices with reprogrammable band structure, J. Phys. Condens. Matter 26 (2014) 123202.

[41] A.V. Chumak, A.A. Serga, B. Hillebrands, Magnonic crystals for data processing, J. Phys. D Appl. Phys. 50 (2017) 244001.

[42] A.A. Serga, A.V. Chumak, B. Hillebrands, YIG magnonics, J. Phys. D Appl. Phys. 43 (2010) 264002.

[43] A.V. Chumak, A.A. Serga, B. Hillebrands, Magnon transistor for all-magnon data processing, Nat. Commun. 5 (2014) 4700.

[44] A.V. Chumak, V.I. Vasyuchka, A.A. Serga, B. Hillebrands, Nat. Phys. 11 (2014) 453.

[45] M.A. Morozova, A.Yu. Sharaevskaya, A.V. Sadovnikov, S.V. Grishin, D.V. Romanenko, E.N. Beginin, Y.P. Sharaevskii, S.A. Nikitov, Band gap formation and control in coupled periodic ferromagnetic structures, J. Appl. Phys. 120 (2016) 223901.

[46] V.L. Zhang, H.S. Lim, S.C. Ng, M.H. Kuok, X. Zhou, A.O. Adeyeye, AIP Adv. 6 (2016) 115106.

[47] H. Al-Wahsh, A. Akjouj, B. Djafari-Rouhani, J.O. Vasseur, L. Dobrzynski, P.A. Deymier, Large magnonic band gaps and defect modes in one-dimensional comblike structures, Phys. Rev. B 59 (1999) 8709.

[48] L. Dobrzynski, Interface response theory of continuous composite materials, Surf. Sci. 180 (1987) 489.

[49] L. Dobrzynski, Interface response theory of electromagnetism in composite dielectric materials, Surf. Sci. 180 (1987) 505.

[50] L. Dobrzynski, Interface response theory of discrete composite systems, Surf. Sci. Rep. 6 (1986) 119.

[51] L. Dobrzynski, Interface response theory of continuous composite systems, Surf. Sci. Rep. 11 (1990) 139.

[52] H. Al-Wahsh, A. Akjouj, B. Djafari-Rouhani, L. Dobrzynski, Magnonic circuits and crystals, Surf. Sci. Rep. 66 (2011) 29.

[53] A. Mir, H. Al-Wahsh, A. Akjouj, B. Djafari-Rouhani, L. Dobrzynski, J.O. Vasseur, Spin-wave transport in serial loop structures, Phys. Rev. B 64 (2001) 224403.

[54] A. Mir, H. Al-Wahsh, A. Akjouj, B. Djafari-Rouhani, L. Dobrzynski, J.O. Vasseur, Magnonic spectral gaps and discrete transmission in serial loop structures, J. Phys. Condens. Matter. 14 (2002) 637.

[55] J.B. Xia, Quantum waveguide theory for mesoscopic structures, Phys. Rev. B 45 (1992) 3593.

[56] H. Al-Wahsh, A. Mir, A. Akjouj, B. Djafari-Rouhani, L. Dobrzynski, Magneto-transport in asymmetric serial loop structures, Phys. Lett. A 291 (2001) 333.

[57] H. Al-Wahsh, Stop bands and defect modes in a magnonic chain of cells showing single-cell spectral gaps, Phys. Rev. B 69 (2004) 012405.

[58] G.A. Melkov, V.I. Vasyuchka, A.V. Chumak, V.S. Tiberkevich, A.N. Slavin, Wave front reversal of nonreciprocal surface dipolar spin waves, J. Appl. Phys. 99 (2006) 08P513.

[59] S. Choi, K.-S. Lee, S.-K. Kim, Spin-wave interference, Appl. Phys. Lett. 89 (2006) 062501.

[60] J. Podbielski, F. Giesen, D. Grundler, Spin-wave interference in microscopic rings, Phys. Rev. Lett. 96 (2006) 167207.

[61] S.V. Vasiliev, V.V. Kruglyak, M.L. Sokolovskii, A.N. Kuchko, Spin wave interferometer employing a local nonuniformity of the effective magnetic field, J. Appl. Phys. 101 (2007) 113919.

[62] R. Hertel, W. Wulfhekel, J. Kirschner, Domain-wall induced phase shifts in spin waves, Phys. Rev. Lett. 93 (2004) 257202.

[63] D.D. Stancil, Progress in electromagnetic research symposium 2005, Hangzhou, August 23–26, 2005 (unpublished).

[64] M. Bauer, C. Mathieu, S.O. Demokritov, B. Hillebrands, P.A. Kolodin, S. Sure, H. Dotsch, V. Grimalsky, Y. Rapoport, A.N. Slavin, Direct observation of two-dimensional self-focusing of spin waves in magnetic films, Phys. Rev. B 56 (1997) R8483.

[65] V. Veerakumar, R.E. Camley, Magnon focusing in thin ferromagnetic films, Phys. Rev. B 74 (2006) 214401.

[66] A. Akjouj, A. Mir, B. Djafari-Rouhani, J.O. Vasseur, L. Dobrzynski, H. Al-Wahsh, P.A. Deymier, Giant magnonic band gaps and defect modes in serial stub structures: application to the tunneling between two wires, Surf. Sci. 482 (2001) 1062.

[67] P. Zielinski, A. Kulak, L. Dobrzynski, B. Djafari-Rouhani, Propagation of waves and chaos in transmission line with strongly anharmonic dangling resonator, Eur. Phys. J. B 32 (2003) 73.

Chapter 4

Magnons in Nanometric Discrete Structures

Housni Al-Wahsh*,†, Abdellatif Akjouj*, Leonard Dobrzyński*, Bahram Djafari-Rouhani* and El Houssaine El Boudouti*,‡

*Department of Physics, Faculty of Sciences and Technologies, Institute of Electronics, Microelectronics and Nanotechnology, UMR CNRS 8520, Lille University, Villeneuve d'Ascq Cedex, France †Faculty of Engineering, Benha University, Cairo, Egypt ‡LPMR, Department of Physics, Faculty of Sciences, University Mohammed I, Oujda, Morocco

Chapter Outline

4.1 Introduction	**143**	
4.2 Interface Response Theory	**146**	
4.3 Effects of Coupling Infinite Linear Chain of Nano-Particles to Three Local Resonators	**148**	
4.3.1 Model and Calculations	148	
4.3.2 Results and Discussion	155	
4.4 Quasibox Structures	**160**	
4.4.1 Model and Calculations	160	
4.4.2 Applications and Discussion of the Results	163	
4.4.3 Summary	167	

4.5 Magnon Nanometric Multiplexer in Cluster Chains	**167**
4.5.1 Introduction	167
4.5.2 Calculations	169
4.5.3 Applications and Discussion of the Results	171
4.5.4 Summary	177
4.6 General Conclusions and Prospectives	**177**
Acknowledgments	**178**
References	**178**

4.1 INTRODUCTION

Low-dimensional spin systems, that is, magnetic configuration with a dimensionality less than three, have attracted enthusiastic attention in the past few years [1–4]. This was related to both the fundamental interest and the potential applications of spin-transport devices, and is supported by the advanced progress in nanofabrication technology [5, 6]. Arrays of very long ferromagnetic nanowires of, for example, Ni, permalloy, and Co, with diameters in the range of 30–500 nm have been created [7, 8]. They are very uniform in cross-section, with lengths in the range of 20 μm. Thus they are realizations of nanowires

Magnonics. https://doi.org/10.1016/B978-0-12-813366-8.00004-2

which one can reasonably view as infinite in length, to excellent approximation. In addition to the static and spin-transport properties of magnetic nanowire arrays, the dynamical properties of magnetic nanostructure are also of considerable interest in both fundamental research as well as applied research [8]. The study of spin waves is a powerful tool for probing the dynamic properties of magnetic media in general and those of laterally patterned magnetic structures in particular [9]. On the other hand, due to the possible use of electron spin for storage and information transfer in quantum computers [10, 11], there have been many recent studies on spin transport in semiconductor nanostructures [12].

Spin systems which have a regular distribution of scattering centers have been seen to possess a distinct and interesting array of magnonic properties, perhaps most strikingly frequency band gaps within which magnons cannot propagate through the structure-a so-called magnonic band gap [13–26]. The interest in these band gaps is related to a number of advantages, the magnonic crystals have in comparison with photonic crystals. The wavelength of a spin wave and, hence, the properties of such crystals depend on the external magnetic field and can be controlled by this field [16, 27]. The wavelength of propagating spin waves for a wide class of ferromagnetic materials in the microwave range is on the order of tens or even hundreds of micrometers. The phase and group velocities of spin waves are also functions of the structure size and the applied external field and may vary over a wide range [17]. As a rule, the velocity of spin waves is several orders of magnitude smaller than the velocity of electromagnetic waves in a given medium. Thus, it is possible to obtain crystals with a magnon band gap whose wavelength width is on the order of several millimeters. Such crystals may have a planar geometry, which can be extremely important for designing integrated devices such as narrow-frequency optical or microwave filters and high-speed switches [27–30].

Two- and three-dimensional composite systems constituted by periodic inclusions of at least two magnetic materials in a host matrix can exhibit an absolute magnonic band gap where the propagation of spin waves is inhibited in any direction of the space [20, 21, 23, 31–34]. These magnonic band gap materials can have many practical applications such as spin injection into devices [10, 11]. Studies of lower dimensional systems, such as 1D periodic layered media [18, 19, 35–39] and periodic waveguide systems with different geometries [40–46], are conducted as analogs of 2D and 3D systems and for applications in their own right. These structures are attractive because their production is more feasible and they require only simple analytical and numerical calculations.

There is a growing realization of the enormous potential of assemblies of supported nanoclusters in the production of high-performance magnetic materials and devices (e.g., [47]). The recent technological advances and the development of highly elaborated experimental techniques have made possible nowadays to investigate certain magnetic properties of clusters directly [48]. Magnetic force microscopy and magnetization measurements have been used to investigate the arrangement of the magnetic moments within the particles [49–51]. The magnetization and resonance frequencies of submicron Fe magnetic dot arrays were investigated by Brillouin light scattering technics and magneto-optic Kerr effects [52]. The dependence of the magnetic moments of the free clusters on size and temperature was investigated using a Stern-Gerlach magnet and time-of-flight mass spectrometer [53]. The X-ray magnetic circular dichroism techniques have been used to investigate the orbital and spin magnetic moments of supported clusters [54, 55]. Brillouin scattering is also extensively used to study spin excitations in magnetic systems including nanoarrays [56–59]. Let us also mention that coherent inelastic light scattering from a microwave-excited array of magnetic particles were used by Grimsditch et al. to show that it should be possible to drive specific spin modes into nonlinear regimes [60]. Information about the uniaxial anisotropy experienced by the cluster spins is reported in Refs. [61–69]. It turns out that the uniaxial anisotropy experienced by the magnetic cluster spins (such as Mn_{12}, Fe_8, and Mn_4) is very strong. Consequently, the electronic spin-lattice relaxation time becomes very long at low temperatures and the cluster spins become frozen at temperatures of the order of 1 K, that is, much higher than the ordering temperatures $T_c \sim 0.1$ K expected on the basis of the intercluster dipolar couplings [62, 63]. However, for the magnetic cluster compound $Mn_6O_4Br_4(Et_2dbm)_6$, the anisotropy experienced by the total spin of each cluster is so small that spin-lattice relaxation remains fast down to the lowest temperatures, thus enabling dipolar order to occur within experimental times at $T_c \sim 0.16$ K [61]. Although quantum tunneling of these cluster spins has been observed [61–69], and could in principle provide a relaxation path toward the magnetically ordered equilibrium state [62, 63], the associated rates in zero field are extremely small (<100 Hz). For these systems, tunneling becomes effective only when strong transverse fields are applied to increase the tunneling rate. On the other hand, the determination of the internal magnetic structures of nanoparticles is also of fundamental importance for practical applications. In this sense, a new technique was proposed by d'Albuquerque e Castro et al. to investigate the phase diagrams of nanosized systems [70, 71]. Their method is based on a scaling relation when starting from the phase diagrams of much smaller systems. A developed method was reported by Khater and Abou Ghantous to calculate the spin mode energies of a magnetic cluster on the surface of a magnetic substrate [72, 73]. They also investigated the presence of magnetic excitations localized on clusters supported on the surface of a magnetically ordered substrate. These recent developments encouraged us to investigate magnetic excitations in networks composed of nanometric magnetic cluster chains. The motivation behind this investigation is to demonstrate the control of the zeros and ones of transmission which have been substantiated through the transmission spectra of magnons obtained within the interface response theory [74–79].

One aim of the present work is to give a comprehensive review of a simple theoretical approach to the propagation properties of magnons in 1D mono-mode nanometric circuits [80–83]. The interest in these structures is related to quantum effects, which become pronounced for particle sizes in the nanometer range, and is one of the reasons why nanostructured materials are often considered to be the key for future technologies [84, 85]. On the other hand, the progress in nanofabrication technology have made possible the preparation of regular arrays of magnetic particles of different shapes and sizes [49, 50, 86, 87]. These recent developments might make possible the production of ultrahigh-density magnetic storage devices, in which each bit is represented by exactly one magnetic nanoparticle. We will give here the details necessary for undergraduate students to understand this new and fast-growing field.

In Section 4.3, we review first the appealing possibility of designing a magnon filter exhibiting narrow peaks or narrow dips in the magnon transmission spectrum. This structure is composed of a linear chain of nanoparticles bearing a permanent magnetic moment and three additional resonators coupled with the chain. This study is conducted within the frame of the Interface Response Theory [74–79] which we recall briefly in Section 4.2. This theory allows the calculation of the Green's functions of a network structure in terms of the Green's functions of its elementary constituents. Two other network structures are then considered, namely a quasibox structure, and a multiplexer structure composed of a nanometric resonant structure coupled with three semiinfinite linear chains. The first one is shown in Section 4.4 to give rise to sharp peaks (dips) and absolute gaps in the transmission spectra of magnons. The second one is investigated in Section 4.5, where we demonstrate the possibility of designing a simple multiplexing device that can transfer one magnon frequency from one chain to the other.

4.2 INTERFACE RESPONSE THEORY

We will address the propagation of spin waves in composite systems composed of one-dimensional cluster chains adsorbed (grafted) on different substrates. This study is performed with the help of the Interface Response (Discrete) Theory [74–79], which permits to calculate the Green's function of any composite material. In what follows, we present the basic concepts and the fundamental equations of this theory (Discrete version).

Let us consider any composite material contained in its space of definition D and formed out of N different homogeneous pieces situated in their domains $D_i(1 \leq i \leq N)$. Each piece is bounded by an interface M_i, adjacent in general to J other pieces through subinterface domains M_{ij} ($1 \leq j \leq J$). The ensemble of all these interface spaces M_i is called the interface space M of the composite material.

Here we consider the discrete theory designed for problems using matrix formulations for linear operators H_i. The starting point is an infinite homogenous material i described by an infinite matrix $[(E + j\varepsilon)I - H_i]$, where E is the eigenenergy, I is the identity matrix, $j = \sqrt{-1}$, and ε is an infinitesimally positive small number. The inverse of this matrix is called the corresponding Green's function G_i and

$$[(E + j\varepsilon)I - H_i]G_i = I. \tag{4.1}$$

One cuts out of this medium a finite one with free surfaces in its space D_i with the help of a cleavage operator V_i in the interface space M_i. One defines A_{si} as the truncated part within D_i of $A_i = V_i G_i$. In the same manner one constructs also G_{si} out of the truncated part of the G_i. One then find the block diagonal matrices G and A_s by juxtaposition of, respectively, all the G_{si} and A_{si} defined for N different homogenous materials i. A composite material is then constructed by assembling such finite media with the help of a coupling operator V_I defined in the whole interface space M. One defines then in the whole space D of the composite the matrices:

$$A = A_s + V_I G \tag{4.2}$$

and in the interface space M:

$$\Delta(MM) = I(MM) + A(MM). \tag{4.3}$$

The elements of the Green's function $g(DD)$ of any composite material can be obtained from [74–76]:

$$g(DD) = G(DD) - G(DM)[\Delta(MM)]^{-1}A(MD). \tag{4.4}$$

The new interface states can be calculated from [74, 75]:

$$\det[\Delta(MM)] = 0 \tag{4.5}$$

showing that, if one is interested in calculating the interface states of a composite, one only needs to know $\Delta(MM)$ in the interface space M.

The density of states n_i corresponding to H_i can then be obtained from the imaginary part of the trace of G_i, namely:

$$n_i(E) = -\frac{1}{\pi} Im Tr G_i(E). \tag{4.6}$$

Moreover, if $U(D)$ [79] represents an eigenvector of the reference system formed by all the infinite materials i, Eq. (4.4) enables one to calculate the eigenvectors $u(D)$ of the composite material:

$$u(D) = U(D) - U(M)[\Delta(MM)]^{-1}A(MD). \tag{4.7}$$

In Eq. (4.7), $U(D)$, $U(M)$, and $u(D)$ are row-vectors. Eq. (4.7) enables also to calculate all the waves reflected and transmitted by the interfaces as well as the reflection and the transmission coefficients of the composite system. In this case, $U(D)$ must be replaced by a bulk wave launched in one homogeneous piece of the composite material [79].

4.3 EFFECTS OF COUPLING INFINITE LINEAR CHAIN OF NANO-PARTICLES TO THREE LOCAL RESONATORS

4.3.1 Model and Calculations

In Ref. [88], Rivkin et al. studied theoretically the magnon propagation along linear chain of dipole-coupled clusters. The object of this section is to investigate the magnon propagation in a nanometric magnetic cluster chain and focus on the effects of a few additional clusters near the chain. We take into account the dipole-dipole interactions between the nearest-neighbor cluster local magnetic moments. We show that an appropriate choice of the geometrical or magnetic parameters of the additional clusters can lead either to narrow peaks or to narrow dips in the transmission spectrum of magnons along the cluster chain. Such a device can be useful as a selecting or rejecting magnon filter.

We consider such a linear chain of nanoparticles bearing a permanent magnetic moment (as depicted in Fig. 4.1A) and study the effect on the magnon transmission spectrum of three additional clusters coupled with the chain (as depicted in Fig. 4.2). The effect of coupling the infinite wire to such a local resonator is to induce peaks and dips (or zeros) in the transmission coefficient. The main purpose of this section is to discuss the possibility of narrow peaks or narrow dips in the magnon transmission by selecting appropriately the geometrical or magnetic parameters of the problem [80, 81]. As shown in Fig. 4.2, we call d the distance between two clusters in the infinite chain and in the three cluster chain and d_1 the distance between the additional clusters and the wire.

Our strategy to model the magnetic properties of this device is similar to the one Rivkin et al. [88] used for another problem. The model is based on simply summing the magnetic field from discrete, point dipoles associated with each cluster. Thus the field on a dipole i is given by summing over the fields produced by the remaining dipoles, l, which is given by the usual form

$$\mathbf{h}_i^d = \sum_{l \neq i} \mathbf{m}_l \left[\nabla \nabla \left(\frac{1}{r_{il}} \right) \right] = \sum_{l \neq i} \left[\frac{3\mathbf{r}_{il}(\mathbf{m}_l \cdot \mathbf{r}_{il})}{r_{il}^5} - \frac{\mathbf{m}_l}{r_{il}^3} \right], \qquad (4.8)$$

where $\mathbf{r}_{il}$ is the vector joining the sites i and l and $\mathbf{m}_l$ the magnetic moment on cluster l. Because of the r_{il}^{-3} dependance of $\mathbf{h}_i^d$, nearest-neighbor interactions are dominant [80, 81]. Depending on the relative importance of the magnetic dipole-dipole and exchange interactions, different models for the magnetic behavior need to be employed. For instance, for small enough values of the excitation wave vector (typically less than about 10^7 m^{-1} in a ferromagnet) dipolar effects are dominant for the dynamics, and magnetostatic modes are the resulting excitations that propagate in such structures [89]. On the other hand, at large enough wave vectors, typically greater than 10^8 m^{-1}, the exchange interaction, which thus provide the restoring force for spin waves, will be dominant [89]. In this work, we adopt the nearest-neighbor interaction model for a chain with

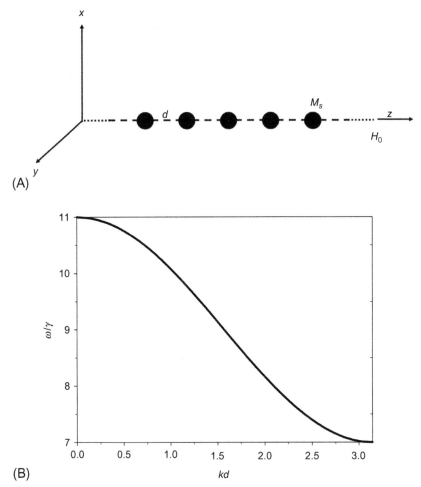

FIG. 4.1 (A) Sketch of an infinite linear cluster chain. The distance between two neighboring clusters is d, the static magnetic moments is $\mathbf{M}_s$, and the external magnetic field is $\mathbf{H}_0$, along the z-axis. (B) Dispersion relation (Eq. 4.16) of an infinite linear chain.

additional clusters, in order to discuss qualitatively the corresponding main physical features. We neglect the exchange interactions whose effects should be in general small (although this may not always be true for atom size clusters) although there is also interest in the intermediate (or "dipole-exchange") case, where both types of interactions influence the dynamical behavior. This is a first step toward more sophisticated calculations, in particular on the effects of longer-ranged interactions, which are expected to produce some changes in the position and shape of the transmission peaks and dips whose existence is mainly reported here.

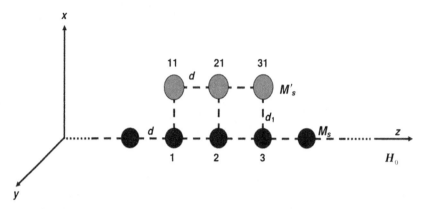

FIG. 4.2 Sketch of the geometry of the considered nanometric device. It consists of one cluster chain and three other clusters. The distances between these clusters are, respectively, d and d_1 as indicated on the figure. The static magnetic moments are $\mathbf{M}_s$ (for the chain clusters) and $\mathbf{M}'_s$ (for the three additional clusters) and the external magnetic field is $\mathbf{H}_0$, along the z-axis.

The interest in pure dipolar magnets comes from very old times; Luttinger and Tisza [90] have written in 1946: "It may certainly be said that dipole ferromagnetism, if it exists, has a character essentially different from exchange ferromagnetism." Recently it has been proved that so-called molecular magnets (such as Mn_{12} and Fe_8) are ordered by dipolar interactions only [62, 63, 91]; it is believed that they may have important future applications in magnetic memory and quantum computing. On the other hand, the small magnetic clusters conventionally used in signal processing devices mostly have nonellipsoidal shape. In such small magnetic particles the internal field nonuniformity due to the particle shape has to be taken into account, and, of course, the dominant mechanism in the creation of such a nonhomogeneity are long-range dipolar interactions. However, in the work of Prakash and Henley [92], who studied "order from disorder" phenomena, it was shown that for a model of magnetic moments arranged on a square lattice, *with dipolar interactions truncated at the nearest-neighbor distance*, the ordered structure has moments aligned along the square lattice axis of the film. Subsequent work [93, 94] has confirmed the qualitative features of this work apply to a system with full long-range dipolar interactions. This justifies our interest in studying the spin wave excitations of pure nearest-neighbor dipolar systems.

The total field on the cluster i is given by

$$\mathbf{h}_i = \mathbf{h}_i^d + \mathbf{H}_0, \tag{4.9}$$

where $\mathbf{H}_0$ is the external field. We assume here that this field $\mathbf{H}_0$ and the cluster chain are along the z-axis. The resonant magnon modes follow from the solution of the Larmor equation

$$\frac{d\mathbf{m}_i}{dt} = -\gamma \mathbf{m}_i \times \mathbf{h}_i, \tag{4.10}$$

where γ is the gyromagnetic ratio, such that $\gamma = ge/2mc$. Here g is the Landé factor, m and e, respectively, the mass and charge of the electron, and c the speed of light.

We linearize the problem by writing

$$\mathbf{m}_i = \mathbf{m}_i^{(0)} + \mathbf{m}_i^{(1)}, \tag{4.11}$$

where the $\mathbf{m}_i^{(0)}$ have only one nonzero component due to the permanent static magnetic moment $\mathbf{M}_s$ (for the chain clusters) and $\mathbf{M}_s'$ (for the three additional clusters) along the z-axis. The $\mathbf{m}_i^{(1)}$ have only nonzero components m_{ix} and m_{iy} along the x- and y-axes. Inserting Eqs. (4.9), (4.11) into Eq. (4.10) and assuming a solution of the form

$$\mathbf{m}_i^{(1)}(t) = \mathbf{m}_i^{(1)} \exp(-j\omega t), \tag{4.12}$$

where $j = \sqrt{-1}$, leads to

$$j\omega \mathbf{m}_i^{(1)} = \gamma \mathbf{m}_i^{(0)} \times \sum_{l \neq i} \left[\frac{3\mathbf{r}_{il}(\mathbf{m}_l^{(1)} \cdot \mathbf{r}_{il})}{r_{il}^5} - \frac{\mathbf{m}_l^{(1)}}{r_{il}^3} \right] + \gamma \mathbf{m}_i^{(1)} \times \mathbf{h}_i. \tag{4.13}$$

This equation has the structure of a vector eigenvalue problem. It can be solved by using the standard transformation

$$m_i^{\pm} = m_{ix} \pm i m_{iy}. \tag{4.14}$$

For the infinite cluster chain, one can use the Bloch theorem to write

$$m_i^{\pm} = m_i^{\pm(0)} e^{jkd}, \tag{4.15}$$

where k is the wave vector. Then, from Eq. (4.13), one obtains [88] a positive-frequency solution

$$\frac{\omega}{\gamma} = \pm H_0 \pm \frac{2M_s}{d^3}(2 + \cos(kd)) \tag{4.16}$$

for m_i^- (associated with the positive sign in Eq. 4.16) and the same solution but with the opposite sign for m_i^+ (associated with the negative sign in Eq. 4.16). When $\mathbf{H}_0$ is taken to be perpendicular to the cluster chain, then one obtains another dispersion relation [88] with frequencies below those given by Eq. (4.16). However, this geometry is out of scope of the present work and in the following $\mathbf{H}_0$ is assumed to be parallel to the cluster chain and strong enough in order to have the static magnetization of the clusters parallel to the z-axis (see Fig. 4.2). From Eq. (4.16) one can also note that the transmission band can be modulated with the applied magnetic field. For the photonic or acoustic crystals, the transmission band is determined by the material properties and microstructure. In contrast, as for the present structure, the magnon energy band depends not only on the material anisotropy and microstructure, but also on the magnetic configuration. So with applied transverse magnetic field, the reconstructing magnetic configuration will result in the modulation of the transmission band.

The corresponding Green's function elements associated with $m_i^{\pm}$ and the negative $(+)$ and positive $(-)$ frequency bands, respectively, are [105]

$$G_{\pm}(n, n') = \frac{t_{\pm}^{|n-n'|}}{F_{\pm}}, \tag{4.17}$$

where

$$t_{\pm} = \begin{cases} \zeta_{\pm} - \sqrt{\zeta_{\pm}^2 - 1}, & (\zeta_{\pm} > 1) \\ \zeta_{\pm} \pm j\sqrt{1 - \zeta_{\pm}^2}, & (-1 < \zeta_{\pm} < 1) \\ \zeta_{\pm} + \sqrt{\zeta_{\pm}^2 - 1}, & (\zeta_{\pm} < -1), \end{cases} \tag{4.18}$$

$$F_{\pm} = \pm \frac{M_s}{d^3}(t_{\pm} - t_{\pm}^{-1}), \tag{4.19}$$

and

$$\zeta_{\pm} = -2 \mp \frac{d^3}{2M_s}\left(\frac{\omega}{\gamma} \pm H_0\right). \tag{4.20}$$

Our aim now is to create a magnon filter with the help of additional clusters having magnetic moments $\mathbf{M}_s'$ and to couple it to the cluster chain. If one wants this device to create zeros of transmission, its magnon eigenmodes have to fall inside the bulk magnon band given by Eq. (4.16). In order to achieve this goal, it is helpful to remember that a chain where the dipole-dipole interactions are perpendicular to $\mathbf{H}_0$ has eigenmode frequencies below those of a chain where the dipole-dipole interactions are parallel to $\mathbf{H}_0$ [88]. Then, it is necessary to construct a system with a resonator having dipolar interactions parallel to those of the chain. It is easy then to check that one and two clusters do not have eigenmodes falling inside the chain bulk band. Then, it becomes obvious that the simplest system which achieves this goal is the three clusters one given in Fig. 4.2. Of course more than three clusters can be used also and in such cases more peaks and dips are expected to appear in the transmission spectra. Thanks to the geometry of Fig. 4.2 and to the consideration of nearest-neighbor interactions only, all the dipole-dipole interactions, which enter this model, are either parallel or perpendicular to the external magnetic field $\mathbf{H}_0$ and therefore this model avoids the complications due to the anisotropic feature of the dipole-dipole interactions.

Consider now as the reference system the infinite cluster chain and the three additional clusters 11, 21, and 31 having a magnetic moment $\mathbf{M}'_s$ and situated like depicted in Fig. 4.2, but without any dipole interactions between themselves, nor with the chain clusters 1, 2, and 3 in front of which they are deposited. This enables us to obtain from the earlier results the Green's function elements $G(MM)$ in the interface space M as a (12×12) matrix whose rows and columns are labeled by $(m_1^+, m_2^+, m_3^+, m_1^-, m_2^-, m_3^-, m_{11}^+, m_{21}^+, m_{31}^+, m_{11}^-, m_{21}^-, m_{31}^-)$. The inverse of this matrix is easily found to be

$$[G(MM)]^{-1}$$

$$= \begin{pmatrix} Y_+ & -Y_+t_+ & 0 & 0 & 0 & 0 & 0 & 0 & 0 & 0 & 0 & 0 \\ -Y_+t_+ & Y_+(1+t_+^2) & -Y_+t_+ & 0 & 0 & 0 & 0 & 0 & 0 & 0 & 0 & 0 \\ 0 & -Y_+t_+ & Y_+ & 0 & 0 & 0 & 0 & 0 & 0 & 0 & 0 & 0 \\ 0 & 0 & 0 & Y_- & -Y_-t_- & 0 & 0 & 0 & 0 & 0 & 0 & 0 \\ 0 & 0 & 0 & -Y_-t_- & Y_-(1+t_-^2) & -Y_-t_- & 0 & 0 & 0 & 0 & 0 & 0 \\ 0 & 0 & 0 & 0 & -Y_-t_- & Y_- & 0 & 0 & 0 & 0 & 0 & 0 \\ 0 & 0 & 0 & 0 & 0 & 0 & \Omega_+ & 0 & 0 & 0 & 0 & 0 \\ 0 & 0 & 0 & 0 & 0 & 0 & 0 & \Omega_+ & 0 & 0 & 0 & 0 \\ 0 & 0 & 0 & 0 & 0 & 0 & 0 & 0 & \Omega_+ & 0 & 0 & 0 \\ 0 & 0 & 0 & 0 & 0 & 0 & 0 & 0 & 0 & \Omega_- & 0 & 0 \\ 0 & 0 & 0 & 0 & 0 & 0 & 0 & 0 & 0 & 0 & \Omega_- & 0 \\ 0 & 0 & 0 & 0 & 0 & 0 & 0 & 0 & 0 & 0 & 0 & \Omega_- \end{pmatrix},$$
$$(4.21)$$

where

$$Y_+ = \frac{F_+}{1 - t_+^2} \tag{4.22}$$

$$Y_- = \frac{F_-}{1 - t_-^2} \tag{4.23}$$

$$\Omega_+ = \frac{\omega}{\gamma} + H_0 \tag{4.24}$$

and

$$\Omega_- = \frac{\omega}{\gamma} - H_0. \tag{4.25}$$

Now we switch on the dipolar interactions between the three additional clusters themselves and their nearest neighbors in the chain. So six clusters are perturbed by the interactions we switch on. But as to each cluster is attached two unknown magnetic moment components m_i^+ and m_i^-, this leads to 12 unknowns, to an interface space of dimension 12 and to the following (12×12) interaction matrix

$V_I(MM)$

$$= \begin{pmatrix}
-\frac{M'_s}{d_1^3} & 0 & 0 & 0 & 0 & 0 & -\frac{M_s}{2d_1^3} & 0 & 0 & -\frac{3M_s}{2d_1^3} & 0 & 0 \\
0 & -\frac{M'_s}{d_1^3} & 0 & 0 & 0 & 0 & 0 & -\frac{M_s}{2d_1^3} & 0 & 0 & -\frac{3M_s}{2d_1^3} & 0 \\
0 & 0 & -\frac{M'_s}{d_1^3} & 0 & 0 & 0 & 0 & 0 & -\frac{M_s}{2d_1^3} & 0 & 0 & -\frac{3M_s}{2d_1^3} \\
0 & 0 & 0 & \frac{M'_s}{d_1^3} & 0 & 0 & \frac{3M_s}{2d_1^3} & 0 & 0 & \frac{M_s}{2d_1^3} & 0 & 0 \\
0 & 0 & 0 & 0 & \frac{M'_s}{d_1^3} & 0 & 0 & \frac{3M_s}{2d_1^3} & 0 & 0 & \frac{M_s}{2d_1^3} & 0 \\
0 & 0 & 0 & 0 & 0 & \frac{M'_s}{d_1^3} & 0 & 0 & \frac{3M_s}{2d_1^3} & 0 & 0 & \frac{M_s}{2d_1^3} \\
-\frac{M'_s}{2d_1^3} & 0 & 0 & -\frac{3M'_s}{2d_1^3} & 0 & 0 & A & \frac{M'_s}{d^3} & 0 & 0 & 0 & 0 \\
0 & -\frac{M'_s}{2d_1^3} & 0 & 0 & -\frac{3M'_s}{2d_1^3} & 0 & \frac{M'_s}{d^3} & B & \frac{M'_s}{d^3} & 0 & 0 & 0 \\
0 & 0 & -\frac{M'_s}{2d_1^3} & 0 & 0 & -\frac{3M'_s}{2d_1^3} & 0 & \frac{M'_s}{d^3} & A & 0 & 0 & 0 \\
\frac{3M'_s}{2d_1^3} & 0 & 0 & \frac{M'_s}{d_1^3} & 0 & 0 & 0 & 0 & 0 & -A & -\frac{M'_s}{d^3} & 0 \\
0 & \frac{3M_s}{2d_1^3} & 0 & 0 & \frac{M'_s}{2d_1^3} & 0 & 0 & 0 & 0 & -\frac{M'_s}{d^3} & -B & -\frac{M'_s}{d^3} \\
0 & 0 & \frac{3M'_s}{2d_1^3} & 0 & 0 & \frac{M'_s}{2d_1^3} & 0 & 0 & 0 & 0 & -\frac{M'_s}{d^3} & -A
\end{pmatrix}, \tag{4.26}$$

where

$$A = \frac{2M'_s}{d^3} - \frac{M_s}{d_1^3} \tag{4.27}$$

and

$$B = \frac{4M'_s}{d^3} - \frac{M_s}{d_1^3}. \tag{4.28}$$

All the physical information about the model system depicted in Fig. 4.2 can then be obtained from the sum of the two previous matrices, namely from [106]

$$[g(MM)]^{-1} = [G(MM)]^{-1} + V_I(MM). \tag{4.29}$$

In particular when a progressive wave of positive frequency ω is launched from the left along the linear chain, the transmitted amplitude is given by

$$t = F_- g(1_-, 3_-) \tag{4.30}$$

and the reflected amplitude becomes

$$r = F_- g(1_-, 1_-) - 1. \tag{4.31}$$

In the earlier equations, $g(1_-, 3_-)$ and $g(1_-, 1_-)$ are Green's function elements between clusters 1 and 3 and on the cluster 1 for the m_i^- magnetic moment. These elements are obtained by inversion of the matrix given by Eq. (4.29).

4.3.2 Results and Discussion

In what follows, we illustrate the previous results by assuming for simplicity [88], $\gamma = 1$, $M_s = 1$, $d = 1$, and $H_0 = 5$. The dispersion curve (frequency as a function of reduced wave vector, Eq. 4.16) is given in Fig. 4.1B (positive kd only).

Fig. 4.3A presents the intensity transmission coefficient $T = |t|^2$ in function of ω/γ, for $d_1 = 1.5$ and $M_s' = 1$. Note the sharp zero of transmission a little above the middle of the bulk band. Such a narrow stop band could be useful to construct a rejecting signal device. Fig. 4.3C shows the variation of the phase in function of (ω/γ) for the same parameters given in Fig. 4.3A. The variation of the phase shows an abrupt change in π at the transmission zero induced by the additional clusters. The geometrical origin of this phase jump can clearly be seen in Fig. 4.3B which shows the frequency evolution of the real and imaginary parts of the transmission function $t(\omega)$. This result demonstrates that the evolution has cusp-like behavior in the vicinity of the transmission zero. Furthermore, at the origin of the coordinates $(Re(t) = 0, Im(t) = 0)$, one can easily see that the value of the transmission function $t(\omega)|_{\omega_0} = 0$. This result agrees with the

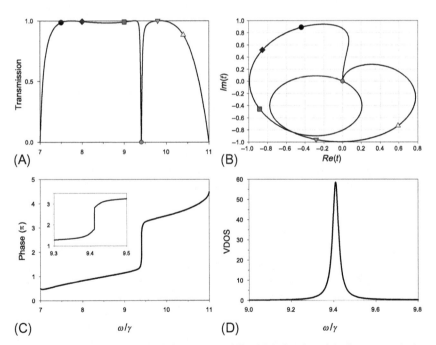

FIG. 4.3 (A) Transmission through the structure of Fig. 4.2 in function of the frequency ω/γ for $H_0 = 5$, $M_s = M_s' = 1$, $d = 1$, and $d_1 = 1.5$. (B) Evolution of real and imaginary parts of the transmission function $t(\omega)$. The parameters have the same values as in (A). (C) The same as in (A) but for the variation of the transmission phase. (D) Variation of the density of states (VDOS) versus ω/γ.

general statement given in Ref. [98] that a sharp jump of the phase by π do appear when the transmission function intersects the origin.

In addition to the information that can be deduced from the amplitude and the phase of the transmission function, one can also use the variation of the density of states (VDOS) or the difference in the DOS due to the presence of scatterer. This quantity is given by [104]

$$\Delta n(\omega) = \pi^{-1}\frac{d\eta(\omega)}{d\omega}, \qquad (4.32)$$

where

$$\eta(\omega) = -\arg(\det([g(MM)]^{-1}G(MM))) \qquad (4.33)$$

is the state phase shift. The derivative of this state phase has, within the bulk bands, the same value as the derivatives of Friedel phase [96, 103] and the transmission phase. This state phase includes also the $+\pi$ or $-\pi$ discontinuities due to the discrete states and to the localized states of the reference and final systems and enables then to check the state conservation between these two systems (see Figs. 4.4 and 4.6). The Friedel phase is related to these discontinuities indirectly when going to the Friedel sum rule, enabling to check, in particular for electrons, the charge conservation. It is worth mentioning that the difference between the Friedel phase and the phase of the transmission amplitude has been reported in several works devoted to electronic transport in mesoscopic structures using the Aharonov-Bohm systems [96, 97]. These studies are related to the investigation of the electronic states of quantum dots as well as to the understanding [95–97] of the transmission phase jumps by π between two adjacent resonances in relation with the experiments of Yacoby et al. [99]. An experimental confirmation of this analogy has been performed by some of us [100, 101] for electromagnetic waves in coaxial cables. The analogy between scattering properties of electrons, phonons, photons, and magnons suggests that this type of feature can also appear in magnonic systems [102]. However, to our knowledge, such a study has not been performed yet in magnonic crystals.

Fig. 4.3D gives the VDOS, between the final and reference systems, versus the frequency. This quantity shows the existence of a resonance, blocking the transmission, at the frequency of the zero of transmission, for which $g(1_-, 3_-)$ vanishes, see Eq. (4.30). The system acts as a rejecting filter. At this frequency, $g(1_-, 2_-)$ vanishes also. So to an incident wave on cluster 1, the clusters 1, 11, 12, and 13 respond, but not the clusters 2 and 3, so the wave is reflected. This response of the clusters 11, 12, and 13 is consistent with the positive VDOS showed in Fig. 4.3D.

In Fig. 4.4, we present the state phase $\eta(\omega)$ as a function of frequency. The plot is given for the same parameters considered in Fig. 4.3A. In the positive frequency range displayed in this figure one can remark the 3π phase drop at $(\omega/\gamma) = 5$ due to the localized states of the three isolated clusters, the π phase jump at all the positions of the localized states of the final system, and finally

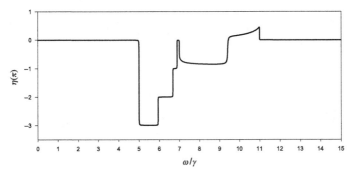

FIG. 4.4 Variation of the state phase $\eta(\omega)$ in function of the frequency ω/γ. The other parameters have the same values as in Fig. 4.3A.

the conservation of the number of states. For the negative frequency range the situation is similar due to the symmetry of the problem.

For $M'_s = 2.5$ and $d_1 = 0.9$, we obtain Fig. 4.5A which presents interesting transmission peak. This feature could be used as a transmission filter for a magnon signal. This is also shown in the plot describing the variation of the phase (Fig. 4.5B). The VDOS is given in Fig. 4.5C. It shows a resonance peak situated at the frequency for which the module of $F_{-}g(1_-, 3_-)$ is unity (see Eq. 4.30) and the transmission is unity. The system acts as a selecting filter.

The conservation of the number of states in the case of the transmission peak is reported in Fig. 4.6 where we present the state phase $\eta(\omega)$ as a function of frequency and for the same parameters considered in Fig. 4.5A.

For other values of the parameters, namely $d_1 = 1$ and $M'_s = 3.15$, it is even possible to obtain two sharp transmission peaks (see Fig. 4.7). These results indicate how the magnon ejection frequency may be controlled by selecting the appropriate parameters of the structure.

It is interesting to underline that the sharp zeros of transmission are obtained when the resonator clusters are weakly coupled with the chain and can be related to the eigenfrequencies of the isolated three cluster resonator. The sharp transmission peaks appear when the resonator clusters are strongly coupled with the chain. In that case the transmission features can be interpreted as mainly due to the widening of the frequencies falling inside the bulk chain band of the eigenmodes of the ensemble of the six perturbed clusters.

In the limit of a weak coupling between the resonator and the infinite chain, the following comment is useful to establish the relationship between the frequencies of the zeros of transmission and the eigenfrequencies of the resonator clusters alone. Indeed, the latter are the solutions of the following determinantal equation:

$$
\begin{vmatrix}
\Omega_- - \dfrac{2M'_s}{d^3} & -\dfrac{M'_s}{d^3} & 0 \\[2mm]
-\dfrac{M'_s}{d^3} & \Omega_- - \dfrac{4M'_s}{d^3} & -\dfrac{M'_s}{d^3} \\[2mm]
0 & -\dfrac{M'_s}{d^3} & \Omega_- - \dfrac{2M'_s}{d^3}
\end{vmatrix} = 0,
\tag{4.34}
$$

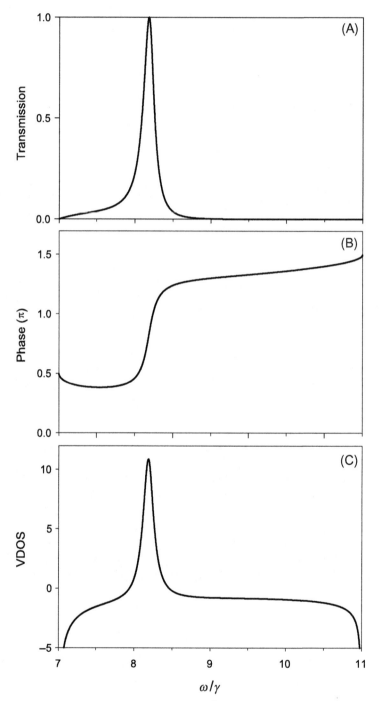

FIG. 4.5 (A) Transmission in function of ω/γ for $M_s' = 2.5$ and $d_1 = 0.9$. The other parameters have the same values as in Fig. 4.3A. (B) The same as in (A) but for the variation of the phase. (C) The same as in (A) but for the VDOS.

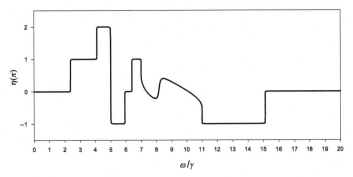

FIG. 4.6 Variation of the state phase $\eta(\omega)$ in function of the frequency ω/γ. The other parameters have the same values as in Fig. 4.5A.

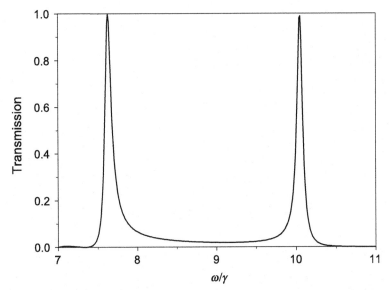

FIG. 4.7 Transmission in function of ω/γ for $H_0 = 5$, $M_s = 1$, $M'_s = 3.15$, $d = 1$, and $d = d_1 = 1$.

which gives the three solutions $\Omega_- \equiv \omega/\gamma - H_0 = 2; 3 + \sqrt{3}; 3 - \sqrt{3}$ in units of M'_s/d^3. We have indeed checked that in the mathematical limit of $d_1 \gg d$, the zeros of transmission coincide with the above solutions (let us emphasize that, depending on the value of M'_s/d^3, only zero, one, or at most two of these solutions fall inside the bulk band of the infinite chain).

However, this mathematical result should not be taken too seriously from the practical point of view. Indeed, we have assumed only nearest-neighbor interactions in our model which obviously fails to be valid in the limit of $d_1 \gg d$. When d_1 is of the order of d, the positions of the zeros of transmission are also more or less significantly affected by the interaction between the resonator clusters and the infinite linear chain.

For all the results displayed here, we checked that the system was stable by calculating the localized magnon states and checking that the positive frequency ones were not falling below the $\omega = 0$ instability limit.

In summary, Figs. 4.3–4.7 show clearly that the device presented in this section can be tuned in order to be a good magnon filter as well by transmission as by reflection. In order to achieve this, the resonant structure must provide zeros and ones of transmission inside the bulk band. This was realized with the three adsorbed cluster device reported here. In order to get sharp transmission peaks one must tune the system parameters such that the transmission ones fall in between two zeros of transmission close one to the other.

4.4 QUASIBOX STRUCTURES

4.4.1 Model and Calculations

In this section we propose a new geometry as depicted in Fig. 4.8B and C. In Fig. 4.8B, there are six additional clusters, three above and three below the infinite chain. We call d the distance between two clusters in the infinite chain and in the three cluster chains, d_1 the distance between the above (or below) three additional clusters and the wire, and d_2 the distance between the below (or above) three additional clusters and the wire. The static magnetic moments are $\mathbf{M}_s$ for the chain clusters and $\mathbf{M}_s'$ for the additional clusters, and the external magnetic field is $\mathbf{H_0}$ along the z-axis. The quasibox structure shown in Fig. 4.8C is obtained from the structure shown in Fig. 4.8B by removing cluster 2 in the wire. The geometrical parameters are considered to be the same as in Fig. 4.8B. Such structures (Fig. 4.8B and C) may exhibit new features, in comparison with the structure shown in Fig. 4.8A (which is exactly the same as Fig. 4.2 and redrawn for the sake of comparison) [80]. For example, the appearance of sharp peaks (dips) in the transmission spectrum achieves almost complete transmission gap by changing the geometrical parameters of the structure. Furthermore, very localized modes appear. These new features (which could be of potential interest in waveguide structures) are essentially due to the three additional clusters adsorbed below the chain and the asymmetry ($d_1 \neq d_2$) of the structures shown in Fig. 4.8B and C, which is quite different from the case shown in Fig. 4.8A [80].

When a progressive wave is launched from the left along the infinite chain, the equation that relates the transmission amplitude t (the reflection amplitude r) to the pulsation of the magnon signal ω is given by Eq. (4.30) (Eq. 4.31).

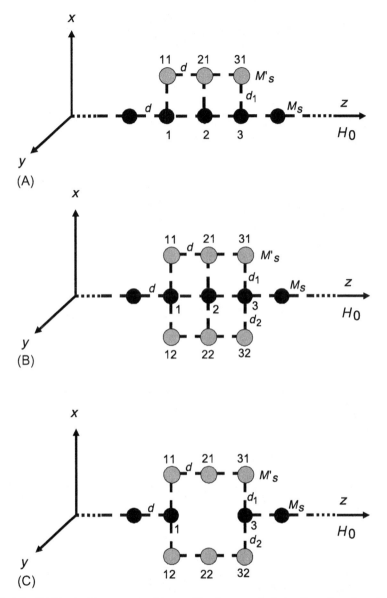

FIG. 4.8 (A) Sketch of the geometry of the considered nanometric device. It consists of one cluster chain and three other clusters. The distances between these clusters are, respectively, d and d_1 as indicated on the figure. The static magnetic moments are $\mathbf{M}_s$ for the chain clusters and $\mathbf{M}'_s$ for the additional clusters, and the external magnetic field is $\mathbf{H_0}$ along the z-axis. (B) The same as in (A) but with additional three adsorbed clusters (at a distance d_2) below the chain. (C) The quasibox structure discussed in the text. The magnetic and geometrical parameters are the same as in (B).

The Green's function elements can be obtained from the inverse of the matrix $[g(MM)]^{-1} = [G(MM)]^{-1} + V_I(MM)$ (see Eq. 4.29). Here $G(MM)$ is the matrix of the Green's function elements in the interface cluster space $M(= 1, 2, 3, 11, 21, 31, 12, 22, 32)$. Its rows and columns are labeled by $(m_1^+, m_2^+, m_3^+, m_1^-, m_2^-, m_3^-, m_{11}^+, m_{21}^+, m_{31}^+, m_{11}^-, m_{21}^-, m_{31}^-, m_{12}^+, m_{22}^+, m_{32}^+, m_{12}^-, m_{22}^-, m_{32}^-)$. $V_I(MM)$ is the interaction matrix in the interface space M [77, 80]. For the structure shown in Fig. 4.8B, $[G(MM)]^{-1}$ and $V_I(MM)$ are given by:

$$[G(MM)]^{-1} =$$

$$\begin{pmatrix}
Y_+ & A_+ & 0 & 0 & 0 & 0 & 0 & 0 & 0 & 0 & 0 & 0 & 0 & 0 & 0 & 0 & 0 & 0 \\
A_+ & A & A_+ & 0 & 0 & 0 & 0 & 0 & 0 & 0 & 0 & 0 & 0 & 0 & 0 & 0 & 0 & 0 \\
0 & A_+ & Y_+ & 0 & 0 & 0 & 0 & 0 & 0 & 0 & 0 & 0 & 0 & 0 & 0 & 0 & 0 & 0 \\
0 & 0 & 0 & Y_- & B_- & 0 & 0 & 0 & 0 & 0 & 0 & 0 & 0 & 0 & 0 & 0 & 0 & 0 \\
0 & 0 & 0 & B_- & B & B_- & 0 & 0 & 0 & 0 & 0 & 0 & 0 & 0 & 0 & 0 & 0 & 0 \\
0 & 0 & 0 & 0 & B_- & Y_- & 0 & 0 & 0 & 0 & 0 & 0 & 0 & 0 & 0 & 0 & 0 & 0 \\
0 & 0 & 0 & 0 & 0 & 0 & \Omega_+ & 0 & 0 & 0 & 0 & 0 & 0 & 0 & 0 & 0 & 0 & 0 \\
0 & 0 & 0 & 0 & 0 & 0 & 0 & \Omega_+ & 0 & 0 & 0 & 0 & 0 & 0 & 0 & 0 & 0 & 0 \\
0 & 0 & 0 & 0 & 0 & 0 & 0 & 0 & \Omega_+ & 0 & 0 & 0 & 0 & 0 & 0 & 0 & 0 & 0 \\
0 & 0 & 0 & 0 & 0 & 0 & 0 & 0 & 0 & \Omega_- & 0 & 0 & 0 & 0 & 0 & 0 & 0 & 0 \\
0 & 0 & 0 & 0 & 0 & 0 & 0 & 0 & 0 & 0 & \Omega_- & 0 & 0 & 0 & 0 & 0 & 0 & 0 \\
0 & 0 & 0 & 0 & 0 & 0 & 0 & 0 & 0 & 0 & 0 & \Omega_- & 0 & 0 & 0 & 0 & 0 & 0 \\
0 & 0 & 0 & 0 & 0 & 0 & 0 & 0 & 0 & 0 & 0 & 0 & \Omega_+ & 0 & 0 & 0 & 0 & 0 \\
0 & 0 & 0 & 0 & 0 & 0 & 0 & 0 & 0 & 0 & 0 & 0 & 0 & \Omega_+ & 0 & 0 & 0 & 0 \\
0 & 0 & 0 & 0 & 0 & 0 & 0 & 0 & 0 & 0 & 0 & 0 & 0 & 0 & \Omega_+ & 0 & 0 & 0 \\
0 & 0 & 0 & 0 & 0 & 0 & 0 & 0 & 0 & 0 & 0 & 0 & 0 & 0 & 0 & \Omega_- & 0 & 0 \\
0 & 0 & 0 & 0 & 0 & 0 & 0 & 0 & 0 & 0 & 0 & 0 & 0 & 0 & 0 & 0 & \Omega_- & 0 \\
0 & 0 & 0 & 0 & 0 & 0 & 0 & 0 & 0 & 0 & 0 & 0 & 0 & 0 & 0 & 0 & 0 & \Omega_-
\end{pmatrix}$$

$$(4.35)$$

where

$$Y_+ = \frac{F_+}{1 - t_+^2}, \quad Y_- = \frac{F_-}{1 - t_-^2} \tag{4.36}$$

$$A = Y_+(1 + t_+^2), \quad A_+ = -Y_+ t_+ \tag{4.37}$$

$$B = Y_-(1 + t_-^2), \quad B_- = -Y_- t_- \tag{4.38}$$

and

$$\Omega_+ = \frac{\omega}{\gamma} + H_0, \quad \Omega_- = \frac{\omega}{\gamma} - H_0. \tag{4.39}$$

$t_\pm$ and $F_\pm$ are given by Eqs. (4.18), (4.19), respectively.

$V_I(MM) =$

$$
\begin{pmatrix}
-V_1 & 0 & 0 & 0 & 0 & 0 & \frac{-P}{2} & 0 & 0 & \frac{-3P}{2} & 0 & 0 & \frac{-E}{2} & 0 & 0 & \frac{-3E}{2} & 0 & 0 \\
0 & -V_1 & 0 & 0 & 0 & 0 & 0 & \frac{-P}{2} & 0 & 0 & \frac{-3P}{2} & 0 & 0 & \frac{-E}{2} & 0 & 0 & \frac{-3E}{2} & 0 \\
0 & 0 & -V_1 & 0 & 0 & 0 & 0 & 0 & \frac{-P}{2} & 0 & 0 & \frac{-3P}{2} & 0 & 0 & \frac{-E}{2} & 0 & 0 & \frac{-3E}{2} \\
0 & 0 & 0 & V_1 & 0 & 0 & \frac{3P}{2} & 0 & 0 & \frac{P}{2} & 0 & 0 & \frac{3E}{2} & 0 & 0 & \frac{E}{2} & 0 & 0 \\
0 & 0 & 0 & 0 & V_1 & 0 & 0 & \frac{3P}{2} & 0 & 0 & \frac{P}{2} & 0 & 0 & \frac{3E}{2} & 0 & 0 & \frac{E}{2} & 0 \\
0 & 0 & 0 & 0 & 0 & V_1 & 0 & 0 & \frac{3P}{2} & 0 & 0 & \frac{P}{2} & 0 & 0 & \frac{3E}{2} & 0 & 0 & \frac{E}{2} \\
\frac{-C}{2} & 0 & 0 & \frac{-3C}{2} & 0 & 0 & V_2 & Q & 0 & 0 & 0 & 0 & 0 & 0 & 0 & 0 & 0 & 0 \\
0 & \frac{-C}{2} & 0 & 0 & \frac{-3C}{2} & 0 & Q & V_3 & Q & 0 & 0 & 0 & 0 & 0 & 0 & 0 & 0 & 0 \\
0 & 0 & \frac{-C}{2} & 0 & 0 & \frac{-3C}{2} & 0 & Q & V_2 & 0 & 0 & 0 & 0 & 0 & 0 & 0 & 0 & 0 \\
\frac{3C}{2} & 0 & 0 & \frac{C}{2} & 0 & 0 & 0 & 0 & 0 & -V_2 & -Q & 0 & 0 & 0 & 0 & 0 & 0 & 0 \\
0 & \frac{3C}{2} & 0 & 0 & \frac{C}{2} & 0 & 0 & 0 & 0 & -Q & -V_3 & -Q & 0 & 0 & 0 & 0 & 0 & 0 \\
0 & 0 & \frac{3C}{2} & 0 & 0 & \frac{C}{2} & 0 & 0 & 0 & -Q & -V_2 & 0 & 0 & 0 & 0 & 0 & 0 & 0 \\
\frac{-D}{2} & 0 & 0 & \frac{-3D}{2} & 0 & 0 & 0 & 0 & 0 & 0 & 0 & 0 & V_4 & Q & 0 & 0 & 0 & 0 \\
0 & \frac{-D}{2} & 0 & 0 & \frac{-3D}{2} & 0 & 0 & 0 & 0 & 0 & 0 & 0 & Q & V_5 & Q & 0 & 0 & 0 \\
0 & 0 & \frac{-D}{2} & 0 & 0 & \frac{-3D}{2} & 0 & 0 & 0 & 0 & 0 & 0 & 0 & Q & V_4 & 0 & 0 & 0 \\
\frac{3D}{2} & 0 & 0 & \frac{D}{2} & 0 & 0 & 0 & 0 & 0 & 0 & 0 & 0 & 0 & 0 & 0 & -V_4 & -Q & 0 \\
0 & \frac{3D}{2} & 0 & 0 & \frac{D}{2} & 0 & 0 & 0 & 0 & 0 & 0 & 0 & 0 & 0 & 0 & -Q & -V_5 & -Q \\
0 & 0 & \frac{3D}{2} & 0 & 0 & \frac{D}{2} & 0 & 0 & 0 & 0 & 0 & 0 & 0 & 0 & 0 & 0 & -Q & -V_4
\end{pmatrix},
$$

$$(4.40)$$

where $V_1 = C + D$, $V_2 = -P + 2Q$, $V_3 = -P + 4Q$, $V_4 = -E + 2Q$, $V_5 = -E + 4Q$, and

$$
C = \frac{M_s'}{d_1^3}; \quad D = \frac{M_s'}{d_2^3}; \quad P = \frac{M_s}{d_1^3}; \quad E = \frac{M_s}{d_2^3}; \quad Q = \frac{M_s'}{d^3}. \tag{4.41}
$$

For the quasibox structure shown in Fig. 4.8C, $[G(MM)]^{-1}$ and $V_I(MM)$ are of dimensions 16×16. They have the same structure as the matrices given in Eqs. (4.35), (4.40) but without the rows and the columns labeled by (m_2^+, m_2^-) [77]. It is worth mentioning that the Green's function method employed in this series can work for any first-neighbor coupled systems; however, the equations reported in this chapter are specific of the discrete dipole approximation used in the calculation [88].

4.4.2 Applications and Discussion of the Results

Now we turn to discussing our numerical results on stop bands and transmission spectrum. Fig. 4.9A displays the intensity transmission coefficient $T = |t|^2$ as a function of ω/γ for the structure shown in Fig. 4.8A. The parameters are $H_0 = 5$, $M_s = M_s' = 1$, $d = 1$, and $d_1 = 1.5$ for the dashed line and 2 for the solid line, respectively. In this plot one can see that the stop band created by the structure shown in Fig. 4.8A with $d_1 = 2$ presents a more narrow dip than the one with $d_1 = 1.5$ (see the inset of Fig. 4.9A). This is attributed to the weaker coupling (in the case of $d_1 = 2$) between the resonator clusters and the chain. What is interesting in Fig. 4.9B is that the two stop bands shown in Fig. 4.9A can be created from the structure of Fig. 4.8B by taking, respectively, $d_1 = 1.5$ and $d_2 = 2$ (or $d_1 = 2$, $d_2 = 1.5$, asymmetric case). The other parameters have the

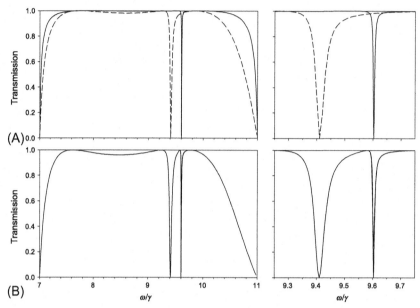

FIG. 4.9 (A) Transmission in function of ω/γ for the structure shown in Fig. 4.8A. The parameters are $H_0 = 5$, $M_s = M'_s = 1$, $d = 1$, and $d_1 = 1.5$ for the *dashed line* and $d_1 = 2$ for the *solid line*, respectively. The *inset* gives a close picture for the two sharp zeros of transmission shown in the figure. (B) The same as in (A) but for the structure shown in Fig. 4.8B. The parameters are $H_0 = 5$, $M_s = M'_s = 1$, $d = 1$, $d_1 = 1.5$, and $d_2 = 2$ (asymmetric case).

same values as in Fig. 4.9A. These results show that the results of the geometry shown in Fig. 4.8B can (in general) be understood as a superposition (with a good precision) of those of the geometry shown in Fig. 4.8A with, respectively, $d_1 = 1.5$ and $d_2 = 2$ (or vice versa). This asymmetric system could be useful to construct a reflector for magnons in analogy to photons in photonic band gap crystals.

Fig. 4.10 depicts the effect of the strong coupling between the six-cluster resonator and the chain on the transmission factor. We show the ω/γ dependence of the transmission for the structure shown in Fig. 4.8B. The parameters are $H_0 = 5$, $M_s = M'_s = 1$, $d = 1$, and $d_2 = d_1 = 0.95$ for the solid line and 0.8 for the dashed line, respectively (symmetric case). The curve associated with the three-cluster resonator coupled with the chain (Fig. 4.8A) is redrawn (see Ref. [80]) for the sake of comparison (dashed-dotted line). The parameters for this plot are $H_0 = 5$, $M_s = M'_s = 1$, $d = 1$, and $d_1 = 0.8$. In this figure one can notice that the new geometry (presented in Fig. 4.8B) gives a more sharp peaks in comparison with the old geometry (Fig. 4.8A). One can also see that the resonance given by the solid line lies between two zeros of transmission, induced by the surrounding six-cluster resonator. This resonance shifts toward the middle of the bulk band when d_1 is slightly increased.

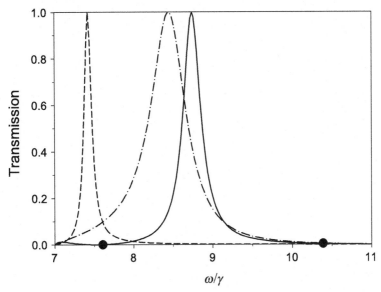

FIG. 4.10 Transmission in function of ω/γ for the structure shown in Fig. 4.8A (*dashed-dotted line*) and Fig. 4.8B (*dashed and solid lines*). The parameters for the *dashed-dotted line* are $H_0 = 5$, $M_s = M_s' = 1$, $d = 1$, and $d_1 = 0.8$. For the *dashed line*, $d_1 = d_2 = 0.8$ and for the *solid line*, $d_1 = d_2 = 0.95$, respectively (symmetric case). The rest of the parameters have the same values as for the *dashed-dotted line*. *Solid circles* on the abscissa axis indicate the positions of the transmission zeros induced by the resonators on both sides of the resonance.

Finally we turn to study the transmission probability through the quasibox structure shown in Fig. 4.8C. The dashed line shown in Fig. 4.11A is redrawn (see Ref. [80]) for the sake of comparison. It represents the transmission through the infinite chain coupled with three adsorbed clusters (Fig. 4.8A). The plot is given as the transmission T versus ω/γ and for the parameters $H_0 = 5$, $M_s = 1$, $M_s' = 3.15$, and $d = d_1 = 1$. One can see two sharp transmission peaks at $\omega/\gamma \simeq 7.6$ and 9.1, and surrounding a region ($\omega/\gamma \simeq 7.9 \longrightarrow 9.9$) where the transmission factor has observable nonzero value. However, when one removes cluster 2 from the chain and adds three adsorbed clusters below the chain and at the same distance like the three clusters above, we obtain the solid-line plot depicted in Fig. 4.11A. The parameters for this plot have the same values as the parameters of the dashed-line plot, namely $H_0 = 5$, $M_s = 1$, $M_s' = 3.15$, and $d = d_1 = d_2 = 1$. This plot presents now interesting two sharp resonances surrounding a region ($\omega/\gamma \simeq 7.9 \longrightarrow 8.5$) where $T \ll 0.005$ (i.e., of practically stop band). Moreover, we increased the value of M_s' and left the other parameters intact. The results (not shown) exhibit a wider (and shifted toward the middle of the bulk band) stop band.

Now if the resonators above and below the chain are not at the same distance ($d_1 \neq d_2$ in Fig. 4.8C) a very sharp mode placed between two zeros of

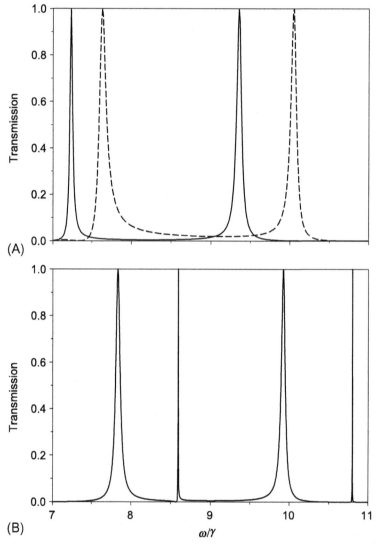

FIG. 4.11 (A) Transmission in function of ω/γ for the structures shown in Fig. 4.8A (*dashed line*) and C (*solid line*). For both plots $H_0 = 5$, $M_s = 1$, $M_s' = 3.15$, and $d = 1$. For the *dashed line*, $d_1 = 1$ and for the *solid line*, $d_1 = d_2 = 1$ (symmetric case). Note the low transmission region ($\omega/\gamma \simeq 7.9 \longrightarrow 8.5$) created by the quasibox structure. (B) Transmission in function of ω/γ for the quasibox structure (asymmetric case). Here $d_1 = 1$, $d_2 = 1.3$, and the rest of the parameters have the same values as in (A). Note the very sharp resonances created at $\omega/\gamma \simeq 8.6$ and 10.8.

transmission is created at $\omega/\gamma \simeq 10.8$ and another very sharp mode in a vicinity of a zero of transmission is created at $\omega/\gamma \simeq 8.6$, see Fig. 4.11B. Note also the two sharp peaks appearing at $\omega/\gamma \simeq 7.8$ and 9.9. The plot is given for $d_1 = 1$, $d_2 = 1.3$ (or $d_1 = 1.3$, $d_2 = 1$), and the rest of the parameters have the same values as in Fig. 4.11A. It is also worth mentioning that slight increase (decrease) in the length of d_1 (or d_2) shifts the positions of the peaks to the right (left) border of the bulk band. The appearance of four peaks in Fig. 4.11B rather than two peaks in Fig. 4.11A is related to the asymmetry of the structure, which in turn has more distinct resonating modes than the corresponding symmetric one. This also helps to get more sharp peaks between two close zeros.

4.4.3 Summary

In summary, we have considered new nanometric resonant structures exhibiting sharp peaks (dips) in the transmission spectra of magnons. A theoretical investigation of the magnon transport through these nanometric structures using a Green's function method is presented. Compared to the infinite chain coupled with three adsorbed clusters, absolute gaps exist indeed in the bulk transmission band of the quasibox structure. Numerical results on very localized modes in nonperturbed structure were also reported. These localized states appear as resonances of strong amplitude near the vicinity of a transmission zero or placed between two transmission zeros. These numerical results confirm that the proposed simple nanostructure resonantly coupling two semiinfinite magnetic cluster chains indeed acts as a magnon filter, selectively transferring magnons of a given frequency from one chain to the other, while letting all the other magnons to be reflected along the input chain. Moreover, the earlier-derived closed-form expressions enable us to find easily the optimal parameters for the device desired, enabling one to engineer it at will for specific applications.

Our theoretical predictions on the generality of the approach must be accompanied by a warning about its limitations. We have assumed only nearest-neighbor interactions in our model which obviously fails to be valid in the limit of $d_1(d_2) \gg d$. When $d_1(d_2)$ is of the order of d, the positions of the zeros of transmission are also more or less significantly affected by the interaction between the resonator clusters and the infinite linear chain. In other words, the dipole-dipole interactions, which entered to this model, are either parallel or perpendicular to the external magnetic field $\mathbf{H}_0$ and therefore this model avoids the complications due to the anisotropic feature of the dipole-dipole interactions which will have to be taken into account in future more sophisticated models.

4.5 MAGNON NANOMETRIC MULTIPLEXER IN CLUSTER CHAINS

4.5.1 Introduction

Complete channel drop tunneling between one-dimensional continua has been extensively studied for electrons ([107] and references therein) as well as for electromagnetic waves [108–111]. This selective transfer of a propagating state from one continuum to the other, leaving the other neighbor states unaffected, may occur when the continua are coupled through a coupling element that supports localized resonant states [108–112]. On the other hand, a directional ejection of quasiparticles (e.g., plasmons, photons, or phonons) from one nanocluster waveguide to another one is now intensively investigated as such a transfer process is particularly important in wavelength multiplexing and telecommunication router devices [108–115].

In the present section we describe a simple system, which under certain conditions realizes the directional transfer of magnons with a good selectivity. The system is depicted in Fig. 4.12A and B. It consists of three semiinfinite nanometric magnetic cluster chain and six (or nine) additional clusters (which play the role of the coupling elements). The distance between two clusters in the semiinfinite chains is considered to be d, and the distance between the additional clusters and the three semiinfinite chains is considered to be d_1.

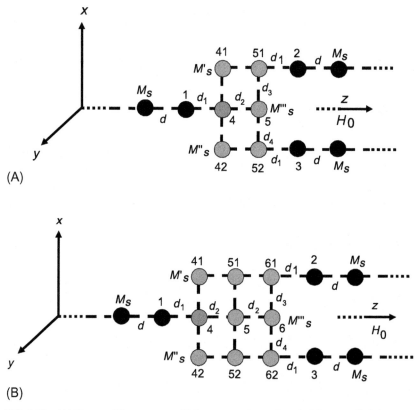

FIG. 4.12 (A) One possible magnon multiplexer geometry made out of three semiinfinite nanometric cluster chains and six other clusters. The distances between these clusters are, respectively, d, d_1, d_2, d_3, and d_4 as indicated on the figure. The static magnetic moments are $\mathbf{M}_s$ (for the chain clusters), $\mathbf{M}_s'$ (for the clusters 41, 51), $\mathbf{M}_s''$ (for the clusters 42, 52), and $\mathbf{M}_s'''$ (for the clusters 4, 5). The external magnetic field is $\mathbf{H_0}$, along the z-axis. (B) The same as in (A) but with additional nine adsorbed clusters.

The horizontal distance between two clusters in the coupling elements is d_2. The vertical distance between clusters 4, 5 and clusters 41, 51 (42, 52) is d_3 (d_4), see Fig. 4.12A. The vertical distances between the coupling elements in Fig. 4.12B is considered to be the same as in Fig. 4.12A. The static magnetic moments are $\mathbf{M}_s$ (for the chain clusters) and $\mathbf{M}_s'$, $\mathbf{M}_s''$, $\mathbf{M}_s'''$ (for the additional clusters shown in Fig. 4.12A and B), along the z-axis. The external magnetic field is $\mathbf{H_0}$, along the z-axis. Finally, the main purpose of this section is to discuss the possibility of narrow peaks (with a high amplitude) in the magnon transmission by selecting appropriately the geometrical or magnetic parameters of the problem.

4.5.2   Calculations

Following the same procedure as in the pervious two sections, the corresponding Green's function elements for the input cluster chain associated with $m_i^\pm$ and, respectively, the negative $(+)$ and positive $(-)$ frequency bands are [105]

$$G_\pm(n, n') = \frac{t_\pm^{|n-n'|} + t_\pm^{3-n-n'}}{F_\pm}, \quad (n, n' \le 1), \tag{4.42}$$

where $t_\pm$ and $F_\pm$ are given by Eqs. (4.18), (4.19), respectively.

For the upper (lower) output cluster chain, the corresponding Green's function elements, respectively, are

$$G_\pm(n, n') = \frac{t_\pm^{|n-n'|} + t_\pm^{n+n'-3}}{F_\pm}, \quad (n, n' \ge 2) \tag{4.43}$$

$$G_\pm(n, n') = \frac{t_\pm^{|n-n'|} + t_\pm^{n+n'-5}}{F_\pm}, \quad (n, n' \ge 3). \tag{4.44}$$

To create a magnon multiplexer out of the coupling elements (having magnetic moments $\mathbf{M}_s'$, $\mathbf{M}_s''$, $\mathbf{M}_s'''$) and the three semiinfinite cluster chains, one has to take into account that the magnon eigenmodes of such a device have to fall inside the bulk magnon band given by Eq. (4.16). Consider now as the reference system the three semiinfinite cluster chains and the additional coupling clusters having magnetic moments $\mathbf{M}_s'$, $\mathbf{M}_s''$, $\mathbf{M}_s'''$ and situated like depicted in Fig. 4.12A and B. When a progressive wave is launched from the left along the semiinfinite input chain, the equation that relates the intensity transmission coefficient in the upper (lower) cluster chain T_{12} (T_{13}), and the intensity reflection coefficient R, to the pulsation of the magnon signal ω can be derived using the Green's function method [77]. It can be written as

$$T_{12} = |F_- g(1_-, 2_-)|^2 \tag{4.45}$$

$$T_{13} = |F_- g(1_-, 3_-)|^2 \tag{4.46}$$

and

$$R = |F_- g(1_-, 1_-) - 1|^2. \tag{4.47}$$

In Eqs. (4.45), (4.46), (4.47), $g(1_-, 2_-)$, $g(1_-, 3_-)$, and $g(1_-, 1_-)$ are, respectively, Green's function elements between clusters 1 and 2, and between clusters 1 and 3, and on the cluster 1 for the magnetic moment m_i^-. The Green's function elements can be obtained from the inverse of the matrix $[g(MM)]^{-1} (= [G(MM)]^{-1} + V_I(MM))$ [106]. Here $G(MM)$ is the matrix of the Green's function elements in the interface cluster space $M(= 1, 2, 3, 42, 52, 4, 5, 41, 51)$, see Fig. 4.12A. Its rows and columns are labeled by $(m_1^+, m_2^+, m_3^+, m_1^-, m_2^-, m_3^-, m_{42}^+, m_{52}^+, m_4^+, m_5^+, m_{41}^+, m_{51}^+, m_{42}^-, m_{52}^-, m_4^-, m_5^-, m_{41}^-, m_{51}^-)$. $V_I(MM)$ is the interaction matrix in the interface space M [77, 80]. For the structure shown in Fig. 4.12A, $[G(MM)]^{-1}$ and $V_I(MM)$ are given by

$$[G(MM)]^{-1} = \mathrm{diag}(Y_+, Y_+, Y_+, Y_-, Y_-, Y_-, \Omega_+, \Omega_+, \Omega_+, \Omega_+, \Omega_+, \Omega_-,$$
$$\Omega_-, \Omega_-, \Omega_-, \Omega_-, \Omega_-), \tag{4.48}$$

where

$$Y_+ = \frac{F_+}{1 + t_+}, \quad Y_- = \frac{F_-}{1 + t_-} \tag{4.49}$$

and

$$\Omega_+ = \frac{\omega}{\gamma} + H_0, \quad \Omega_- = \frac{\omega}{\gamma} - H_0, \tag{4.50}$$

$V_I(MM) =$

$$\begin{pmatrix}
2B_1 & 0 & 0 & 0 & 0 & 0 & 0 & 0 & M_4 & 0 & 0 & 0 & 0 & 0 & 0 & 0 & 0 & 0 \\
0 & 2M_1 & 0 & 0 & 0 & 0 & 0 & 0 & 0 & 0 & 0 & M_4 & 0 & 0 & 0 & 0 & 0 & 0 \\
0 & 0 & 2M_9 & 0 & 0 & 0 & 0 & 0 & M_6 & 0 & 0 & 0 & 0 & 0 & 0 & 0 & 0 & 0 \\
0 & 0 & 0 & -2B_1 & 0 & 0 & 0 & 0 & 0 & 0 & 0 & 0 & 0 & -M_4 & 0 & 0 & 0 & 0 \\
0 & 0 & 0 & 0 & -2M_1 & 0 & 0 & 0 & 0 & 0 & 0 & 0 & 0 & 0 & 0 & 0 & 0 & -M_4 \\
0 & 0 & 0 & 0 & 0 & -2M_9 & 0 & 0 & 0 & 0 & 0 & 0 & 0 & -M_6 & 0 & 0 & 0 & 0 \\
0 & 0 & 0 & 0 & 0 & 0 & A & M_7 & -\frac{M_8}{2} & 0 & 0 & 0 & 0 & 0 & -\frac{3M_8}{2} & 0 & 0 & 0 \\
0 & 0 & M_9 & 0 & 0 & 0 & M_7 & B & 0 & -\frac{M_8}{2} & 0 & 0 & 0 & 0 & 0 & -\frac{3M_8}{2} & 0 & 0 \\
B_1 & 0 & 0 & 0 & 0 & 0 & -\frac{B_4}{2} & 0 & C & B_2 & -\frac{B_3}{2} & 0 & -\frac{3B_4}{2} & 0 & 0 & 0 & -\frac{3B_3}{2} & 0 \\
0 & 0 & 0 & 0 & 0 & 0 & 0 & -\frac{B_4}{2} & B_2 & D & 0 & -\frac{B_3}{2} & 0 & -\frac{3B_4}{2} & 0 & 0 & 0 & -\frac{3B_3}{2} \\
0 & 0 & 0 & 0 & 0 & 0 & 0 & 0 & -\frac{M_3}{2} & 0 & E & M_2 & 0 & 0 & -\frac{3M_3}{2} & 0 & 0 & 0 \\
0 & M_1 & 0 & 0 & 0 & 0 & 0 & 0 & 0 & -\frac{M_3}{2} & M_2 & F & 0 & 0 & 0 & -\frac{3M_3}{2} & 0 & 0 \\
0 & 0 & 0 & 0 & 0 & 0 & 0 & 0 & \frac{3M_8}{2} & 0 & 0 & 0 & -A & -M_7 & \frac{M_8}{2} & 0 & 0 & 0 \\
0 & 0 & 0 & 0 & 0 & -M_9 & 0 & 0 & 0 & \frac{3M_8}{2} & 0 & 0 & -M_7 & -B & 0 & \frac{M_8}{2} & 0 & 0 \\
0 & 0 & 0 & -B_1 & 0 & 0 & \frac{3B_4}{2} & 0 & 0 & 0 & \frac{3B_3}{2} & 0 & \frac{B_4}{2} & 0 & -C & -B_2 & \frac{B_3}{2} & 0 \\
0 & 0 & 0 & 0 & 0 & 0 & 0 & \frac{3B_4}{2} & 0 & 0 & 0 & \frac{3B_3}{2} & 0 & \frac{B_4}{2} & -B_2 & -D & 0 & \frac{B_3}{2} \\
0 & 0 & 0 & 0 & 0 & 0 & 0 & 0 & \frac{3M_3}{2} & 0 & 0 & 0 & 0 & 0 & \frac{M_3}{2} & 0 & -E & -M_2 \\
0 & 0 & 0 & 0 & -M_1 & 0 & 0 & 0 & 0 & \frac{3M_3}{2} & 0 & 0 & 0 & 0 & 0 & \frac{M_3}{2} & -M_2 & -F
\end{pmatrix},$$

$$\tag{4.51}$$

where $A = 2M_7 - B_4$, $B = 2M_7 + 2M_6 - B_4$, $C = 2B_2 + 2M_4 - M_3 - M_8$, $D = 2B_2 - M_3 - M_8$, $E = 2M_2 - B_3$, $F = 2M_2 + 2M_4 - B_3$, $M = \frac{M_8}{d^3}$, $M_1 = \frac{M_8'}{d_1^3}$,

$$M_2 = \frac{M_s'}{d_2^3}, M_3 = \frac{M_s'}{d_3^3}, M_4 = \frac{M_s}{d_1^3}, M_5 = \frac{M_s'}{d_5^3}, M_6 = \frac{M_s}{d_5^3}, M_7 = \frac{M_s''}{d_2^3}, M_8 = \frac{M_s''}{d_4^3},$$

$$M_9 = \frac{M_s''}{d_5^3}, M_{10} = \frac{M_s'}{d_4^3} \text{ and } B_1 = \frac{M_s'''}{d_1^3}, B_2 = \frac{M_s'''}{d_2^3}, B_3 = \frac{M_s'''}{d_3^3}, B_4 = \frac{M_s'''}{d_4^3}.$$

For the structure shown in Fig. 4.12B, $[G(MM)]^{-1}$ and $V_I(MM)$ are of dimensions 24 × 24. They have similar structure as the matrices given in Eqs. (4.48), (4.51) [77].

4.5.3 Applications and Discussion of the Results

In order to illustrate the results of the previous model, we present in Fig. 4.13, the variations of the intensity transmission coefficients T_{12} and T_{13} versus ω/γ for the structure shown in Fig. 4.12A. Fig. 4.13 shows together the direct transmission T_{12} from site 1 to site 2 (solid line) and T_{13} from site 1 to site 3 (dashed line). For the parameters $d_1 = 1.1$, $d_2 = d_3 = d_4 = d = 1$, $Ms = 1$, $M_s' = 1.7$, $M_s'' = 1.3$, $M_s''' = 1.55$, $H_0 = 5$, we obtain (Fig. 4.13A) a magnonic signal in the upper cluster chain centered at $\omega/\gamma = 8.2$, and another signal (with lower amplitude) in the lower cluster chain centered at $\omega/\gamma = 9.2$. When we increase M_s''' to 1.8 and d_3 to 1.1 and decrease d_2 to 0.8, leaving all the other parameters as in Fig. 4.13A, we obtain Fig. 4.13B which presents now an interesting two well separated (i.e., two different values for ω/γ, namely $\omega/\gamma = 7.53; 10.77$) transmission peaks in the bulk transmission band. This feature could be used as a transmission filter for a magnon signal. However, the transmission energy is less than 50% of the input signal. For other values of the parameters, namely $M_s''' = 1.8$, $d_1 = d_2 = d_3 = 1$ (the rest of the parameters have the same values as in Fig. 4.13A) it is possible to obtain transmission peaks in the upper and lower channels with amplitude more than 50% of the complete transmission (see Fig. 4.13C). However, for the set of parameters used in Fig. 4.13C, and for other set of parameters (not shown), the transmission factor shows together with a relatively strong peak in one channel, a small transmission in the other channel.

To increase the amplitude of the transmission factor in the output channels (and at the same time keep the peaks separated) we increased the coupling resonators to nine clusters (see Fig. 4.12B). Fig. 4.14 depicts the effect of coupling between the nine cluster resonators and the semiinfinite chains on the transmission factor. We show the ω/γ dependence of the transmission factors, T_{12} (solid line) and T_{13} (dashed line), for the set of parameters $d = d_1 = d_2 = 1$, $d_3 = 1.5$, $d_4 = 1.25$, $M_s = 1$, $M_s' = 1.15$, $M_s'' = 1.1$, $M_s''' = 1.25$, and $H_0 = 5$. In this figure one can notice that the nine clusters' geometry (presented in Fig. 4.12B) gives (within the bulk transmission band) multiple peaks with amplitude more than 90% of the complete transmission. However, there is some overlap (of order 10%–18% of the transmission one) between peaks corresponding to T_{12} and T_{13}.

Finally we present in Fig. 4.15 the effect of the strong coupling (i.e., a smaller value for the distance d_1, see Fig. 4.12) between the nine cluster

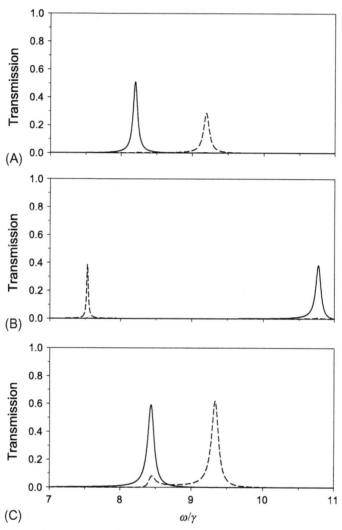

FIG. 4.13 (A) Transmission in function of ω/γ for the structure shown in Fig. 4.12A. The *solid (dashed) line* corresponds to the direct transmission T_{12} (T_{13}) from site 1 to site 2 (from site 1 to site 3). The parameters are $d_1 = 1.1$, $d_2 = d_3 = d_4 = d = 1$, $Ms = 1$, $M'_s = 1.7$, $M''_s = 1.3$, $M'''_s = 1.55$, $H_0 = 5$. (B) The same as in (A) but for the parameters $M'''_s = 1.8$, $d_3 = 1.1$, and $d_2 = 0.8$. The rest of the parameters have the same values as in (A). (C) The same as in (A) but for the parameters $M'''_s = 1.8$ and $d_1 = d_2 = d_3 = 1$. The rest of the parameters have the same values as in (A).

resonators and the three semiinfinite chains on the transmission factor. The plot is given as the transmission T_{12} (solid line) and T_{13} (dashed line) versus ω/γ and for the parameters $M_s = 1$, $M'_s = 1.15$, $M''_s = 1.2$, $M'''_s = 1.25$, $H_0 = 5$, $d_1 = 0.7$, $d = 1$, $d_2 = 1.15$, and $d_3 = d_4 = 1.2$. This plot presents

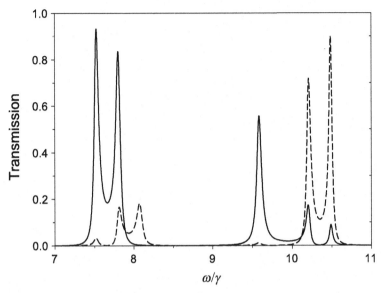

FIG. 4.14 Transmission (through the structure shown in Fig. 4.12B) in function of ω/γ. The *solid (dashed) line* corresponds to T_{12} (T_{13}). The parameters are $d = d_1 = d_2 = 1, d_3 = 1.5, d_4 = 1.25$, $M_s = 1, M_s' = 1.15, M_s'' = 1.1, M_s''' = 1.25$, and $H_0 = 5$.

now interesting two sharp (and well separated) resonances. If we consider the transmission in each of the output channels separately, one can see that the resonance (in each channel) placed between two zeros of transmission, induced by the nine cluster resonators [80, 81].

Here we study the effect of the geometrical parameters on the transmission factor in both channels. Before going to the detailed discussions, let us mention that a first step in the choice of the parameters is done on the basis of our previous calculation where a resonator (made of several clusters) is attached to a linear chain [80]. Then, in a second step, the parameters have to be adjusted around these values in order to increase the transmission in one or both channels. In order to study the effect of d_3 (or d_4 because of the symmetry of the devices shown in Fig. 4.12) on the intensity of transmission, we have kept $H_0 = 5$, all other parameters equal to 1, and slightly changing the value of d_3 around 1. Fig. 4.16A–E shows the ω/γ dependence of the transmission factors (through the structure shown in Fig. 4.12B), T_{12} (solid line) and T_{13} (dashed line), for $H_0 = 5$ and $d_3 = 0.75, 0.85, 1, 1.15$, and 1.25, respectively. For $d_3 = 1$ the results exhibit two sharp peaks appearing at $\omega/\gamma \simeq 7.6$ and 9.13 (see Fig. 4.16C). The amplitude of the first peak is order 0.5 of transmission one while the amplitude of the second one is of order 0.25 of transmission one. For $d_3 < 1$ (which represents a configuration where coupling between clusters 4, 5, 6 and clusters 41, 51, 61 is stronger) the intensity of transmission in the lower channel is strongly reduced while a strong peaks appear in the upper channel

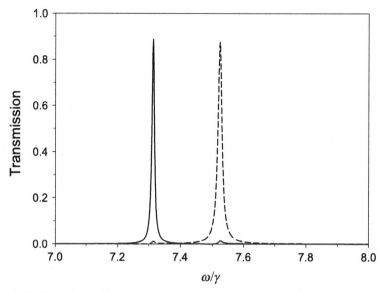

FIG. 4.15 Transmission (through the structure shown in Fig. 4.12B) in function of ω/γ. The *solid (dashed) line* corresponds to T_{12} (T_{13}). The parameters are $M_s = 1$, $M'_s = 1.15$, $M''_s = 1.2$, $M'''_s = 1.25$, $H_0 = 5$, $d_1 = 0.7$, $d = 1$, $d_2 = 1.15$, and $d_3 = d_4 = 1.2$.

with amplitude reaches $\simeq 0.7$ of transmission one (see Fig. 4.16A and B). On the other hand, when $d_3 > 1$ the amplitude of transmission in the upper channel is strongly reduced while in the lower one strong peaks occurs with amplitude greater than 0.7 (see Fig. 4.16D and E). In other words, keeping $H_0 = 5$, all other parameters equal to 1, and slightly increasing (decreasing) the length of d_1 or d_2 (from the value 1) shifts the positions of the peaks to the left (right) border of the bulk band. In such case, identical peaks in both channels occur for all values of d_1 or d_2 (see Fig. 4.16C). Let us finally mention that appearance of multiple (and well separated) resonances in Figs. 4.14 and 4.15 rather than two peaks in Fig. 4.13 is related to the nine coupling clusters presented in the structure shown in Fig. 4.12B which in turn has more distinct resonating modes than the corresponding six clusters one. This also helps to get more sharp peaks between two close zeros. These zeros of transmission are obtained when such a six- or nine-cluster structure is inserted between the linear chains [80–82].

Now we turn to study the effect of the magnetic parameters on the transmission factor. To study the effect of M'''_s on the intensity of transmission, we have kept $H_0 = 5$, all other parameters equal to 1, and slightly changing the value of M'''_s around 1. Identical peaks in both channels occur for all values of M'''_s. Fig. 4.17D shows the transmission factors versus the reduced frequency for $M'''_s = 1$ (which is exactly the same as Fig. 4.16C described previously). The results for $M'''_s < 1$ (not shown) show that the frequency position of the peaks is shifted toward the left border of the bulk band, and the amplitude of transmission

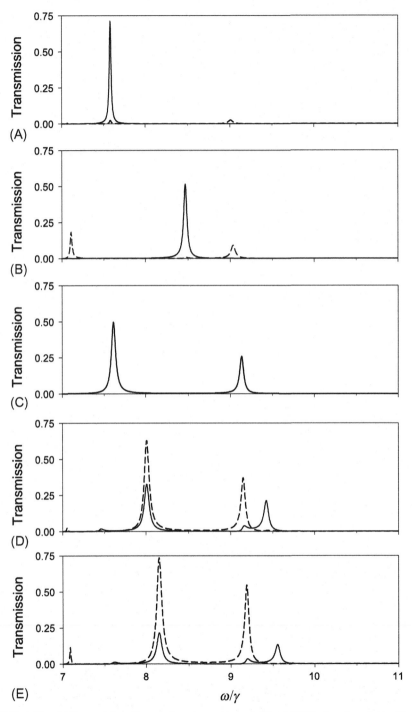

FIG. 4.16 (A) Transmission (through the structure shown in Fig. 4.12B) in function of ω/γ. The *solid (dashed) line* corresponds to T_{12} (T_{13}). $d_3 = 0.75$, $H_0 = 5$, and the rest of parameters have the value 1. (B), (C), (D), and (E) are the same as in (A) but for $d_3 = 0.85$, 1, 1.15, and 1.25, respectively.

is strongly reduced. While for $M_s''' > 1$ the peaks' positions are shifted toward the right border of the bulk band, and the transmission amplitude of both peaks are almost equal and is of order $\simeq 0.45$ of the complete transmission. On the other hand, the effect of M_s' (or M_s'' because of the symmetry of the structures shown in Fig. 4.12) on the transmission factor is different from the effect of M_s'''. Fig. 4.17A–H shows the ω/γ dependence of the transmission factors (through the structure shown in Fig. 4.12B), T_{12} (solid line) and T_{13} (dashed line), for $H_0 = 5$ and $M_s' = 0.7, 0.8, 0.9, 1, 1.1, 1.2, 1.3,$ and 1.4, respectively. The rest of parameters have the value 1. For $M_s' < 1$, the results exhibit multiple peaks with large overlap in both channels. A stronger peak in the upper channel (with amplitude reaches $\simeq 0.63$ of the complete transmission) shows up.

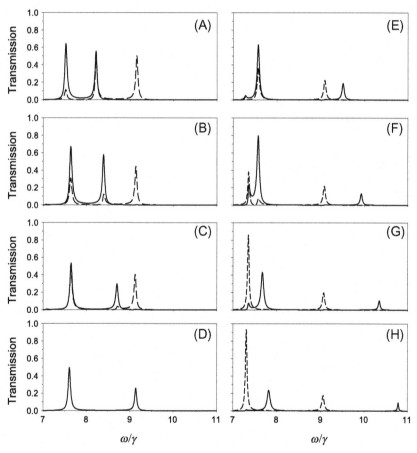

FIG. 4.17 (A) Transmission (through the structure shown in Fig. 4.12B) in function of ω/γ. The *solid (dashed) line* corresponds to T_{12} (T_{13}). $M_s' = 0.7$, $H_0 = 5$, and the rest of parameters have the value 1. (B), (C), (D), (E), (F), (G), and (H) are the same as in (A) but for $M_s' = 0.8, 0.9, 1, 1.1, 1.2, 1.3,$ and 1.4, respectively.

For $M'_s > 1.2$, the transmission in the upper channel is strongly reduced. Instead, a very strong and sharp peak in the lower channel appear at the left border of the transmission band (see Fig. 4.17G and H). It is interesting to underline that the values of the magnetic moments on the coupling clusters M'_s, M''_s, M'''_s, in the results obtained for the structure shown in Fig. 4.12B, are close to the value of M_s (the magnetic moment for the chain clusters). However, for the structure shown in Fig. 4.12A, the previous trend produces transmission peaks with large overlap.

4.5.4 Summary

We have investigated the magnon transfer between two nanometric magnetic cluster chains coupled with a nanometric resonant structure. Inspired by the work of Rivkin et al. [88, 116], we have conceived a model system allowing analytical and numerical study of the conditions for a selective transfer of propagating state from one cluster chain to the other leaving all the other neighbor states unaffected. Our theoretical model enables us to determine completely the geometrical and magnetic parameters of the nanometric structure for a partial magnon transfer at a specific frequency. The frequency domain (transmission bulk band) where the magnon transfer occurs depends on not only the material anisotropy and microstructure, but also the magnetic configuration of the structure. Let us mention that our model system may have potential applications for designing microwave multiplexing devices. For example, a succession of a few nanometric structures with different characteristics may allow us to extract several specific frequencies out of an input signal of large frequency band. However, we have assumed only nearest-neighbor interactions in our model which obviously fails to be valid in the limit of $d_1(d_2, d_3, d_4) \gg d$. When $d_1(d_2, d_3, d_4)$ is of the order of d, the positions (amplitudes) of the peaks are also more or less significantly affected by the interaction between the resonator clusters and the semiinfinite linear chains.

We conclude by stressing that we present here a simple nanometric magnon multiplexer, which may be realized with current production of high-performance magnetic materials and devices using nanofabrication technologies. Of course, as we underline earlier, further model calculations and simulations will be necessary in order to find the appropriate clusters and substrates, before starting to build and test a first prototype.

4.6 GENERAL CONCLUSIONS AND PROSPECTIVES

All particles, bosons (magnon, photon, phonon, plasmon, etc.), and fermions (electrons, etc.) have a wave associated with them. When coherent and mono-mode propagations of such waves is made possible in linear guides, many new circuits and devices can be made out of such guides. We reviewed only a few circuits. But no doubt that many others will be conceived in the future decades, as well with micro as with nanodimensions. One can start and imagine even atomic structures for any specific application.

So this chapter looks like a starting point for many applications. Its backbone is a simple analytical theory for the propagation of coherent spin waves in mono-mode circuits. This simple approach starts to be complemented by more realistic numerical simulations and by experiments for some long-wavelength spin waves. A lot remains to be done for microwaves, especially those used in modern telecommunications. Let us just mention that investigations on nonlinear effects in such structures are also beginning; see, for example, Zielinski et al. [117].

ACKNOWLEDGMENTS

H. Al-Wahsh gratefully acknowledges the hospitality of the Université des Sciences et Technologies de Lille.

REFERENCES

[1] M. Greven, R.J. Birgeneau, U.J. Wiese, Monte Carlo study of correlations in quantum spin ladders, Phys. Rev. Lett. 77 (1996) 1865.

[2] S. Chakravarty, Dimensional crossover in quantum antiferromagnets, Phys. Rev. Lett. 77 (1996) 4446.

[3] D.G. Shelton, A.A. Nersesyan, A.M. Tsvelik, Antiferromagnetic spin ladders: crossover between spin S=1/2 and S=1 chains, Phys. Rev. B 53 (1996) 8521.

[4] J. Piekarewicz, J.R. Shepard, Dynamic spin response for Heisenberg ladders, Phys. Rev. B 57 (1998) 10260.

[5] P. Rai-Chaudhry (Ed.), The handbook of microlithography, micromachining and microfabrication, in: SPIE, 1996.

[6] X.F. Wang, P. Vasilopoulos, F.M. Peeters, Spin-current modulation and square-wave transmission through periodically stubbed electron waveguides, Phys. Rev. B 65 (2002) 165217.

[7] R. Arias, D.L. Mills, Theory of spin excitations and the microwave response of cylindrical ferromagnetic nanowires, Phys. Rev. B 63 (2001) 134439.

[8] A. Encinas-Oropesa, M. Demand, L. Piraux, I. Huynen, U. Ebels, Dipolar interactions in arrays of nickel nanowires studied by ferromagnetic resonance, Phys. Rev. B 63 (2001) 104415.

[9] J. Jorzick, S.O. Demokritov, C. Mathieu, B. Hillebrands, B. Bartenlian, C. Chappert, F. Rousseaux, A.N. Slavin, Brillouin light scattering from quantized spin waves in micron–size magnetic wires, Phys. Rev. B 60 (1999) 15194.

[10] D.P. DiVincenzo, D. Loss, Superlattice. Microstruct. 23 (1998) 419.

[11] B.E. Kane, A silicon-based nuclear spin quantum computer, Nature (London) 393 (1998) 133.

[12] A.G. Mal'shukov, K.A. Chao, Waveguide diffusion modes and slowdown of D'yakonov-Perel' spin relaxation in narrow two-dimensional semiconductor channels, Phys. Rev. B 61 (2000) R2413.

[13] G. Gubbiotti, R. Silvani, S. Tacchi, M. Madami, G. Carlotti, Z. Yang, A.O. Adeyeye, M. Kostylev, Tailoring the spin waves band structure of 1D magnonic crystals consisting of L-shaped iron/permalloy nanowires, J. Phys. D Appl. Phys. 50 (2017) 105002.

[14] B. Lenk, H. Ulrichs, F. Garbs, M. Münzenberg, The building blocks of magnonics, Phys. Rep. 507 (2011) 107.

[15] S.O. Demokritov, A.N. Slavin (Eds.), Magnonics: From Fundamentals to Applications Topics in Applied Physics, vol. 125, Springer, Berlin, 2013.

[16] M. Krawczyk, D. Grundler, Review and prospects of magnonic crystals and devices with reprogrammable band structure, J. Phys. Condens. Matter. 26 (2014) 123202.

[17] S.L. Vysotskiĭ, S.A. Nikitov, Y.A. Filimonov, Magnetostatic spin waves in two-dimensional periodic structures (magnetophoton crystals), J. Exp. Theor. Phys. 101 (2005) 547.

[18] S.A. Nikitov, P. Tailhades, C.S. Tsai, Spin waves in periodic magnetic structures—magnonic crystals, J. Magn. Magn. Mater. 236 (2001) 320.

[19] Y.V. Gulyaev, S.A. Nikitov, Magnonic crystals and spin waves in periodic structures, Dokl. Akad. Nauk. 380 (2001) 469.

[20] J.O. Vasseur, L. Dobrzynski, B. Djafari-Rouhani, H. Puszkarski, Magnon band structure of periodic composites, Phys. Rev. B 54 (1996) 1043.

[21] M. Krawczyk, J.C. Levy, D. Mercier, H. Puszkarski, Forbidden frequency gaps in magnonic spectra of ferromagnetic layered composites, Phys. Lett. A 282 (2001) 186.

[22] A. Figotin, I. Vitebsky, Nonreciprocal magnetic photonic crystals, Phys. Rev. E 63 (2001) 066609.

[23] D.S. Deng, X.F. Jin, R. Tao, Magnon energy gap in a periodic anisotropic magnetic superlattice, Phys. Rev. B 66 (2002) 104435.

[24] L. Berger, Emission of spin waves by a magnetic multilayer traversed by a current, Phys. Rev. B 54 (1996) 9353.

[25] J.C. Slonczewski, Current-driven excitation of magnetic multilayers, J. Magn. Magn. Mater. 159 (1996) L1.

[26] G.A. Prinz, Phys. Today 48 (4) (1995) 58.

[27] V.L. Zhang, H.S. Lim, S.C. Ng, M.H. Kuok, X. Zhou, A.O. Adeyeye, Spin-wave dispersion of nanostructured magnonic crystals with periodic defects, AIP Adv. 6 (2016) 115106.

[28] A.V. Chumak, A.A. Serg, B. Hillebrands, Magnon transistor for all-magnon data processing, Nat. Commun. 5 (2014) 4700.

[29] A.V. Chumak, V.I. Vasyuchka, A.A. Serga, B. Hillebrands, Nat. Phys. 11 (2014) 453.

[30] M.A. Morozova, A.Y. Sharaevskaya, A.V. Sadovnikov, S.V. Grishin, D.V. Romanenko, E.N. Beginin, Y.P. Sharaevskii, S.A. Nikitov, Band gap formation and control in coupled periodic ferromagnetic structures, J. Appl. Phys. 120 (2016) 223901.

[31] M. Krawczyk, H. Puszkarski, Magnonic excitations versus three-dimensional structural periodicity in magnetic composites, Cryst. Res. Technol. 41 (6) (2006) 547.

[32] M. Krawczyk, H. Puszkarski, Magnonic spectra of ferromagnetic composites versus magnetization contrast, Acta Phys. Pol. A 93 (1998) 805.

[33] H. Puszkarski, M. Krawczyk, Magnonic crystals—the magnetic counterpart of photonic crystals, Solid State Phenom. 94 (2003) 125.

[34] Y.V. Gulyaev, S.A. Nikitov, L.V. Zhivotovskii, A.A. Klimov, P. Tailhades, L. Presmanes, C. Bonningue, C.S. Tsai, S.L. Vysotskii, Y.A. Filimonov, Ferromagnetic films with magnon bandgap periodic structures: magnon crystals, JETP Lett. 77 (2003) 670.

[35] H. Puszkarski, M. Krawczyk, On the multiplicity of the surface boundary condition in composite materials, Phys. Lett. A 282 (2001) 106.

[36] H. Puszkarski, M. Krawczyk, J.C.S. Levy, D. Mercier, Ferromagnetic layered composites. Transfer matrix approach, Acta Phys. Pol. A 100 (2001) 195.

[37] V.V. Kruglyak, A.N. Kuchko, Phys. Metals Metallogr. 93 (2002) 511.

[38] V.V. Kruglyak, A.N. Kuchko, Spectrum of spin waves propagating in a periodic magnetic structure, Physica B 339 (2003) 130.

[39] V.V. Kruglyak, A.N. Kuchko, Damping of spin waves in a real magnonic crystal, J. Magn. Magn. Mater. 272 (2004) 302.

[40] H. Al-Wahsh, A. Akjouj, B. Djafari-Rouhani, J.O. Vasseur, L. Dobrzynski, P.A. Deymier, Large magnonic band gaps and defect modes in one-dimensional comblike structures, Phys. Rev. B 59 (1999) 8709.

[41] A. Akjouj, A. Mir, B. Djafari-Rouhani, J.O. Vasseur, L. Dobrzynski, H. Al-Wahsh, P.A. Deymier, Giant magnonic band gaps and defect modes in serial stub structures: application to the tunneling between two wires, Surf. Sci. 482 (2001) 1062.

[42] H. Al-Wahsh, A. Akjouj, B. Djafari-Rouhani, A. Mir, L. Dobrzynski, Effect of pinning fields on the spin wave band gaps in comblike structures, Eur. Phys. J. B 37 (2004) 499.

[43] A. Mir, H. Al-Wahsh, A. Akjouj, B. Djafari-Rouhani, L. Dobrzynski, J.O. Vasseur, Spin-wave transport in serial loop structures, Phys. Rev. B 64 (2001) 224403.

[44] A. Mir, H. Al-Wahsh, A. Akjouj, B. Djafari-Rouhani, L. Dobrzynski, J.O. Vasseur, Magnonic spectral gaps and discrete transmission in serial loop structures, J. Phys. Condens. Matter. 14 (2002) 637.

[45] H. Al-Wahsh, A. Mir, A. Akjouj, B. Djafari-Rouhani, L. Dobrzynski, Magneto-transport in asymmetric serial loop structures, Phys. Lett. A 291 (2001) 333.

[46] H. Al-Wahsh, Stop bands and defect modes in a magnonic chain of cells showing single-cell spectral gaps, Phys. Rev. B 69 (2004) 012405.

[47] J. Bansmann, S.H. Baker, C. Binns, J.A. Blackman, J.P. Bucher, J. Dorantes-Dávila, V. Dupuis, L. Favre, D. Kechrakos, A. Kleibert, K.H. Meiwes-Broer, G.M. Pastor, A. Perez, O. Toulemonde, K.N. Trohidou, J. Tuaillon, Y. Xie, Magnetic and structural properties of isolated and assembled clusters, Surf. Sci. Rep. 56 (2005), 189.

[48] G. Rollmann, S. Sahoo, P. Entel, Nanoscale materials: from science to technology, in: Proceedings of the Indo-US Workshop: Structure and magnetism in iron clusters, Puri, India, 2004.

[49] R.P. Cowburn, D.K. Koltsov, A.O. Adeyeye, M.E. Welland, D.M. Tricker, Single-domain circular nanomagnets, Phys. Rev. Lett. 83 (1999) 1042.

[50] A.V. Jausovec, G. Xiong, R.P. Cowburn, Cycle-by-cycle observation of single-domain-to-vortex transitions in magnetic nanodisks, Appl. Phys. Lett. 88 (2006) 052501.

[51] A. Lebib, S.P. Li, M. Natali, Y. Chen, Size and thickness dependencies of magnetization reversal in Co dot arrays, J. Appl. Phys. 89 (2001) 3892.

[52] M. Grimsditch, Y. Jaccard, I.K. Schuller, Magnetic anisotropies in dot arrays: shape anisotropy versus coupling, Phys. Rev. B 58 (1998) 11539.

[53] I.M.L. Billas, J.A. Becker, A. Châtelain, W.A. de Heer, Magnetic moments of iron clusters with 25 to 700 atoms and their dependence on temperature, Phys. Rev. Lett. 71 (1993) 4067.

[54] J.T. Lau, A. Fröhlisch, R. Nietubyc, M. Reif, W. Wirth, Size-dependent magnetism of deposited small iron clusters studied by X-ray magnetic circular dichroism, Phys. Rev. Lett. 89 (2002) 057201.

[55] P. Gambardella, S. Rusponi, M. Veronese, S.S. Dhesi, C. Grazioli, A. Dallmeyer, I. Cabria, R. Zeller, P.H. Dederichs, K. Kern, C. Carbone, H. Brune, Giant magnetic anisotropy of single cobalt atoms and nanoparticles, Science 300 (2003) 1130.

[56] G. Gubbiotti, M. Conti, G. Carlotti, P. Candeloro, E. Di Fabrizio, K. Guslienko, A. Andre, C. Bayer, A. Slavin, Magnetic field dependence of quantized and localized spin wave modes in thin rectangular magnetic dots, J. Phys. Condens. Matter. 16 (2004) 7709.

[57] K. Perzlmaier, M. Buess, C. Back, V. Demidov, B. Hillebrands, S. Demokritov, Spin-wave eigenmodes of permalloy squares with a closure domain structure, Phys. Rev. Lett. 94 (2005) 057202.

[58] N. Segreeva, S. Cherif, A. Stashkevich, M. Kostylev, J. Ben Youssef, Spin waves quantization in patterned exchange-coupled double layers, J. Magn. Magn. Mater. 288 (2005) 250.

[59] L. Giovannini, F. Montoncello, F. Nizzoli, G. Gubbiotti, G. Carlotti, T. Okuno, T. Shinjo, M. Grimsditch, Spin excitations of nanometric cylindrical dots in vortex and saturated magnetic states, Phys. Rev. B 70 (2004) 172404.

[60] M. Grimsditch, F.Y. Fradin, Y. Ji, A. Hoffmann, R.E. Camley, V. Metlushko, V. Novosad, Coherent inelastic light scattering from a microwave-excited array of magnetic particles, Phys. Rev. Lett. 96 (2006) 47401.

[61] A. Morello, F.L. Mettes, F. Luis, J.F. Fernández, J. Krzystek, G. Aromí, G. Christou, L.J. de Jongh, Long-range ferromagnetic dipolar ordering of high-spin molecular clusters, Phys. Rev. Lett. 90 (2003) 017206.

[62] J.F. Fernández, J.J. Alonso, Ordering of dipolar Ising crystals, Phys. Rev. B 62 (2000) 53.

[63] J.F. Fernández, Dipolar ordering of molecular magnets, Phys. Rev. B 66 (2002) 064423.

[64] C. Sangregorio, T. Ohm, C. Paulsen, R. Sessoli, D. Gatteschi, Quantum tunneling of the magnetization in an iron cluster nanomagnet, Phys. Rev. Lett. 78 (1997) 4645.

[65] S.M. Aubin, N.R. Dilley, L. Pardi, J. Krzystek, M.W. Wemple, L.C. Brunel, M.B. Maple, G. Christou, D.N.J. Hendrickson, Am. Chem. Soc. 120 (1998) 4991.

[66] L. Thomas, A. Caneschi, B. Barbara, Nonexponential dynamic scaling of the magnetization relaxation in Mn_{12} acetate, Phys. Rev. Lett. 83 (1999) 2398.

[67] F.L. Mettes, F. Luis, L.J. de Jongh, Quantum relaxation and quantum coherence in mesoscopic molecular magnets, Phys. Rev. B 64 (2001) 174411.

[68] F. Luis, F.L. Mettes, J. Tejada, D. Gatteschi, L.J. de Jongh, Observation of quantum coherence in mesoscopic molecular magnets, Phys. Rev. Lett. 85 (2000) 4377.

[69] F.L. Mettes, G. Aromí, F. Luis, M. Evangelisti, G. Christou, D. Hendrickson, L.J. de Jongh, Experimental observation of quantum coherence in molecular magnetic clusters with half-integer spin, Polyhedron 20 (2001) 1459.

[70] J. d'Albuquerque e Castro, D. Altbir, J.C. Retamal, P. Vargas, Scaling approach to the magnetic phase diagram of nanosized systems, Phys. Rev. Lett. 88 (2002) 237202.

[71] P. Vargas, D. Altbir, J. d'Albuquerque e Castro, Fast Monte Carlo method for magnetic nanoparticles, Phys. Rev. B 73 (2006) 092417.

[72] A. Khater, M. Abou Ghantous, Calculation of spin modes on magnetic surface clusters, Surf. Sci. 498 (2002) L97.

[73] M. Abou Ghantous, A. Khater, Localised spin modes of decorated magnetic clusters on a magnetic surface, J. Magn. Magn. Mater. 248 (2002) 85.

[74] L. Dobrzynski, Interface response theory of continuous composite materials, Surf. Sci. 180 (1987) 489.

[75] L. Dobrzynski, Interface response theory of electromagnetism in composite dielectric materials, Surf. Sci. 180 (1987) 505.

[76] L. Dobrzynski, Interface response theory of discrete composite systems, Surf. Sci. Rep. 6 (1986) 119.

[77] L. Dobrzynski, Interface response theory of continuous composite systems, Surf. Sci. Rep. 11 (1990) 139.

[78] L. Dobrzynski, Interface-response theory of electromagnetism in dielectric superlattices, Phys. Rev. B 37 (1988) 8027.

[79] L. Dobrzynski, H. Puszkarski, Eigenvectors of composite systems. I. General theory, J. Phys. Condens. Matter. 1 (1989) 1239.

[80] H. Al-Wahsh, L. Dobrzynski, B. Djafari-Rouhani, G. Hernández-Cocoletzi, A. Akjouj, Magnon propagation in a nanometric magnetic cluster chain: effects of additional clusters near the chain, Surf. Sci. 600 (2006) 4883.

[81] H. Al-Wahsh, L. Dobrzynski, B. Djafari-Rouhani, A. Akjouj, Magnon nanometric filters in quasi-one-dimensional cluster chains, Surf. Sci. 601 (2007) 4801.

[82] H. Al-Wahsh, B. Djafari-Rouhani, L. Dobrzynski, A. Akjouj, Magnon nanometric multiplexer in quasi-one-dimensional cluster chains, Surf. Sci. 602 (2008) 1795.

[83] H. Al-Wahsh, Simple nanometric magnon multiplexer, Eur. Phys. J. B 76 (2010) 445.

[84] W.P. Halperin, Quantum size effects in metal particles, Rev. Mod. Phys. 58 (1986) 533.

[85] A.S. Edelstein, R.C. Cammarata, Nanomaterials: Synthesis, Properties and Applications, Institute of Physics Publishing, Bristol, 1996.

[86] C.A. Ross, M. Hwang, M. Shima, J.Y. Cheng, M. Farhoud, T.A. Savas, I. Smith Henry, W. Schwarzacher, F.M. Ross, M. Redjdal, F.B. Humphrey, Micromagnetic behavior of electrodeposited cylinder arrays, Phys. Rev. B 65 (2002) 144417.

[87] W. Xu, D.B. Watkins, L.E. Delong, K. Rivkin, J.B. Ketterson, V.V. Metlushko, Ferromagnetic resonance study of nanoscale ferromagnetic ring lattices, J. Appl. Phys. 95 (2004) 6645.

[88] K. Rivkin, A. Heifetz, P.R. Sievert, J.B. Ketterson, Resonant modes of dipole-coupled lattices, Phys. Rev. B 70 (2004) 184410.

[89] E.L. Albuquerque, M.G. Cottam, Theory of elementary excitations in quasiperiodic structures, Phys. Rep. 376 (2003) 225.

[90] J.M. Luttinger, L. Tisza, Theory of dipole interaction in crystals, Phys. Rev. 70 (1946) 954.

[91] J. Liu, B. Wu, L. Fu, R.B. Diener, Q. Niu, Quantum step heights in hysteresis loops of molecular magnets, Phys. Rev. B 65 (2002) 224401.

[92] S. Prakash, C.L. Henley, Ordering due to disorder in dipolar magnets on two-dimensional lattices, Phys. Rev. B 42 (1990) 6574.

[93] S.M. Patchedjiev, J.P. Whitehead, K. DéBell, Effects of dilution on the magnetic ordering of a two-dimensional lattice of dipolar magnets, J. Phys. Condens. Matter. 17 (2005) 2137.

[94] K. DéBell, A.B. MacIsaac, I.N. Booth, J.P. Whitehead, Dipolar-induced planar anisotropy in ultrathin magnetic films, Phys. Rev. B 55 (1997) 15108.

[95] H. Al-Wahsh, E.H. El Boudouti, B. Djafari-Rouhani, A. Akjouj, L. Dobrzynski, Transmission gaps and sharp resonant states in the electronic transport through a simple mesoscopic device, Phys. Rev. B 75 (2007) 125313.

[96] T. Taniguchi, M. Buttiker, Friedel phases and phases of transmission amplitudes in quantum scattering systems, Phys. Rev. B 60 (1999) 13814.

[97] H.W. Lee, Generic transmission zeros and in-phase resonances in time-reversal symmetric single channel transport, Phys. Rev. Lett. 82 (1999) 2358.

[98] I. Rotter, A.F. Sadreev, Zeros in single-channel transmission through double quantum dots, Phys. Rev. E 71 (2005) 046204; H. Barkay, E. Narevicius, N. Moiseyev, Non-Hermitian scattering theory: resonant tunneling probability amplitude in a quantum dot, Phys. Rev. B 67 (2003) 045322.

[99] A. Yacoby, M. Heiblum, D. Mahalu, H. Shtrikman, Coherence and phase sensitive measurements in a quantum dot, Phys. Rev. Lett. 74 (1995) 4047; R. Schuster, E. Buks, M. Heiblum, D. Mahalu, V. Umansky, H. Shtrikman, Phase measurement in a quantum dot via a double-slit interference experiment, Nature 385 (1997) 417.

[100] A. Mouadili, E.H. El Boudouti, A. Soltani, A. Talbi, A. Akjouj, B. Djafari-Rouhani, Theoretical and experimental evidence of Fano-like resonances in simple monomode photonic circuits, J. Appl. Phys. 113 (2013) 164101.

[101] A. Mouadili, E.H. El Boudouti, A. Soltani, A. Talbi, A. Akjouj, B. Djafari-Rouhani, Electromagnetically induced absorption in detuned stub waveguides: a simple analytical and experimental model, J. Phys. Condens. Matter. 26 (2014) 505901.

[102] J.O. Vasseur, A. Akjouj, L. Dobrzynski, B. Djafari-Rouhani, E.H. El Boudouti, Photon, electron, magnon, phonon and plasmon mono-mode circuits, Surf. Sci. Rep. 54 (2004) 1.

[103] J. Friedel, XIV. The distribution of electrons round impurities in monovalent metals, Philos. Mag. 43 (1952) 153.

[104] L. Dobrzynski, A. Akjouj, E. El Boudouti, et al., in: Interface Transmission Tutorial, Book Series: Phononics, Elsevier, 2017.

[105] D.L. Mills, A.A. Maradudin, Some thermodynamic properties of a semi-infinite Heisenberg ferromagnet, J. Phys. Chem. Solids 28 (1967) 1855.

[106] L. Dobrzynski, V.R. Velasco, F. Garcia-Moliner, Response functions for single interfaces and layered structures, Phys. Rev. B 35 (1987) 5872.

[107] C.C. Eugster, J.A. del Alamo, Tunneling spectroscopy of an electron waveguide, Phys. Rev. Lett. 67 (1991) 3586.

[108] S. Fan, P.R. Villeneuve, J.D. Joannopoulos, H.A. Haus, Channel drop tunneling through localized states, Phys. Rev. Lett. 80 (1998) 960 (and references therein).

[109] K. Sakoda, Optical Properties of Photonic Crystals, vol. 80, Springer Ser. Opt. Sci., Springer, Berlin, 2001.

[110] S.G. Johnson, J.D. Joannopoulos, Photonic Crystals: The Road From Theory to Practice, Kluwer, Boston, MA, 2002.

[111] H.A. Haus, Y. Lai, Narrow-band optical channel-dropping filter, J. Lightwave Technol. 10 (1992) 57.

[112] L. Dobrzynski, B. Djafari-Rouhani, A. Akjouj, J.O. Vasseur, J. Zemmouri, Resonant tunneling between two continua, Phys. Rev. B 60 (1999) 10628.

[113] S.S. Orlov, A. Yariv, S.V. Essen, Coupled-mode analysis of fiber-optic add-drop filters for dense wavelength-division multiplexing, Opt. Lett. 22 (1997) 688.

[114] L. Dobrzynski, A. Akjouj, B. Djafari-Rouhani, J.O. Vasseur, M. Bouazaoui, J.P. Vilcot, H. Al-Wahsh, P. Zielinski, J. Vigneron, Simple nanometric plasmon multiplexer, Phys. Rev. E 69 (2004) 035601(R).

[115] L. Dobrzynski, A. Akjouj, B. Djafari-Rouhani, P. Zielinski, H. Al-Wahsh, A simple phonon multiplexer, Europhys. Lett. 65 (2004) 791.

[116] K. Rivkin, W. Saslow, L.E. De Long, J.B. Ketterson, Dynamic magnetic response of infinite arrays of ferromagnetic particles, Phys. Rev. B 75 (2007) 174408.

[117] P. Zielinski, A. Kulak, L. Dobrzynski, B. Djafari-Rouhani, Propagation of waves and chaos in transmission line with strongly anharmonic dangling resonator, Eur. Phys. J. B 32 (2003) 73.

Chapter 5

Surface, Interface, and Confined Slab Magnons

Leonard Dobrzyński*, Abdellatif Akjouj*, Housni Al-Wahsh†
and Bahram Djafari-Rouhani*
*Department of Physics, Faculty of Sciences and Technologies, Institute of Electronics, Microelectronics and Nanotechnology, UMR CNRS 8520, Lille University, Villeneuve d'Ascq Cedex, France †Faculty of Engineering, Benha University, Cairo, Egypt

Chapter Outline

5.1 Introduction	185	5.7.4 The Localized Modes 198
5.2 Bulk Heisenberg Model	186	5.7.5 Discussion 199
5.3 Bulk Response Function	187	5.8 Surface Reconstruction and Soft
5.4 Planar Defect Magnons	188	Surface Magnons 200
5.5 Surface Magnons	189	5.9 The Effect of Surface-Pinning
5.5.1 Model With First Nearest-		Fields on the Thermodynamic
Neighbor Interactions	189	Properties of a Ferromagnet 203
5.5.2 Model With First and		5.9.1 Motivations 203
Second Nearest-Neighbor		5.9.2 Elementary
Interactions	190	Considerations 205
5.6 Interface Magnons	190	5.9.3 Surface-Specific Heat in
5.7 Confined Slab Magnons	192	Presence of Finite Surface-
5.7.1 Introduction to Slabs	192	Pinning Field 213
5.7.2 The Surface Response		5.10 Summary 218
Operators	192	References 219
5.7.3 The Response Function	193	

5.1 INTRODUCTION

A disturbance in local magnetic ordering can propagate in a magnetic material. Such a wave is first reported by Bloch [1] and is named a spin wave or magnon as it is related to the collective excitations of the electron spin system in ferromagnetic crystals. Surface magnons are magnons localized at surfaces of magnetic materials. Interface magnons are magnons localized at the interface

Magnonics. https://doi.org/10.1016/B978-0-12-813366-8.00005-4

between two magnetic materials; confined slab magnons are magnons localized within a magnetic slab confined between two other magnetic materials. An introduction to surface magnons may be found, for example, in Cottam and Tilley book [2]. This section provides some tutorial examples of: localized magnons at surfaces, interfaces, planar defects, and within confined slabs [3–10]. Examples of soft surface magnons leading to surface superstructures [5, 8] and of some surface thermodynamic magnetic properties [11] will be also given at the end of this chapter.

5.2 BULK HEISENBERG MODEL

Let us define the bulk Heisenberg model for an infinite crystal, used in this chapter. We start from an infinite simple-cubic lattice of atoms, described by the Heisenberg Hamiltonian

$$\mathbf{H}_0 = - \sum_{l,l'} J(l, l') \mathbf{S}_l \cdot \mathbf{S}_{l'}, \tag{5.1}$$

where we retain only the exchange interactions, respectively, J_1 and J_2 between the spins $\mathbf{S}_l$ and $\mathbf{S}_{l'}$ situated on first and second nearest-neighbor atoms. The linearized Holstein-Primakoff transformation [12] enables us to rewrite this Hamiltonian with $b_l^\dagger$ and b_l the creation and annihilation operators and S the spin amplitude.

The equations of motion of the $b_l^\dagger$ operators can then be written in a matrix form.

In an infinite matrix form, all these equations of motion can be written as

$$(\omega \mathbf{I} - \mathbf{H}_0) \mathbf{b} = 0, \tag{5.2}$$

where $\mathbf{b}$ is a column vector having as many rows as we have atoms in the crystal and representing the ensemble of the operators $b_l^\dagger$. So the Heisenberg Hamiltonian $\mathbf{H}_0$ in this matrix form has matrix elements only between first and second nearest-neighbor atoms.

The diagonalization of this Hamiltonian provides the bulk magnon dispersion relation

$$\omega = 4SJ_1(3 - \cos k_1 a_0 - \cos k_2 a_0 - \cos k_3 a_0)$$
$$+ 8SJ_2(3 - \cos k_1 a_0 \cos k_2 a_0 - \cos k_2 a_0 \cos k_3 a_0 - \cos k_1 a_0 \cos k_3 a_0), \tag{5.3}$$

where $\mathbf{k}$ is the propagation vector and a_0 the distance between first nearest-neighbor atoms.

Let us note that the previous frequencies ω have to be positive in order for this spin lattice to keep its ferromagnetic structure. However, we shall see in Section 5.8 that negative values of ω may appear at some boundaries of the Brillouin zone such as for ($k_1 a_0 = k_2 a_0 = 0$, $k_3 a_0 = \pi$) and for ($k_1 a_0 = k_2 a_0 = \pi$, $k_3 a_0 = 0$), leading this crystal to become antiferromagnetic. In order to keep its ferromagnetic structure, the previous two cases lead, respectively, to the following conditions $J_1 + 4J_2 > 0$ and $J_1 > 0$.

Since we shall choose in what follows, surfaces and interfaces parallel to the (001) plane of the cubic crystals, it is convenient to rewrite this dispersion relation as

$$\omega = \Gamma - 2\beta \cos k_3 a_0, \tag{5.4}$$

where

$$\Gamma = 4SJ_1(3 - \cos k_1 a_0 - \cos k_2 a_0) + 8SJ_2(3 - \cos k_1 a_0 \cos k_2 a_0), \tag{5.5}$$

$$\beta = 2SJ_1 + 4SJ_2(\cos k_1 a_0 + \cos k_2 a_0). \tag{5.6}$$

5.3 BULK RESPONSE FUNCTION

The bulk magnetic properties of the above Heisenberg ferromagnet can be studied with the help of its bulk response function defined by

$$(\omega \mathbf{I} - \mathbf{H}_0) \cdot \mathbf{G}_0 = \mathbf{I}, \tag{5.7}$$

where $\mathbf{I}$ is the unit matrix.

Taking advantage now of the periodicity of the system in directions parallel to the (001) planes, we introduce a two-dimensional space vector:

$$\mathbf{x}_\|(l) = a_0(l_1 \hat{x}_1 + l_2 \hat{x}_2), \tag{5.8}$$

a two-dimensional wave vector parallel to the surfaces:

$$\mathbf{k}_\| = k_1 \hat{x}_1 + k_2 \hat{x}_2, \tag{5.9}$$

and a Fourier transformation of the response function:

$$\mathbf{G}_0(l, l'; \omega) = \frac{1}{N^2} \sum_{\mathbf{k}_\|} \mathbf{G}_0(\mathbf{k}_\|, \omega; l_3, l_3') \exp\{j\mathbf{k}_\| \cdot [\mathbf{x}_\|(l) - \mathbf{x}_\|(l')]\}, \tag{5.10}$$

where N^2 is the number of atoms in a (001) plane.

The corresponding bulk response function may be readily obtained in a closed analytic form for the present model [11]

$$\mathbf{G}_0(\mathbf{k}_\|, \omega; l_3, l_3') = \frac{1}{\beta} \frac{t^{|l_3 - l_3'| + 1}}{t^2 - 1}, \tag{5.11}$$

where

$$t = \begin{cases} \zeta - (\zeta^2 - 1)^{1/2}, & \zeta > 1 \\ \zeta + j(1 - \zeta^2)^{1/2}, & -1 < \zeta < 1 \\ \zeta + (\zeta^2 - 1)^{1/2}, & \zeta < -1 \end{cases} \tag{5.12}$$

and

$$\zeta = \frac{\Gamma - \omega}{2\beta}. \tag{5.13}$$

Let us remark that t is essentially equivalent to $\exp(q)$, with $q = jk_3a_0$ inside the bulk bands, with $j^2 = -1$. When $-1 < \zeta < 1$, we obtain $|t| = 1$ which corresponds to the propagation waves in the ferromagnetic crystal and hence the bulk band for magnons. By contrast, for $\zeta < 1$ or $\zeta > 1$, we obtain $|t| < 1$ and the waves are evanescent.

5.4 PLANAR DEFECT MAGNONS

Create a planar defect by changing the exchange interactions J_1 and J_2 between the atomic planes $l_3 = 0$ and $l_3 = 1$ to new values J_{I1} and J_{I2}. Define

$$\alpha = 2S(J_1 + 2J_2). \tag{5.14}$$

The corresponding perturbation to the bulk Hamiltonian $\mathbf{H}_0$ is

$$\mathbf{V}_D(\mathbf{k}_{\parallel}; l_3, l_3') = (\alpha_I - \alpha)(\delta_{l_3,0}\delta_{l_3',0} + \delta_{l_3,1}\delta_{l_3',1}) + (\beta - \beta_I)(\delta_{l_3,0}\delta_{l_3',1} + \delta_{l_3,1}\delta_{l_3',0}), \tag{5.15}$$

where α_I and β_I have the same expressions as the α and β but with J_1 and J_2 replaced, respectively, by J_{I1} and J_{I2}. It is possible now to construct the response function associated with this ferromagnet with the earlier defined planar defect. If one is just interested in the search of localized modes near the planar defect it is enough to calculate the kernel of this Green's function, namely

$$\det(\mathbf{I} - \mathbf{V}_D\mathbf{G}_0) = \frac{[(\alpha - \alpha_I + \beta_I)t - \beta][(\alpha - \alpha_I - \beta_I)t - \beta]}{\beta^2(1 - t^2)}. \tag{5.16}$$

The localized modes exist outside the bulk magnon band when the previous kernel vanishes. This may happen when

$$t_+ = \frac{\beta}{\alpha - \alpha_I + \beta_I} \tag{5.17}$$

or

$$t_- = \frac{\beta}{\alpha - \alpha_I - \beta_I}. \tag{5.18}$$

These solutions correspond to localized modes when

$$|t| \leq 1. \tag{5.19}$$

The frequencies of these modes are then deduced from Eq. (5.13) when noticing from Eq. (5.12) that

$$\zeta = \frac{1}{2}\left(t + \frac{1}{t}\right). \tag{5.20}$$

In this model, it is possible to get either 0, 1, or 2 localized modes, near the planar defect depending on the values of the different parameters. This is discussed in details in Ref. [8]. Here we give only, in Fig. 5.1, one case for which two localized planar defect modes (I_1) and (I_2) exist below the bulk magnon band.

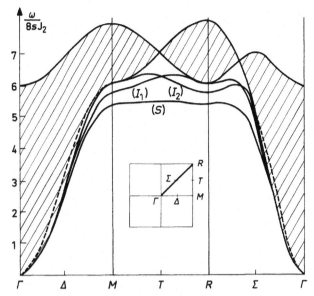

FIG. 5.1 Dispersion of the localized modes (I_1) and (I_2) at a planar defect (001) along the high symmetry directions of the two-dimensional Brillouin zone. The Brillouin zone is sketched inside this figure. The parameters corresponding to these examples are $J_{i1} = 2J_{i2}$ and $\epsilon = J_{I1}/J_{i1} = J_{I2}/J_{I2} = 1/2$. (S) represents the free surface mode. When one computes the variation of the density of states between this planar defect and two free surface semiinfinite crystals, one obtains also resonant modes inside the bulk band represented by the *dotted curve* on this figure.

5.5 SURFACE MAGNONS

In this section we present a simple example of surface localized magnons with the help of the above model of a Heisenberg ferromagnetic crystal with first or first and second nearest-neighbor interactions.

5.5.1 Model With First Nearest-Neighbor Interactions

Let us now create a semiinfinite crystal with a free surface by removing in the infinite crystal all interactions between the atoms situated in the $l_3 = 0$ and $l_3 = 1$ planes. The corresponding cleavage operator $\mathbf{V}_0$, which when added to $\mathbf{H}_0$, gives the Hamiltonian $\mathbf{h}_0$, of the two semiinfinite crystals, is

$$\mathbf{V}_0(l_3, l_3') = -2SJ_1(\delta_{l_3,0}\delta_{l_3',1} + \delta_{l_3,1}\delta_{l_3',0} - \delta_{l_3,0}\delta_{l_3',0} - \delta_{l_3,1}\delta_{l_3',1}). \tag{5.21}$$

With the help of the result given by Eq. (5.11), the interface response theory enables us to derive easily the surface response function [9]

$$\mathbf{G}_s(k_\parallel, \omega; l_3, l_3') = \frac{1}{2SJ_1}\frac{t^{|l_3-l_3'|+1}}{(t^2-1)} + \frac{1}{2SJ_1}\frac{t^{2-(l_3+l_3')}}{(t^2-1)}, \quad l_3, l_3' \le 0. \tag{5.22}$$

The surface localized magnons are given by the poles of the surface Green's function. One sees at once that such poles exist only for $t = +1$ or $t = -1$, which corresponds to the extremities of the bulk bands. So in this very simple model there is no localized surface magnon.

5.5.2 Model With First and Second Nearest-Neighbor Interactions

However, if one adds second nearest-neighbor interactions J_2 to the first nearest ones J_1, one obtains the result for the bulk magnon band dispersion relation given by Eq. (5.3).

The Green's function of such a semiinfinite ferromagnet with a free (001) surface, such that l_3, $l_3' \geq 0$ is

$$\mathbf{G}_s(\mathbf{k}_\parallel, \omega; l_3, l_3') = \frac{1}{\beta(t^2 - 1)} \left(t^{1+|l_3-l_3'|} + \frac{t^{l_3+l_3'+2}(\beta t - \alpha)}{\alpha t - \beta} \right). \tag{5.23}$$

Let us give also explicitly the expression of this Green's function for the complementary semiinfinite, such that l_3, $l_3' \leq -1$

$$\mathbf{G}_s(\mathbf{k}_\parallel, \omega; l_3, l_3') = \frac{1}{\beta_i(t^2 - 1)} \left(t^{1+|l_3-l_3'|} + \frac{t^{-l_3-l_3'}(\beta t - \alpha)}{\alpha t - \beta} \right). \tag{5.24}$$

The localized surface magnons, poles of the previous Green's function have the following analytic expression [3]

$$\omega_s = 8S(J_1 + 4J_2) \left(\sin^2 \frac{k_1 a_0}{2} + \sin^2 \frac{k_2 a_0}{2} \right) - 32SJ_2 \sin^2 \frac{k_1 a_0}{2} \sin^2 \frac{k_2 a_0}{2}$$

$$- \frac{32SJ_2^2}{J_1 + 4J_2} \left(\sin^2 \frac{k_1 a_0}{2} + \sin^2 \frac{k_2 a_0}{2} \right)^2. \tag{5.25}$$

The previous expression for the surface magnons can also be obtained from the results given previously for the planar defect by taking the limit for which the interactions J_{I1} and J_{I2} vanish. An example of such a surface magnon branch S is given in Fig. 5.1.

5.6 INTERFACE MAGNONS

Interface magnons are localized magnons at the interface between two different ferromagnetic materials. One example of such an interface is that between two different ferromagnets $i = 1$ and $i = 2$ (where i is an extra index enabling to distinguish them) such as defined just above with, respectively, interactions α_i and β_i given by Eqs. (5.6), (5.14). At the (001) interface, these interactions are supposed to be α_I and β_I. The operator coupling the two semiinfinite ferromagnets and creating this interface is

$$\mathbf{V}_I(\mathbf{k}_\parallel, \omega; l_3, l_3') = \alpha_I(\delta_{l_3,0}\delta_{l_3',0} + \delta_{l_3,-1}\delta_{l_3',-1}) - \beta_I(\delta_{l_3,0}\delta_{l_3',-1} + \delta_{l_3,-1}\delta_{l_3',0}). \tag{5.26}$$

In order to search for interface magnons, let us construct

$$\Delta = \det(\mathbf{I} - \mathbf{V}_I\mathbf{G})$$

$$= \left(1 - \frac{\alpha_I t_1}{\alpha_1 t_1 - \beta_1}\right)\left(1 - \frac{\alpha_I t_2}{\alpha_2 t_2 - \beta_2}\right) - \frac{\beta_I{}^2 t_1 t_2}{(\alpha_1 t_1 - \beta_1)(\alpha_2 t_2 - \beta_2)}, \tag{5.27}$$

where $\mathbf{G}$ is the reference Green's function constituted from $\mathbf{G}_{s1}$ for $l_3, l_3' \geq 0$ (Eq. 5.23) and $\mathbf{G}_{s2}$ for $l_3, l_3' \leq -1$ (Eq. 5.24).

The interface localized magnons are obtained outside the bulk bands of the two ferromagnets when the real and imaginary part of Δ vanish. Resonant interface states may exist inside the bulk bands of the ferromagnets when only the real part of Δ vanish. Such a resonant state appears as a peak in the density of state variation between the final system and the reference one constituted by the two independent semiinfinite ferromagnets. The different possible resonant and localized interface magnons are discussed in details in Ref. [8]. An example of such resonant and localized magnons is given in Fig. 5.2.

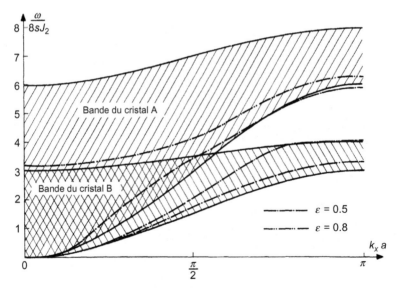

FIG. 5.2 Dispersion of the localized and resonant interface magnons at an (001) interface for $k_2 = 0$ and along the k_1 axes of the Brillouin zone for different values of $\epsilon = J_{I1}/J_{11} = J_{I2}/J_{12}$. In this example, the exchange interactions are chosen to be such that $J_{11}/J_{12} = J_{21}/J_{22} = J_{11}/J_{21} = 2$.

5.7 CONFINED SLAB MAGNONS

5.7.1 Introduction to Slabs

An exciting development in materials science is the appearance of new composite materials prepared by deposition of thin layers of one solid on another one. Most particularly, by means of a sputtering technique one may achieve a precise control of the thickness from a few angstroms to several hundred angstroms of the thin layers. Superlattices made out of alternating thin slabs of two solids having different magnetic properties have been studied experimentally and theoretically, see, for example, Refs. [9, 13–20].

In this example, we address ourselves to a "magnetic quantum well," made from a ferromagnetic slab confined between two other ferromagnetic semiinfinite materials (see Fig. 5.3). When taking different limits of this model, we obtain as well the properties of a thin slab deposited on a substrate, of an interface between two semiinfinite ferromagnets, of a slab with two free surfaces, and of a surface of a semiinfinite ferromagnet.

We study these systems within an atomic model, the Heisenberg one, including exchange effects between first nearest neighbors and neglecting dipolar and Zeeman energies. The ferromagnetic quantum well under study here is built up from L atomic (001) planes of one simple-cubic lattice confined between two other simple-cubic lattices having the same lattice parameter a_0. The three different ferromagnets are characterized by their Heisenberg exchange interactions (J_1, J_2, and J_3) between first nearest-neighbor atoms. The interface atoms are bound together by exchange interactions J_I and J_{II} between first nearest-neighbor atoms. This simple model enables one to obtain in closed form the response function for this composite system. The knowledge of this function enables us to study all the magnetic properties of this ferromagnetic quantum well, and in particular the localized interface magnons. In the next section, we describe more precisely the bulk Heisenberg model used in this study and give the surface response operators necessary for the investigation of the quantum well ferromagnet performed in the next sections.

5.7.2 The Surface Response Operators

Let us now create a slab ($i = 2$) by removing in the infinite crystal all interactions between the atoms situated in the $l_3 = 0$ and $l_3 = 1$ planes and also between those situated in the $l_3 = L$ and $l_3 = L + 1$ planes. The corresponding

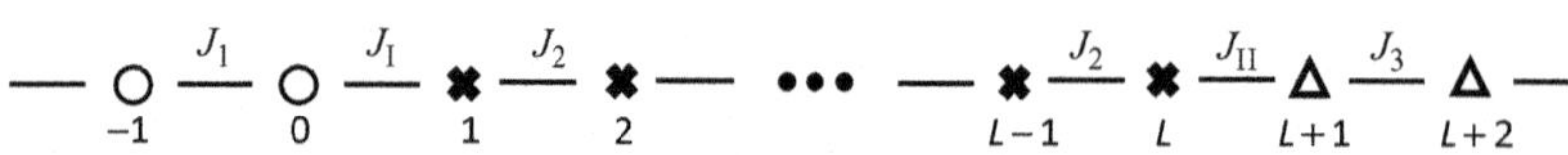

FIG. 5.3 Geometry of the film ($1 \leq l_3 \leq L$) sandwiched between two other ferromagnets. The *circles*, *crosses*, and *triangles* represent the (001) atomic planes of the simple-cubic crystals. J_1, J_2, and J_3 represent the bulk interactions and J_I and J_{II} the interface interactions.

cleavage operator $\mathbf{V}_{02}$, which when added to $\mathbf{H}_{02}$, gives the Hamiltonian $\mathbf{h}_{02}$, of the slab and of the two semiinfinite crystals, is

$$\mathbf{V}_{02}(l_3, l_3') = -2SJ_2(\delta_{l_3,0}\delta_{l_3',1} + \delta_{l_3,1}\delta_{l_3',0} - \delta_{l_3,0}\delta_{l_3',0} - \delta_{l_3,1}\delta_{l_3',1})$$
$$- 2SJ_2(\delta_{l_3,L}\delta_{l_3',L+1} + \delta_{l_3,L+1}\delta_{l_3',L} - \delta_{l_3,L}\delta_{l_3',L} - \delta_{l_3,L+1}\delta_{l_3',L+1}).$$

$$(5.28)$$

The surface response operator $\mathbf{A}_{s2}$ associated with this slab is formed by definition out of the elements of

$$\mathbf{A}_{02} = \mathbf{V}_{02} \cdot \mathbf{G}_{02}, \qquad (5.29)$$

belonging only to the slab, and is easily found to be

$$\mathbf{A}_{s2}(l_3, l_3') = -\frac{1}{t_2 + 1}\left(\delta_{l_3,1}t_2^{l_3'} + \delta_{l_3,L}t_2^{L-l_3'+1}\right), \qquad 1 \le l_3, l_3' \le L. \quad (5.30)$$

In the same manner, one obtains the surface response operators for the semiinfinite crystals

$$\mathbf{A}_{s1}(l_3, l_3') = -\delta_{l_3,0}\frac{t_1^{1-l_3'}}{t_1 + 1}, \qquad l_3, l_3' \le 0 \qquad (5.31)$$

and

$$\mathbf{A}_{s3}(l_3, l_3') = -\delta_{l_3,L+1}\frac{t_3^{l_3'-L}}{t_3 + 1}, \qquad l_3, l_3' \ge L + 1. \qquad (5.32)$$

With the help of the results given by Eqs. (5.11), (5.30)–(5.32), the interface response theory enables us to derive easily the surface and slab response functions. We do not give them here. Their explicit forms are given for the semiinfinite ferromagnet by Eq. (5.22) and for the ferromagnetic slab in Ref. [19]. We need only the surface response operators given by Eqs. (5.30)–(5.32) and not the whole surface and slab response functions, for what follows.

5.7.3 The Response Function

5.7.3.1 The Ferromagnetic Quantum Well

The ferromagnetic quantum well under study here is formed out of a ferromagnetic slab ($i = 2$, $1 \le l_3 \le L$) bounded by two different semiinfinite ferromagnets ($i = 1$, $l_3 \le 0$) and ($i = 3$, $l_3 \ge L + 1$).

Following the interface response theory, we define first the reference response function $\mathbf{G}_s$ for the composite system under study here as a block diagonal matrix whose nonzero elements are given by

$$\mathbf{G}_s(l_3, l_3'; \mathbf{k}_\parallel, \omega^2) = \begin{cases} \mathbf{G}_{01}(l_3, l_3'; \mathbf{k}_\parallel, \omega^2), & l_3, l_3' \le 0 \\ \mathbf{G}_{02}(l_3, l_3'; \mathbf{k}_\parallel, \omega^2), & 1 \le l_3, l_3' \le L \\ \mathbf{G}_{03}(l_3, l_3'; \mathbf{k}_\parallel, \omega^2), & l_3, l_3' \ge L + 1. \end{cases} \qquad (5.33)$$

The interface coupling operator, which binds the three different ferromagnets together through the interactions J_I and J_{II} between the interface atoms, is

$$\mathbf{V}_I(l_3, l_3') = 2SJ_I(\delta_{l_3,0}\delta_{l_3',1} + \delta_{l_3,1}\delta_{l_3',0} - \delta_{l_3,0}\delta_{l_3',0} - \delta_{l_3,1}\delta_{l_3',1})$$
$$+ 2SJ_{II}(\delta_{l_3,L}\delta_{l_3',L+1} + \delta_{l_3,L+1}\delta_{l_3',L} - \delta_{l_3,L}\delta_{l_3',L} - \delta_{l_3,L+1}\delta_{l_3',L+1}).$$
$$\tag{5.34}$$

The interface response operator $\mathbf{A}$ is defined as

$$\mathbf{A} = \mathbf{A}_s + \mathbf{V}_I \cdot \mathbf{G}_s, \tag{5.35}$$

where the nonzero elements of the surface response operator $\mathbf{A}_s$ are, with the help of Eqs. (5.30)–(5.32)

$$\mathbf{A}_s(l_3, l_3') = \begin{cases} \mathbf{A}_{s1}(l_3, l_3'), & l_3, l_3' \leq 0 \\ \mathbf{A}_{s2}(l_3, l_3'), & 1 \leq l_3, l_3' \leq L \\ \mathbf{A}_{s3}(l_3, l_3'), & l_3, l_3' \geq L+1. \end{cases} \tag{5.36}$$

The response function $\mathbf{g}$ of the quantum well system can then be calculated from the general equation of the interface response theory

$$\mathbf{g}(\mathbf{I} + \mathbf{A}) = \mathbf{G}_s. \tag{5.37}$$

For the evaluation of the matrix elements of $\mathbf{g}$ this equation becomes

$$\mathbf{g}(l_3, l_3') = \mathbf{G}_s(l_3, l_3') - \sum_{m,m' \in M} \mathbf{G}_s(l_3, m)\Delta^{-1}(m, m')\mathbf{A}(m', l_3'), \tag{5.38}$$

where M stands for the total interface space, namely $m, m' = 0, 1, L$, and $L+1$ and Δ is defined by

$$\Delta(m, m') = \delta_{m,m'} + \mathbf{A}(m, m'), \quad m, m' \in M. \tag{5.39}$$

In Eqs. (5.35)–(5.39), the dependence on $k_\parallel$ and ω was not explicitly written, for simplicity.

The only difficulty in the calculation of the matrix elements of $\mathbf{g}$ is the inversion of the 4×4 matrix Δ. When calculating the determinant of Δ, it is convenient to define it as being proportional to

$$W(k_\parallel, \omega) = \left[1 - J_I \left\{ \frac{t_1}{J_1(t_1 - 1)} + \frac{t_2}{J_2(t_2 - 1)} \right\} \right]$$

$$\left[1 - J_{II} \left\{ \frac{t_2}{J_2(t_2 - 1)} + \frac{t_3}{J_3(t_3 - 1)} \right\} \right] - t_2^{2L} \left[1 + J_I \left\{ \frac{1}{J_2(t_2 - 1)} - \frac{t_1}{J_1(t_1 - 1)} \right\} \right]$$

$$\left[1 + J_{II} \left\{ \frac{1}{J_2(t_2 - 1)} - \frac{t_3}{J_3(t_3 - 1)} \right\} \right]. \tag{5.40}$$

Finally all the matrix elements of $\mathbf{g}$ are found to be as follows.

1. The source point and observation point are both in the semiinfinite ferromagnet 1

$$
\mathbf{g}(k_\parallel, \omega; l_3, l'_3) = \frac{1}{2SJ_1} \frac{t_1^{|l_3 - l'_3| + 1}}{(t_1^2 - 1)} + \frac{1}{2SJ_1} \frac{t_1^{2 - (l_3 + l'_3)}}{(t_1^2 - 1)} \frac{1}{W(k_\parallel, \omega)}
$$

$$
\times \left[\left[1 - J_{II} \left\{ \frac{t_2}{J_2(t_2 - 1)} + \frac{t_3}{J_3(t_3 - 1)} \right\} \right] \right.
$$

$$
\left[1 + J_I \left\{ \frac{1}{J_1(t_1 - 1)} - \frac{1}{J_2(t_2 - 1)} \right\} \right]
$$

$$
- t_2^{2L} \left[1 + J_{II} \left\{ \frac{1}{J_2(t_2 - 1)} - \frac{t_3}{J_3(t_3 - 1)} \right\} \right]
$$

$$
\left. \left[1 + J_I \left\{ \frac{1}{J_2(t_2 - 1)} + \frac{1}{J_1(t_1 - 1)} \right\} \right] \right], \quad (5.41)
$$

for $l_3, l'_3 \leq 0$.

2. The source point is in ferromagnet 1 and the observation point is in slab 2

$$
\mathbf{g}(k_\parallel, \omega; l_3, l'_3) = -\frac{J_1}{2SJ_1 J_2} \frac{t_1^{1 - l_3}}{(t_1 - 1)(t_2 - 1)} \frac{1}{W(k_\parallel, \omega)}
$$

$$
\times \left\{ \left[1 - J_{II} \left\{ \frac{t_2}{J_2(t_2 - 1)} + \frac{t_3}{J_3(t_3 - 1)} \right\} \right] t_2^{l'_3} \right.
$$

$$
\left. + \left[1 + J_{II} \left\{ \frac{1}{J_2(t_2 - 1)} - \frac{t_3}{J_3(t_3 - 1)} \right\} \right] t_2^{2L + 1 - l'_3} \right\},
$$

$$
(5.42)
$$

for $l_3 \leq 0$ and $1 \leq l'_3 \leq L$.

3. For $1 \leq l_3 \leq L$ and $l'_3 \leq 0$, $\mathbf{g}(k_\parallel, \omega; l_3, l'_3)$ can be obtained just by interchanging the indices l_3 and l'_3 in the right-hand side of Eq. (5.42).

4. The source point is in ferromagnet 1 and the observation point is in ferromagnet 3

$$
\mathbf{g}(k_\parallel, \omega; l_3, l'_3) = \frac{J_1 J_{II}}{2SJ_1 J_2 J_3} \frac{t_1^{1 - l_3}}{(t_1 - 1)} \frac{(t_2 + 1)t_2^L}{(t_2 - 1)} \frac{t_3^{l'_3 - L}}{(t_3 - 1)} \frac{1}{W(k_\parallel, \omega)}, \quad (5.43)
$$

for $l_3 \leq 0$ and $l'_3 \geq L + 1$.

For $l_3 \geq L + 1$ and $l'_3 \leq 0$, $\mathbf{g}(k_\parallel, \omega; l_3, l'_3)$ can be obtained just by interchanging the indices l_3 and l'_3 in the right-hand side of Eq. (5.43).

5. The source and observation points are in ferromagnet 2

$$
\mathbf{g}(k_\parallel, \omega; l_3, l'_3) = \frac{1}{2SJ_2} \frac{t_2^{|l_3 - l'_3| + 1}}{(t_2^2 - 1)} + \frac{1}{2SJ_2} \frac{1}{(t_2^2 - 1)} \frac{1}{W(k_\parallel, \omega)}
$$

$$
\times \left\{ t_2^{l_3 + l'_3} \left[1 + J_I \left\{ \frac{1}{J_2(t_2 - 1)} - \frac{t_1}{J_1(t_1 - 1)} \right\} \right] \right.
$$

$$\left[1 - J_{\mathrm{II}}\left\{\frac{t_3}{J_3(t_3-1)} + \frac{t_2}{J_2(t_2-1)}\right\}\right]$$

$$+ t_2^{2L+2-l_3-l_3'}\left[1 - J_I\left\{\frac{t_2}{J_2(t_2-1)} + \frac{t_1}{J_1(t_1-1)}\right\}\right]$$

$$\left[1 + J_{\mathrm{II}}\left\{\frac{1}{J_2(t_2-1)} - \frac{t_3}{J_3(t_3-1)}\right\}\right]$$

$$+ t_2^{2L+1}\left(t_2^{l_3-l_3'} + t_2^{-l_3+l_3'}\right)\left[1 + J_I\left\{\frac{1}{J_2(t_2-1)} - \frac{t_1}{J_1(t_1-1)}\right\}\right]$$

$$\left[1 + J_{\mathrm{II}}\left\{\frac{1}{J_2(t_2-1)} - \frac{t_3}{J_3(t_3-1)}\right\}\right]\right\}, \tag{5.44}$$

for $1 \leq l_3, l_3' \leq L$.

6. The source point is in slab 2 and the observation point is in ferromagnet 3

$$\mathbf{g}(k_\parallel, \omega; l_3, l_3') = -\frac{J_{\mathrm{II}}}{2SJ_2J_3}\frac{t_3^{l_3'-L}}{(t_2-1)(t_3-1)}\frac{1}{W(k_\parallel, \omega)}$$

$$\times\left\{\left[1 + J_I\left\{\frac{1}{J_2(t_2-1)} - \frac{t_1}{J_1(t_1-1)}\right\}\right]t_2^{L+l_3}\right.$$

$$\left. + \left[1 - J_I\left\{\frac{t_1}{J_1(t_1-1)} + \frac{t_2}{J_2(t_2-1)}\right\}\right]t_2^{L+1-l_3}\right\}, \tag{5.45}$$

for $1 \leq l_3 \leq L$ and $l_3' \geq L+1$.
For $l_3 \geq L+1$ and $1 \leq l_3' \leq L$, $\mathbf{g}(k_\parallel, \omega; l_3, l_3')$ can be obtained just by interchanging the indices l_3 and l_3' in the right-hand side of Eq. (5.45).
7. The source and observation points are in ferromagnet 3

$$\mathbf{g}(k_\parallel, \omega; l_3, l_3') = \frac{1}{2SJ_3}\frac{t_3^{|l_3-l_3'|+1}}{(t_3^2-1)} + \frac{1}{2SJ_3}\frac{t_3^{l_3+l_3'-2L}}{(t_3^2-1)}\frac{1}{W(k_\parallel, \omega)}$$

$$\times\left\{\left[1 - J_I\left\{\frac{t_1}{J_1(t_1-1)} + \frac{t_2}{J_2(t_2-1)}\right\}\right]\right.$$

$$\left[1 + J_{\mathrm{II}}\left\{\frac{1}{J_3(t_3-1)} - \frac{t_2}{J_2(t_2-1)}\right\}\right]$$

$$- t_2^{2L}\left[1 + J_I\left\{\frac{1}{J_2(t_2-1)} - \frac{t_1}{J_1(t_1-1)}\right\}\right]$$

$$\left[1 + J_{\mathrm{II}}\left\{\frac{1}{J_2(t_2-1)} + \frac{1}{J_3(t_3-1)}\right\}\right]\right\}, \tag{5.46}$$

for $l_3, l_3' \geq L+1$.

5.7.3.2 Particular Limits of the Quantum Well

From the general results given earlier, Eqs. (5.40)–(5.46), it is possible to obtain the response functions of several other interesting systems.

1. *An adsorbed ferromagnetic slab on another semiinfinite ferromagnet*
   The response function of the system formed by L atomic layers $(1 \leq l_3 \leq L)$ of a ferromagnetic slab adsorbed on another semiinfinite ferromagnet $(l_3 \leq 0)$ reads directly by setting $J_{II} = 0$ in Eqs. (5.40)–(5.42), (5.44).
2. *An interface between two semiinfinite ferromagnets*
   The response function for the system formed by a semiinfinite ferromagnet 1 $(l_3 \leq 0)$ coupled with another semiinfinite ferromagnet 2 $(l_3 \geq 1)$ is obtained by putting $J_3 = J_{II} = J_2$ in Eqs. (5.40)–(5.46).
   Let us give here explicitly these results as they do not read directly from the earlier equations

$$\mathbf{g}(k_{\parallel}, \omega; l_3, l_3') = \frac{1}{2SJ_1} \left[\frac{t_1^{|l_3 - l_3'| + 1} + t_1^{2 - l_3 - l_3'}}{(t_1^2 - 1)} + \frac{J_1 J_2}{W_I(k_{\parallel}, \omega)} \frac{(t_2 - 1)}{(t_1 - 1)} t_1^{2 - l_3 - l_3'} \right],$$
(5.47)

for $l_3, l_3' \leq 0$,

$$\mathbf{g}(k_{\parallel}, \omega; l_3, l_3') = \frac{1}{2SJ_2} \left[\frac{t_2^{|l_3 - l_3'| + 1} + t_2^{l_3 + l_3'}}{(t_2^2 - 1)} + J_1 J_1 \frac{(t_1 - 1)}{(t_2 - 1)} \frac{t_2^{l_3 + l_3'}}{W_I(k_{\parallel}, \omega)} \right],$$
(5.48)

for $l_3, l_3' \geq 1$,

$$\mathbf{g}(k_{\parallel}, \omega; l_3, l_3') = -\frac{J_I}{2S} \frac{t_1^{1 - l_3} t_2^{l_3'}}{W_I(k_{\parallel}, \omega)},$$
(5.49)

for $l_3 \leq 0$ and $l_3' \geq 1$,

$$\mathbf{g}(k_{\parallel}, \omega; l_3, l_3') = -\frac{J_I}{2S} \frac{t_1^{1 - l_3'} t_2^{l_3}}{W_I(k_{\parallel}, \omega)},$$
(5.50)

for $l_3 \geq 1$ and $l_3' \leq 0$,
   where

$$W_I(k_{\parallel}, \omega) = J_1 J_2 (t_1 - 1)(t_2 - 1) - J_1 [J_1 t_2 (t_1 - 1) + J_2 t_1 (t_2 - 1)].$$
(5.51)

A similar result was also obtained [21], in a less explicit form.
3. *The free surface ferromagnetic slab*
   The response function obtained [19] for a ferromagnetic slab $(1 \leq l_3 \leq L)$ with free surfaces reads directly from Eqs. (5.40), (5.44) with $J_I = J_{II} = 0$.
4. *The free surface of a semiinfinite ferromagnet*
   The response function, Eq. (5.22), for a semiinfinite ferromagnet situated at $l_3 \leq 0$ can also be obtained from Eqs. (5.40), (5.41) when $J_I = J_{II} = 0$.

5.7.4 The Localized Modes

The knowledge of the previous response functions enables us for the systems to address all the ferromagnetic properties, which can be studied with the help of the Heisenberg model. However, here we will just illustrate their usefulness by a study of the localized magnons within the film confined between two other ferromagnets.

These localized magnons can be obtained from the new poles of the response function, which are the values of $\omega(k_\parallel)$ for which $W(k_\parallel, \omega)$ (Eq. 5.40) vanishes. Note, however, that the value $t_2 = -1$ (edge of the bulk band of medium 2) is not to be retained as the corresponding numerators of the response function vanish also. Another useful expression for the study of these localized modes can be obtained when defining a new variable q_2 by

$$t_2 = e^{q_2}. \tag{5.52}$$

Then the closed-form expression, which gives the localized modes, becomes

$$2 \left[1 - \frac{J_I}{J_1} \frac{t_1}{t_1 - 1} \right] \left[1 - \frac{J_{II}}{J_3} \frac{t_3}{t_3 - 1} \right] \sinh(q_2 L) + \frac{J_I J_{II}}{2 J_2^2} \frac{\sinh[q_2(L-1)]}{\sinh^2(q_2^2)}$$
$$- \frac{J_I J_{II}}{J_2} \left[- \left[\frac{1}{J_I} + \frac{1}{J_{II}} \right] + \frac{1}{J_1} \frac{t_1}{t_1 - 1} + \frac{1}{J_3} \frac{t_3}{t_3 - 1} \right] \frac{\cosh[q_2(L-0.5)]}{\sinh\left(\frac{q_2}{2}\right)} = 0, \tag{5.53}$$

with $q_2 \neq i\pi$.

When the ferromagnets 1 and 3 are identical, then this result separates into solutions, which are, respectively, symmetric or antisymmetric with respect to the middle of the film,

$$\coth[q_2 L/2] \coth[q_2/2] = 1 + 2 J_2 \left[\frac{1}{J_1} \frac{t_1}{t_1 - 1} - \frac{1}{J_I} \right] \tag{5.54}$$

and

$$\tanh[q_2 L/2] \coth[q_2/2] = 1 + 2 J_2 \left[\frac{1}{J_1} \frac{t_1}{t_1 - 1} - \frac{1}{J_I} \right], \tag{5.55}$$

with $q_2 \neq i\pi$.

The localized states due to an adsorbed slab 2 on the ferromagnet 1 are given by

$$2 \left[1 - \frac{J_I}{J_1} \frac{t_1}{t_1 - 1} \right] \sinh(q_2 L) - \frac{J_I}{J_2} \frac{\cosh[q_2(L-0.5)]}{\sinh(q_2/2)} = 0, \tag{5.56}$$

with $q_2 \neq i\pi$.

Note that Eqs. (5.53)–(5.55) outside the bulk bands of ferromagnets 1 and 3 are purely real or imaginary.

The simplicity of these results allow to easily obtain through a simple numerical calculation, a qualitative feeling about the existence of localized magnons. So we will just illustrate here these general analytic results by a few results.

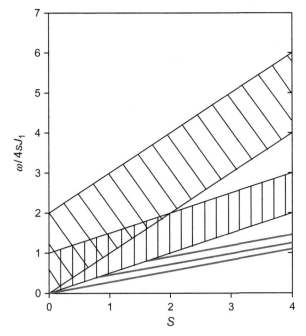

FIG. 5.4 Dispersion [function of $S = 2 - \cos(k_1 a_0) - \cos(k_2 a_0)$] of three localized magnons due to three atomic layers of one ferromagnet sandwiched between two different ones. The *hatched areas* represent the bulk bands of the two semiinfinite ferromagnets, $J_2/J_1 = 0.25$, $J_3/J_1 = 0.5$, $J_I/J_1 = 1$, and $J_{II}/J_1 = 1$.

As a function of the compact quantity $S = 2 - \cos(k_1 a_0) - \cos(k_2 a_0)$, Fig. 5.4 shows the dispersion curves of three localized magnons below the bulk bands of the semiinfinite crystals for a three-layered slab confined between two other magnets and for $J_2/J_1 = 0.25$, $J_3/J_1 = 0.5$, $J_I/J_1 = 1.0$, and $J_{II}/J_1 = 1.0$.

Fig. 5.5 gives another illustration for a three-layered slab where we show as a function of S, the dispersion curves of two localized magnons appearing in a gap between the bulk bands of the two semiinfinite crystals. Here $J_2/J_1 = 0.66$, $J_3/J_1 = 0.5$, $J_I/J_1 = 1$, and $J_{II}/J_1 = 1$. When J_2/J_1 and J_3/J_1 are greater than 1, localized magnons appear also above the bulk bands of the semiinfinite magnets.

Similar results were also calculated for adsorbed slabs. We do not show them here as the position of the localized states due to adsorption was found with a good precision to be similar to those due to the same thin film confined between the same semiinfinite ferromagnets.

5.7.5 Discussion

This section contains the results for the response function of a system of two semiinfinite ferromagnets separated by a ferromagnetic slab. These results are used for the study of the dispersion relations of confined slab magnons. They

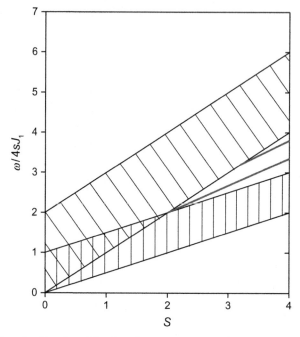

FIG. 5.5 Dispersion [function of $S = 2\cos(k_1 a_0) - \cos(k_2 a_0)$] of two localized magnons due to a three-layered slab sandwiched between two different ferromagnets, $J_2/J_1 = 0.66$, $J_3/J_1 = 0.5$, and $J_1/J_1 = J_{II}/J_1 = 1$.

can be used for the investigations of all the other physical properties of such a system, like, for example, the confined slab magnetization, specific heat, and the response of the system at a given observation point to a magnetic field applied at a given source point.

5.8 SURFACE RECONSTRUCTION AND SOFT SURFACE MAGNONS

Some choices of the exchange parameters J_1 and J_2 (as we have here only one ferromagnet we do not use the index i) may lead to the existence of surface magnon with a negative frequency, a so-called soft magnon. This is reminiscent of an instability at the surface of the ferromagnet. One possibility of rising this instability would be the modification of the magnetic order at the surface, known as a surface reconstruction (or more specifically a superstructure), according to the wave vector where the soft magnon occurs first. In what follows, let us give an example of such an instability. Consider the (110) surface of the simple cubic crystal considered above with interactions J_1 and J_2 between, respectively, first and second nearest-neighbor spins. Let us label by the integer l the atomic planes parallel to this surface. A Fourier transformation in the directions parallel to the (110) surface may be done by using a solution of the form

$$b^\dagger(\mathbf{R}_l) = b^\dagger(l, \mathbf{k}_\parallel)e^{j\mathbf{k}_\parallel \cdot \mathbf{R}_l}, \tag{5.57}$$

where $\mathbf{R}_l$ are the spin positions in the three-dimensional lattice

$$\mathbf{k}_\parallel = k_3\mathbf{z} + k_Y\mathbf{Y} \tag{5.58}$$

and

$$\mathbf{Y} = \frac{1}{\sqrt{2}}(\mathbf{y} - \mathbf{x}). \tag{5.59}$$

One gets then

$$ib^\dagger(l, \mathbf{k}_\parallel) = -2SJ_1\left[(6 - 2\cos k_3 a_0)b^\dagger(l, \mathbf{k}_\parallel) - 2\cos\frac{k_Y a_0}{\sqrt{2}}\sum_{\sigma=\mp 1}b^\dagger(l+\sigma, \mathbf{k}_\parallel) \right]$$

$$- 2SJ_2\left[(12 - 2\cos k_Y a_0\sqrt{2})b^\dagger(l, \mathbf{k}_\parallel) \right.$$

$$\left. -4\cos\frac{k_Y a_0}{\sqrt{2}}\cos k_3 a_0\sum_{\sigma=\mp 1}b^\dagger(l+\sigma, \mathbf{k}_\parallel) - \sum_{\sigma=\mp 2}b^\dagger(l+\sigma, \mathbf{k}_\parallel) \right]. \tag{5.60}$$

Let us chose for all what follows $k_Y a_0 = \pi/\sqrt{2}$ and $k_3 = 0$. Eq. (5.60) becomes

$$ib^\dagger(l, \mathbf{k}_\parallel) = -8SJ_1 b^\dagger(l, \mathbf{k}_\parallel) - 2SJ_2\left[14b^\dagger(l, \mathbf{k}_\parallel) - \sum_{\sigma=\mp 2}b^\dagger(l+\sigma, \mathbf{k}_\parallel) \right]. \tag{5.61}$$

Notice that in this case the equations for the even and odd planes l decouple. One may then address each type of planes independently. The bulk Green's function can be obtained in the same manner as in Eq. (5.11)

$$\mathbf{G}_0(k_3, k_Y, l, l', \omega) = \begin{cases} \frac{1}{J_2}\frac{t^{1+\frac{|l-l'|}{2}}}{t^2-1}, & |l-l'| \quad \text{even} \\ 0, & |l-l'| \quad \text{odd.} \end{cases} \tag{5.62}$$

The definition of t is the same as in Eq. (5.12), with

$$\zeta = 2\frac{J_1}{J_2} + 7 - \frac{\omega}{4SJ_2}. \tag{5.63}$$

Create now two semiinfinite lattices with free (110) surfaces from the infinite lattice. The corresponding perturbation V_s is obtained from the previous equations for the planes $l = -2, -1, 0, 1$. This perturbation has two independent parts for, respectively, the even and odd l planes

$$\mathbf{V}_S(\mathbf{k}_\parallel, l, l') = \mathbf{V}_S^e + \mathbf{V}_S^o, \tag{5.64}$$

$$\mathbf{V}_S^e = -J_2(\delta_{l,-2}\delta_{l',-2} - \delta_{l,0}\delta_{l',-2} - \delta_{l,-2}\delta_{l',0}) - (2J_1 + 5J_2)\delta_{l,0}\delta_{l',0}, \tag{5.65}$$

$$\mathbf{V}_S^o = -J_2(\delta_{l,1}\delta_{l',1} - \delta_{l,1}\delta_{l',-1} - \delta_{l,-1}\delta_{l',1}) - (2J_1 + 5J_2)\delta_{l,-1}\delta_{l',-1}, \tag{5.66}$$

where the indexes e and o stand, respectively, for even and odd.

From Eqs. (5.62), (5.64), one obtains for the kernel of the surface Green's function

$$\Delta = \det(\mathbf{I} - \mathbf{V}_S\mathbf{G}_0) = \Delta_e\Delta_o, \tag{5.67}$$

with

$$\Delta_e = \Delta_o = \frac{-t^2 J_2(2J_1 + 5J_2) + 2t(J_1 + 3J_2)J_2 - J_2^2}{J_2^2(t^2 - 1)}. \tag{5.68}$$

When Δ vanishes, one obtains for the surface magnons the double solution

$$t = \frac{J_2}{2J_1 + 5J_2}. \tag{5.69}$$

This solution is those of a surface mode in agreement with the condition

$$|t| < 1. \tag{5.70}$$

This solution is valid inside all the ferromagnetic region of the (J_1, J_2) plane, see Fig. 5.6.

Using Eqs. (5.63), (5.69) together with

$$\zeta = \frac{1}{2}\left(t + \frac{1}{t}\right), \tag{5.71}$$

one obtains the localized mode frequency

$$\omega_s = \frac{8S(J_1^2 + 7J_1J_2 + 11J_2^2)}{2J_1 + 5J_2}. \tag{5.72}$$

The region of the (J_1, J_2) plane where $\omega_s < 0$ is displayed in Fig. 5.6. It is the hatched region bounded by the lines

$$J_1 + 4J_2 = 0 \tag{5.73}$$

and

$$J_1 + \left(\frac{7 + \sqrt{5}}{2}\right)J_2 = 0. \tag{5.74}$$

This simple example shows that in this region the (110) surface of this ferromagnet is unstable and should lead to a magnetic reconstruction. Note that we explore here only the point $k_Y a_0 = \pi/\sqrt{2}$ and $k_3 = 0$ of the two-dimensional surface Brillouin zone. If other soft surface magnons exist for other points of the Brillouin zone, the instability region of the (J_1, J_2) plane may be larger.

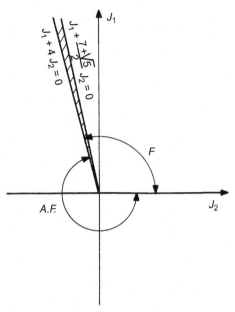

FIG. 5.6 Ferromagnetic and antiferromagnetic stability domains of a simple cubic crystal are, respectively, labeled by F and $A.F.$ in the plane J_1 and J_2, where J_1 and J_2 are the first and second nearest-neighbor exchange interactions. This figure displays also in the *hatched area*, the domain instability of a (110) surface.

Note also that soft localized magnons may exist near other planes like planar defects and interfaces between two different ferromagnets [8].

5.9 THE EFFECT OF SURFACE-PINNING FIELDS ON THE THERMODYNAMIC PROPERTIES OF A FERROMAGNET

5.9.1 Motivations

In a crystal of finite volume, the specific heat, as well as other thermodynamic quantities, will differ from the values computed from a theory that presumes the crystal to have infinite volume. The differences occur because one may have normal modes of the system in which the excitation is localized near the surface, and because the perturbation produced by the presence of the surface alters the distribution in frequency of the bulk modes. The leading corrections to the thermodynamic properties of the crystal from these two sources are proportional to the surface area of the crystal.

A number of theoretical studies of the surface contribution to the specific heat from lattice vibrations have appeared [22, 23]. These studies assume that the surface layer is identical in atomic structure and mass with the appropriate atomic layers in the interior of the crystal. When the temperature is low compared to the Debye temperature, one finds a contribution to the specific heat proportional to the square of the temperature T. In principle, this term should become observable at sufficiently low temperatures, since the bulk phonon-specific heat is proportional to T^3.

There have been a number of theoretical investigations of surface effects on the magnetic degrees of freedom of ferromagnets and antiferromagnets. One finds that surface spin waves appear in the excitation spectrum of these materials under a variety of conditions [24–27]. In the low-temperature region, where spin-wave theory is valid, the mean spin deviation in the vicinity of the surface, and the surface-specific heat of the Heisenberg ferromagnet [28] and antiferromagnet [27] have been studied theoretically. In the Heisenberg ferromagnet, the term in the specific heat proportional to the surface area is found to vary linearly with temperature [28], for low temperatures. Measurements [29] of the specific heat of small YIG particles indicate that a term proportional to T is present in the specific heat. It has been suggested [29] that this excess specific heat is the surface contribution, although the magnitude of this term is larger than the result obtained from the theory [28].

Although the effect on the surface spin-wave frequencies of pinning fields and changes in exchange constants near the surface have been examined [24–27] in some of the investigations, the calculations of the specific heat and mean spin deviation near the surface assume first a perfect semiinfinite array of planes of spins, with no surface-pinning fields present, and exchange constants unaltered near the surface. Thus, the theoretical studies of the effect of surfaces on the thermodynamic properties of crystals have employed severely idealized descriptions of the surface region. While it appears possible to make crystals with nearly perfect surfaces in some instances [30], it is by no means clear that crystals suitable for the study of the size dependence of their thermodynamic properties can be fabricated with surfaces that closely approximate the conditions assumed in the theoretical models. It is thus necessary to examine the influence of surface imperfections on the surface contribution to the thermodynamic properties of the material.

At low temperatures, the surface contribution to the phonon-specific heat and the magnon-specific heat of the antiferromagnet appear insensitive to some kinds of surface imperfections. For example, it has been found [31] that a monolayer of mass defect impurities on the surface of a simple model of a semiinfinite, simple cubic crystal leaves unaltered the coefficient of the leading T^2 term in the phonon contribution to the low-temperature form of the surface-specific heat. This is presumably because at low temperatures, the phonons that make the dominant contribution to the specific heat have wavelengths, which are very long when compared with a lattice constant. These very long-wavelength modes are little affected by a perturbation with spatial extent very much smaller than the wavelength. In the antiferromagnet, the frequency of long-wavelength surface magnons is found to be rather insensitive to changes in the anisotropy field H_A or the exchange field H_E near the surface, in the limit where $H_A \ll H_E$.

The main purpose of this section is to explore the effect of a surface-pinning field on the low-temperature specific heat and mean spin deviation of a simple Heisenberg ferromagnet. We focus our attention on the effect of surface-pinning fields because they can vary over a wide range of values, depending on the nature of the surface and the properties of the magnetic moment bearing ions. If one has a perfect semiinfinite crystal, in which the surface layer has the same atomic structure and lattice constant as a similar crystallographic plane in the interior of the medium, then the crystal field at a site in the surface layer will have lower symmetry than the crystal field at an interior site. This fact, combined with the presence of spin-orbit coupling can give rise to pinning fields in the surface layer [32]. While one might expect this contribution to the pinning field to be small if the magnetic ion is an S-state ion, if orbital degeneracy is present, one may expect rather strong surface-pinning fields to be generated even in a perfect surface. Also, it is possible for magnetically ordered oxide layers to form on the surface. The effect of such layer on the material under study may be phenomenologically represented by introducing an effective pinning field on the ions near the surface.

It is clear from the preceding paragraph that one can encounter effective pinning fields that vary in magnitude over a wide range. Since a pinning field in the surface can strongly affect the spin motion near the boundary, it is clear that the surface contribution to the magnon-specific heat can be severely modified. We shall examine this effect for a wide range of pinning fields in this section.

In what follows, we present a simple treatment of the surface-specific heat and mean sublattice deviation that provides expressions for these quantities in the two limiting cases of zero and infinite pinning fields. We also discuss the intermediate case in a qualitative fashion. Then, we employ a Green's function technique to derive an analytic expression for the surface-specific heat in the presence of a surface-pinning field. We confine our attention to the case of a simple cubic Heisenberg ferromagnet, with nearest-neighbor exchange interactions between the spins, and a (100) surface.

5.9.2 Elementary Considerations

In the studies of the surface-specific heat associated with phonons and spin waves cited earlier, response function techniques were employed in the computations. While the response function techniques are powerful, and allow a large number of diverse problems to be treated within the context of a single formalism, it is often instructive to examine problems with a simple structure by more elementary techniques when this is possible. In this section, we consider a simple cubic array of spins with nearest-neighbor exchange interactions in a film of thickness La, with two (100) surfaces, $L + 1$ atomic layers and a being the lattice parameter. We find that if the pinning field is either 0 or infinite, the surface-specific heat and the mean spin deviation in the vicinity of the surface may be easily computed in a straightforward fashion. The calculation of the surface-specific heat is also carried out for arbitrary pinning fields with a response function.

Before proceeding with these calculations, let us recall briefly some features of the study of the semiinfinite Heisenberg ferromagnet described first by Mills and Maradudin [28] and given earlier in this book chapter. It had been pointed out by Wallis et al. [24], Filippov [25], and Puszkarski [26] that if one considers a semiinfinite simple cubic lattice of spins with a free (100) surface, and with nearest-neighbor and next-nearest-neighbor interactions J_1 and J_2, respectively, then one has surface spin waves associated with the surface region. These surface magnons are not unlike the Rayleigh waves of elasticity theory [33], since they lie below the bulk manifold in frequency. For this particular geometry, the surface wave solutions exist only if the next-nearest-neighbor exchange interaction $J_2 \neq 0$, if the exchange parameters assume their bulk value in the surface region. Mills and Maradudin [28] computed the surface-specific heat, and the spin deviation near the surface for the geometry just described. A response function method was employed in the calculation. Both the quantities computed in the work receive contributions from the thermally excited surface magnons. There are also contributions from the bulk waves, because the frequency distribution of the bulk waves is perturbed by the presence of the surface. A delicate feature of the calculation is a partial cancelation between the perturbed bulk excitation contribution and the surface-wave contribution to the surface-specific heat, and the change in the mean spin deviation near the surface. Although we mentioned earlier that the existence of surface magnons depends on the finiteness of J_2 in an intimate way for this geometry, the magnitude of the surface-specific heat and the functional form of the mean spin deviation near the surface are insensitive to the presence of J_2. The next-nearest-neighbor interaction J_2 enters the result only in the combination $(J_1 + 4J_2)$. Thus, while the calculation of Mills and Maradudin [28] illustrates in detail how one may include the surface-wave portion of the computed quantities, and points out the delicate partial cancelation of the surface wave and bulk wave contributions, the qualitative features of the surface modification of the mean spin deviation near the surface, and the magnon contribution to the specific heat must not depend on whether or not the particular geometry allows the presence of surface magnons.

With these remarks in mind, let us proceed to examine the properties of a simple cubic ferromagnet in the shape of a film of thickness La with (100) surfaces, constructed of n atomic layers. Let there be N_s, spins in each layer, so the total number of spins $N = nN_s$. We are interested in the properties of the film in the limit as $N_s \to \infty$, for fixed n. We assume here that the spins are exchange coupled via nearest-neighbor exchange interactions J, with Hamiltonian[1]

$$H_0 = -J \sum_{ij} S_i S_j. \qquad (5.75)$$

---

1. Since the Hamiltonian of Eq. (5.75) is rotationally invariant, we may choose the magnetization along any convenient axis.

5.9.2.1 Without Surface-Pinning Fields

Thus, we ignore surface-pinning fields for the moment as well as modifications of the exchange coupling near the surface.

We choose a coordinate system with the x_3-axis normal to the film, and the x_1, x_2-axes in the plane of the film. We assume the film placed so it occupies the region $0 < x_3 < La$. For definiteness, let us take the number of film layers to be even.

Now consider the spin-wave spectrum of a material described by the Hamiltonian of Eq. (5.75). For a bulk crystal, the spin-wave dispersion relation is the well-known form, given earlier in this chapter

$$\omega(k_1 k_2 k_3) = 4JS(3 - \cos(k_1 a) - \cos(k_2 a) - \cos(k_3 a)), \tag{5.76}$$

where a is the lattice constant. If we denote the position of a lattice site by the vector $\mathbf{R} = \mathbf{l}a$, where $(\mathbf{l}) = (l_1 l_2 l_3)$ is a set of three integers, then the eigenfunction $|\mathbf{k}\rangle$ for a spin wave in a bulk crystal has the form

$$|\mathbf{k}\rangle = \frac{1}{\sqrt{N}} \sum_\mathbf{l} e^{i\mathbf{k}\mathbf{l}a} a^+(\mathbf{l})|0\rangle, \tag{5.77}$$

where $|0\rangle$ is the ground state and $a^+(\mathbf{l})$ the appropriate boson operator that creates a spin deviation on the site $\mathbf{l}$.

Next consider the film geometry described earlier. The work of Wallis et al. [24], Filippov [25], and Puszkarski [26] shows that when the next-nearest-neighbor exchange $J_2 = 0$, one has no surface spin waves for this geometry, as we mentioned earlier. Thus, we consider the bulk solutions only. Because the full translational symmetry in the x_1 and x_2-directions remains in the presence of the two surfaces, we seek solutions of the Bloch form to be

$$|\mathbf{k}_{//}\rangle = \frac{1}{\sqrt{N_s}} \sum_\mathbf{l} e^{i\mathbf{k}_{//}\mathbf{l}a} f(l_3) a^+(\mathbf{l})|0\rangle, \tag{5.78}$$

where $\mathbf{k}_{//} = k_1 \hat{x}_1 + k_2 \hat{x}_2$, $\hat{x}_1$ is the unit vector in the x_1-direction and $\hat{x}_2$ is the unit vector in the x_2-direction. For the function $f_\alpha(l_3)$, we take a linear combination of the two functions $e^{ik_3 l_3 a}$ and $e^{-ik_3 l_3 a}$, because these are solutions of the bulk equations for the same frequency $\omega(k_1 k_2 k_3)$. Thus, we have

$$f(l_3) = \alpha e^{ik_3 l_3 a} + \beta e^{-ik_3 l_3 a}, \tag{5.79}$$

where $k_3 > 0$. The ratio of the constants α and β, as well as the possible values of k_3, is determined by requiring the solution in Eq. (5.78) to satisfy the equations of motion for spins in the surface layers $l_3 a = 0$ and $l_3 a = La$. This analysis is presented in detail by Maradudin and Mills [28]. For the case considered here, with a simple cubic array of spins coupled by nearest-neighbor interactions only, and free (100) surfaces, a simple result is obtained. One finds that the allowed values of k_3 are:

$$k_3 = 0, \frac{\pi}{La}, \frac{2\pi}{La}, \frac{3\pi}{La}, \ldots. \tag{5.80}$$

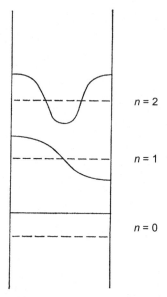

FIG. 5.7 The dependence of the spin deviation with distance from the surfaces of the film for the first few low-lying modes, when the surface-pinning field is zero.

The normalized eigenfunctions reduce to simply

$$f(l_3) = \left(\frac{2}{n}\right)^{1/2} \cos\left[k_3 a \left(l_3 + \frac{1}{2}\right)\right], \tag{5.81}$$

if one has an even number n of layers between $al_3 = 0$ and $al_3 = a(n-1)$ units. The first few lowest-lying eigenfunctions are shown in Fig. 5.7.

Because the eigenfrequencies of the film are given in Eq. (5.76) for the values of k_3 given in Eq. (5.80), we can write down an exact expression for the internal energy of the film, in the context of spin-wave theory

$$U = \frac{A}{(2\pi)^2} \int dk_1 dk_2 \sum_{n=0}^{\infty} \omega\left(k_1 k_2, \frac{n\pi}{La}\right) n_{BE}\left[\omega\left(k_1 k_2, \frac{n\pi}{La}\right)\right]. \tag{5.82}$$

In this expression,

$$n_{BE}(\omega) = \left(e^{\hbar\omega/k_B T} - 1\right)^{-1} \tag{5.83}$$

is the Bose Einstein occupation-number factor, $\hbar$ is the Planck constant, and A is the area of one surface of the film.

Let us next consider the case where the film thickness La is great enough that for fixed k_1 and k_2, the mode spacing is small compared with $k_B T$. We can then replace the sum over the index n in Eq. (5.82) by an integral. Of course, if we carry out this procedure, we recover the well-known expression for the contribution of the spin waves to the internal energy of the crystal. We shall convert the sum into an integral, but retain the first correction term by employing the formula [34] (when $f(\infty) = 0$)

$$\sum_{n=0}^{\infty} f(n) = \int_0^{\infty} dn f(n) + \frac{1}{2} f(0) + \cdots . \tag{5.84}$$

Upon noting that $\omega(k_1 k_2 k_3)$ is an even function of k_3, we obtain

$$U = \frac{ALa}{(2\pi)^2} \int d\mathbf{k} \, \omega(\mathbf{k}) n_{BE}(\omega(\mathbf{k})) + \frac{1}{2} \frac{A}{(2\pi)^2} \int dk_1 dk_2 \omega(k_1 k_2, 0) n_{BE}(\omega(k_1 k_2, 0)). \tag{5.85}$$

The first term of Eq. (5.85) may be recognized as the well-known spin-wave contribution to the internal energy of a large crystal of volume $V = ALa$. The second term is a correction to the internal energy proportional to the surface area of the film. Notice that Eq. (5.85) contains only the first correction to the volume internal energy from the presence of the surfaces. In the limit of very low temperatures, where the finite mode spacing for fixed k_1 and k_2 is important, the internal energy cannot be written simply as the sum of a volume term plus a surface term. We return to this point later in this section.

The integrals in Eq. (5.85) are easily evaluated for temperatures sufficiently low that $\omega(\mathbf{k})$ may be approximated by its long-wavelength form, and the upper limits of the integration may be replaced by infinity. For long wavelengths

$$\omega(\mathbf{k}) = Dk^2, \tag{5.86}$$

where $D = 2SJa^2$.

Upon evaluating the integrals using the low-temperature approximations just mentioned, and differentiating the results to obtain the specific heat

$$C(T) = \frac{\partial U(T)}{\partial T}, \tag{5.87}$$

we find

$$C(T) = k_B \left[V \frac{15}{32\pi^{3/2}} \zeta\left(\frac{5}{2}\right) \left(\frac{k_B T}{D}\right)^{3/2} + S_0 \frac{\zeta(2)}{8\pi} \left(\frac{k_B T}{D}\right) \right], \tag{5.88}$$

where $S_0 = 2A$ is the total surface area of the film.

In Eq. (5.88), the first term is the well-known $T^{3/2}$ spin-wave contribution to the specific heat of a ferromagnet of volume V. The second term is the surface correction to the volume term. The surface term exhibits the linear temperature dependence found in the response function ion calculation of Mills and Maradudin. If one compares the coefficient of the surface term in Eq. (5.88) with the result obtained in Mills and Maradudin [28] for the case $J_2 = 0$, it is found that the surface term in Eq. (5.88) is smaller than the earlier result by a factor of 2. In Mills and Maradudin [28], the surface-specific heat was overestimated by a factor of 2 because one negative contribution to the change in the spin-wave density of states was omitted from the work [35].

For the case in which the next-nearest-neighbor exchange interaction J_2 vanishes, we see that the surface term in the spin-wave-specific heat may be quickly computed by the remarkably simple method described earlier. This is possible because one can solve the boundary value problem for the allowed values of k_3 in a trivial manner for this simple geometry. It appears more difficult to include the effect of next-nearest exchange coupling on the surface-specific heat by this simple approach, although this may readily be done by the response-function method of Mills and Maradudin [28].

Before proceeding with the discussion, we comment on the range of validity of the approximation in Eq. (5.85) to the exact result exhibited in Eq. (5.82). If one examines the most important contribution to the right-hand side of Eq. (5.82), one must have the spacing $\Delta\omega$ between adjacent modes of fixed k_1 and k_2 small compared with k_BT. For fixed k_1 and k_2, the spacing between adjacent modes is given by:

$$\Delta\omega = 2Dk_3\Delta k_3 = (2\pi D/La)k_3. \tag{5.89}$$

The most important values of k_3 are those for which $Dk_3^2 \approx k_BT$. If we require $\Delta\omega$ to be small compared with k_BT for these values of k_3, we find that Eq. (5.85) adequately represents Eq. (5.82) when $La > 2\pi(D/k_BT)^{1/2}$.

In essence, this criterion requires that the film thickness be large compared with the wavelength of a spin wave with energy k_BT. This means that the form given in Eq. (5.85) is a valid representation of the specific heat only when the surface term is a small correction to the volume term. This same criterion for the validity of the decomposition of the magnon-specific heat into a surface and volume term can be seen to apply to the phonon-specific heat [22, 23]. It is not difficult to phrase the preceding argument in general terms.

We next evaluate the mean spin deviation near the lower surface at $l_3 = 0$. This may be quickly done by employing the wave function in Eq. (5.81). If $\Delta_{l_3} = S - \langle S(l_3)\rangle$ is the deviation of a spin in the plane l_3, units from the bottom of the film, we have (with $N = nN_s$, the total number of spins in the film)

$$\Delta(l_3) = \frac{2}{N}\frac{A}{(2\pi)^2}\int dk_1 dk_2 \sum_{n=0}^{\infty} \cos^2\left[\frac{n\pi a}{La}\left(l_3 + \frac{1}{2}\right)\right] n_{BE}\left[\omega\left(k_1 k_2, \frac{n\pi}{La}\right)\right]. \tag{5.90}$$

The spin deviation near the surface of a semiinfinite film at $l_3 = 0$ may be computed by taking the limit as $La \to \infty$ in this result, and replacing the sum over n by an integration. One has then

$$\Delta(l_3) = \frac{2}{N}\frac{ALa}{(2\pi)^2}\int dk_1 dk_2 \int_0^{\infty} dk_3 \times \cos^2\left[k_3 a\left(l_3 + \frac{1}{2}\right)\right] n_{BE}[\omega(k_1 k_2 k_3)]$$

$$= \frac{2a^3}{(2\pi)^3}\int d^3k \cos^2\left[k_3 a\left(l_3 + \frac{1}{2}\right)\right] n_{BE}\left[\omega(\mathbf{k})\right]. \tag{5.91}$$

Now notice

$$2\cos^2\left[k_3 a\left(l_3 + \frac{1}{2}\right)\right] = 1 + \cos[k_3 a(2l_3 + 1)]. \tag{5.92}$$

Thus, we write Eq. (5.91) in the form $\Delta(l_3) = \Delta(\infty) + \delta\Delta(l_3)$, where $\Delta(\infty)$ is the expression for the mean spin deviation in the interior of a bulk crystal, and

$$\delta\Delta(l_3) = \frac{2a^3}{(2\pi)^3}\int d^3k \cos[k_3 a(2l_3 + 1)]n_{BE}[\omega(\mathbf{k})]. \tag{5.93}$$

Eq. (5.93) is precisely the same expression for $\delta\Delta(l_3)$ as the one derived in Mills and Maradudin [28], except that in Mills and Maradudin [28], the spin-wave frequency $\omega(\mathbf{k})$ contained the contribution from the next-nearest-neighbor exchange interaction J_2. The derivative of Eq. (5.93) in the earlier work was quite formidable, because one had to include the contribution from the surface magnons present when $J_2 \neq 0$, and there were cancelations between the surface magnon contribution to $\delta\Delta(l_3)$ and that from the perturbed continuum spin waves.

We briefly recall two properties of the result for $\delta\Delta(l_3)$. In the low-temperature limit, where the thermal magnon wavelengths are long compared with the lattice constant, one easily sees that

$$\delta\Delta(0) = \Delta(\infty)$$

that is, the mean spin deviation at the surface is precisely twice the value in the interior of the crystal. For general values of l_3, the integral in Eq. (5.93) has been evaluated by contour integration [28]. One finds

$$\delta\Delta(l_3) = \frac{1}{4\pi(2l_3 + 1)\alpha}\left\{1 + 2\sum_{m=1}^{\infty}\exp\left[-(2l_3 + 1)\left(\frac{m\pi}{\alpha}\right)^{1/2}\right]\right.$$
$$\left.\times \cos\left[(2l_3 + 1)\left(\frac{m\pi}{\alpha}\right)^{1/2}\right]\right\}$$

where $\alpha = D/k_B T$.

5.9.2.2 With an Infinite Surface-Pinning Field

The preceding discussion applies to crystals in which the surface-pinning field is zero. Let us conclude this section by considering the opposite extreme by supposing the surface-pinning field infinite. For this case, one may write down the spin-wave eigenfunctions at once by noting the boundary condition

$$S^{(+)}(0) = S^{(+)}(La) = 0$$

imposed by the infinite pinning field. The spin-wave eigenfunctions may be write down at once:

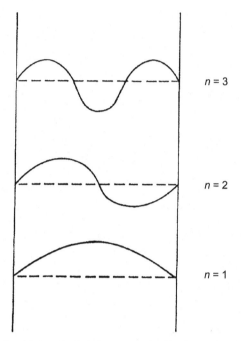

FIG. 5.8   The variation of the spin deviation across the film for the first few low-lying modes associated with a given k_x and k_y, when the pinning field is infinite.

$$|\mathbf{k}_{//}\mathbf{n}\rangle = \sqrt{\frac{2}{N}} \sum_l e^{ik_{//}la} \sin\left(\frac{n\pi l_3}{La}\right) a^+(l)|0\rangle$$

where $n = 1, 2, 3, \ldots$. The spin deviation in the low-lying modes is shown in Fig. 5.8. One has modes characterized by the quantity $k_3 = (n\pi/La))$. For each value of k_1 and k_2, the set k_3 is precisely the same as in the case where the surface pinning was zero, except the mode with $k_3 = 0$ is missing.

This last remark enables one to easily compute the internal energy. We have

$$U = \frac{A}{(2\pi)^2} \int dk_1 dk_2 \sum_{n=1}^{\infty} \omega\left(k_1 k_2, \frac{n\pi}{La}\right) n_{BE}\left[\omega\left(k_1 k_2, \frac{n\pi}{La}\right)\right]$$

$$= \frac{A}{(2\pi)^2} \int dk_1 dk_2 \sum_{n=0}^{\infty} \omega\left(k_1 k_2, \frac{n\pi}{La}\right) n_{BE}\left[\omega\left(k_1 k_2, \frac{n\pi}{La}\right)\right]$$

$$- \frac{A}{(2\pi)^2} \int dk_1 dk_2 \omega(k_1 k_2, 0) n_{BE}[\omega(k_1 k_2, 0)]. \tag{5.94}$$

On the first line of the last equation, we have the internal energy of the film in the absence of the pinning field. The term on the second line is just $2U_s$, where U_s is the surface correction to U for zero pinning field. Upon replacing the sum in the first term by an integral, plus the first correction term as we did earlier, we obtain the simple result

$$U = U_v - U_s,$$ (5.95)

where U_v is the volume contribution to U, and U_s the surface contribution to U for zero pinning field. Thus, when the pinning field is infinite, for the low-temperature form of the specific heat, one has

$$C(T) = k_B \left[V \frac{15}{32\pi^{3/2}} \zeta\left(\frac{5}{2}\right) \left(\frac{k_B T}{D}\right)^{3/2} - S_0 \frac{\zeta(2)}{8\pi} \left(\frac{k_B T}{D}\right) \right]$$

(infinite pinning field). (5.96)

Because the spins near the surface are frozen by the pinning field, the total specific heat is decreased from its value in the bulk. Notice that exact expression for the specific heat is positive, from our earlier discussion, one sees that the form exhibited in Eq. (5.96) is valid only where the surface term is a small correction to the bulk term.

The spin deviation near a surface is easily computed by proceeding as we did earlier, since the eigenfunctions have a simple form. When the temperature is low, and the thermal magnon wavelength is long compared with a lattice constant, one finds $\Delta(l_3)$ is well approximated by the form:

$$\Delta(l_3) = \Delta(\infty) - \delta\Delta(l_3) \quad \text{(infinite-pinning field)}.$$ (5.97)

As one expects, the mean spin deviation in the surface layer is zero because $\delta\Delta(0) = \Delta(\infty)$.

In this section, we have been able to reproduce the principal results of Mills and Maradudin [28] by elementary arguments. Also, we have examined the case where the surface-pinning field is infinite by the same elementary means. We have done this by ignoring all effects associated with the presence of the next-nearest-neighbor exchange J_2, for this geometry. As one sees from the earlier work, inclusion of finite values of J_2 causes a surface magnon branch to split off from below the bulk band, and introduces strong variations in the bulk magnon density of states near the bottom of the band. When one then proceeds with the calculation of the surface-specific heat and mean spin deviation near the surface, the computation becomes nontrivial and is best carried out with response function methods. We find it remarkable that the analysis produces results, which do not differ in structure from the simple formulas exhibited in this section.

5.9.3 Surface-Specific Heat in Presence of Finite Surface-Pinning Field

In this section, we consider the modification of the specific heat by the presence of free surfaces, with pinning fields present in the surface. We confine our attention to a semiinfinite Heisenberg ferromagnet in the form of a simple cubic crystal with free (100) surfaces. We proceed in the manner described by Maradudin and Wallis [23]. We begin with a perfect, infinitely extended crystal with periodic boundary conditions applied to a region of macroscopic volume V. Then one creates two (100) surfaces by "cutting" all interactions that cross a fictitious plane that passes between two (100) atomic planes and then introducing the pinning fields in these two layers. The changes in the state phase shift and in density of states may be obtained from the response function method. We shall see that this response function method yields the first correction term to the volume-specific heat of the spins. This correction term is proportional to the area of the surfaces created by the bond cutting procedure.

5.9.3.1 The State Phase Shift

Let us denote the ferromagnetic ground state of the array by $|0\rangle$. The state with a single spin deviation at the site l will be denoted by $|l\rangle$. Explicitly,

$$|l\rangle = a^+(l)|0\rangle = S^-(l)/(2S)^{1/2}|0\rangle. \qquad (5.98)$$

In the spin-wave approximation, the equations of motion assume the form

$$[\omega I - H]|l\rangle = 0, \qquad (5.99)$$

where, if N is the number of atoms in the crystal, I is the $N * N$ unit matrix, H is an $N * N$ effective Hamiltonian, and the vector $|1\rangle$ is the N component column vector

$$|1\rangle = \begin{pmatrix} |l_1\rangle \\ \vdots \\ |l_i\rangle \\ \vdots \\ |l_N\rangle \end{pmatrix}. \qquad (5.100)$$

In Eq. (5.99), H is the Hamiltonian of the full crystal, in the presence of surface-pinning fields and the "broken" exchange bonds associated with the region between the two surfaces. We write

$$\omega I - H = \omega I - H_0 + V, \qquad (5.101)$$

where H_0 is the Hamiltonian of the perfect, infinitely extend crystal, and V the perturbation that results from the presence of the surfaces. With only nearest-neighbor interactions present, one has

$$\langle l|H_0|l'\rangle = 12JS\delta_{ll'} - 2JS\sum_{\delta}\delta_{l',l+\delta}. \qquad (5.102)$$

In Eq. (5.102), $\delta_{ll'}$ is the Kronecker symbol, and the sum is over the six nearest-neighbor sites that surround the lattice site l.

It is convenient to introduce a new set of basis states that take advantage of the translational symmetry of the perturbed crystal in the 1 and 2 directions. Define

$$|k_1 k_2; l_3\rangle = \frac{1}{N_s^{1/2}} \sum_{l_1 l_2} e^{iak_1 l_1} e^{iak_2 l_2} |l\rangle, \tag{5.103}$$

where k_1 and k_2 are values of the wave vector allowed by the periodic boundary conditions in the 1 and 2 directions.

The corresponding bulk response function is in a closed analytic form for the present model, see Eqs. (5.11)–(5.13)

$$\mathbf{G}(\mathbf{k}_\parallel, \omega; l_3, l_3') = \frac{1}{2SJ} \frac{t^{|l_3 - l_3'|+1}}{t^2 - 1}, \tag{5.104}$$

where

$$t = \begin{cases} \zeta - (\zeta^2 - 1)^{1/2}, & \zeta > 1 \\ \zeta + j(1 - \zeta^2)^{1/2}, & -1 < \zeta < 1 \\ \zeta + (\zeta^2 - 1)^{1/2}, & \zeta < -1 \end{cases} \tag{5.105}$$

and

$$\zeta = \frac{3 - \cos(k_1 a) - \cos(k_2 a) - \omega}{4SJ}. \tag{5.106}$$

The perturbation $\mathbf{V}$ creating two free (001) surfaces, one at $l_3 = 0$ and one at $l_3 = 1$, with a pinning surface field H_S is

$$\mathbf{V} = -\begin{pmatrix} g\mu_B H_S - 2JS & 2JS \\ 2JS & g\mu_B H_S - 2JS \end{pmatrix}, \tag{5.107}$$

where g, μ_B are, respectively, the Landé factor and μ_B the Bohr magneton.

To compute the effect of the surfaces on the spin wave contribution to the internal energy and to the specific heat, one needs to know the change in the density of states of the system produced by the perturbation $\mathbf{V}$. A convenient way of computing the change in density of states is to employ the state phase shift method (e.g., [36]). One forms first the matrix

$$\mathbf{\Delta} = \mathbf{I} + \mathbf{VG}, \tag{5.108}$$

for values of l_3 and l_3' in the interface space $l_3 = 0$ and 1.

With the expressions given here for $\mathbf{V}$ and $\mathbf{G}$, one obtains

$$\det \mathbf{\Delta} = \left[1 - (2 - \epsilon)\frac{t}{t+1}\right]\left[1 - \epsilon\frac{t}{t-1}\right], \tag{5.109}$$

where

$$\epsilon = \frac{g\mu_B H_S}{2SJ} \tag{5.110}$$

is a without dimension parameter measuring the pinning field strength as compared with the exchange interaction J.

The state phase shift due to the creation of the two free surfaces is then

$$\eta(\omega, k_1, k_2) = -\arg \det \boldsymbol{\Delta} = -\frac{\pi}{2} + \tan^{-1}\left[\left(\frac{2}{\epsilon} - 1\right)\left(\frac{1-\zeta}{1+\zeta}\right)^{1/2}\right]. \quad (5.111)$$

5.9.3.2 The Variation of the Density of States

The density of state variation due to the surfaces and to the pinning field can now be obtained (e.g., [36]) from

$$\Delta n(\omega) = \frac{2N_s}{2} \frac{1}{(2\pi)^2} \int dk_1 dk_2 \frac{1}{\pi} \frac{d\eta(\omega, k_1, k_2)}{d\omega}, \quad (5.112)$$

where $2N_S$ is the total number of surface atoms. The integration has to be done inside the bulk band, namely for $-\pi \leq k_1, k_2 \leq \pi$.

For the calculation of the low-temperature variation of the density of states, one needs only to take the contribution of the frequencies such that $\hbar\omega$ very smaller than $k_B T_C$, where T_C is the Curie temperature. In this frequency range, we need only the very small values of k_1 and k_2 in the integration to be done in the previous equation. So we may replace ζ with

$$\zeta \cong 1 + \frac{1}{2}(k_1^2 + k_2^2)a^2 - \frac{\omega}{4SJ}. \quad (5.113)$$

Using the previous three equations, one obtains for the density of states variation due to the two surfaces and the pinning field

$$\Delta n(\omega) = \frac{2N_s}{32\pi SJ}\left[\frac{4}{\pi} \tan^{-1}\left(\left[\frac{1}{\epsilon} - \frac{1}{2}\right]\sqrt{\frac{\omega}{2SJ}}\right) - 1\right]. \quad (5.114)$$

5.9.3.3 The Variation of the Specific Heat

Eq. (5.114) is valid for small ω as compared to the maximum $12SJ$ of the bulk magnon frequencies. One can then deduce, at low temperatures the surface contribution to the internal energy

$$U_S(T) = \int_0^\infty d\omega \, \omega n_{BE}(\omega) \Delta n(\omega), \quad (5.115)$$

where $n_{BE}(\omega)$ is the Bose-Einstein occupation factor of Eq. (5.83).

Then the surface contribution to the specific heat is obtained from

$$C_S(T) = \frac{\partial U_S(T)}{\partial T}. \quad (5.116)$$

It is convenient to calculate first the surface contribution to the specific heat $C_S^0(T)$ for a pinning field $H_S = 0$ and to express $C_S(T)$ in the following form

$$C_S(T) = C_S^0(T)R_C(T),$$ (5.117)

where

$$C_S^0(T) = k_B S_0 \frac{\zeta(2)}{8\pi} \left(\frac{k_B T}{2SJa^2} \right),$$ (5.118)

as obtained before by the elementary considerations, see Eq. (5.88) and

$$R_C(T) = 2J(\alpha) - 1 + \frac{1}{2}\alpha \frac{\partial J(\alpha)}{\partial \alpha},$$ (5.119)

where

$$J(\alpha) = -J(-\alpha) = \frac{2}{\pi \zeta(2)} \int_0^\infty dx x n_{BE}(x) \tan^{-1}(\alpha \sqrt{x}),$$ (5.120)

$$\alpha = \sqrt{\frac{k_B T}{2SJ}} \left(\frac{1}{\epsilon} - \frac{1}{2} \right),$$ (5.121)

and

$$\epsilon = \frac{g\mu_B H_S}{4SJ}.$$ (5.122)

The numerical values of $J(\alpha)$ are obtained by integration of Eq. (5.120) and given in Fig. 5.9.

Also in Fig. 5.10, the specific heat reduction factor $R_C(T)$ is presented as a function of temperature, for various values of the pinning parameter $\epsilon = \frac{g\mu_B H_S}{4SJ}$. The temperature is measured in units of the exchange temperature $T_x = \frac{2SJ}{k_B}$.

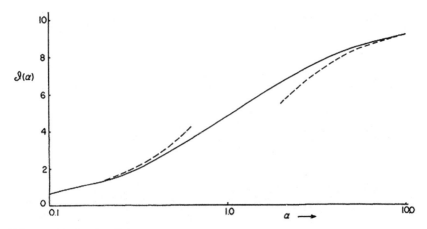

FIG. 5.9 The function $J(\alpha)$.

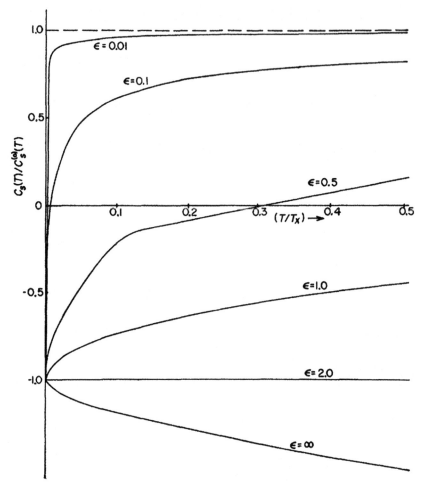

FIG. 5.10 The specific heat reduction factor $R_C(T)$ as a function of temperature, for various values of the pinning parameter $\epsilon = \frac{g\mu_B H_S}{4SJ}$. The temperature is measured in units of the exchange temperature $T_x = \frac{2SJ}{k_B}$.

5.10 SUMMARY

This chapter presents simple examples of surface, interface, and slab magnons. It addresses also magnetic surface reconstructions due to soft surface magnons. The influence of surface-pinning fields on the spin contribution to the thermodynamic properties of a ferromagnet is also presented with a simple model.

REFERENCES

[1] F. Bloch, Zur Theorie des Ferromagnetismus, Z. Phys. 61 (1930) 206–219.

[2] M.G. Cottam, D.R. Tilley, Introduction to Surface and Superlattice Excitations, Cambridge University Press, Cambridge, MA, 1989, p. 85.

[3] R.F. Wallis, A.A. Maradudin, I.P. Ipatova, A.A. Klochitkin, Surface spin waves, Solid State Commun. 5 (1967) 89.

[4] S.E. Trullinger, D.L. Mills, Analogue of surface reconstruction in a Heisenberg ferromagnet, Solid State Commun. 12 (1973) 819.

[5] A. Blandin, Soft surface magnons and magnetic structures of surfaces, Solid State Commun. 13 (1973) 1537.

[6] K. Sattler, H.C. Stegmann, Paramagnetic Sheet at the Surface of the Heisenberg Ferromagnet EuO, Phys. Rev. Lett. 29 (1972) 1565.

[7] A. Herpin, Théorie du Magnétisme, Presses Universitaires, de France, 1968, p. 593.

[8] B. Djafari-Rouhani, L. Dobrzyński, Magnons et surstructures d'interfaces, J. Phys. 36 (1975) 835.

[9] A. Akjouj, B. Sylla, P. Zieliński, L. Dobrzyński, Ferromagnetism of composites with two interfaces, Phys. Rev. B 37 (1988) 5670.

[10] A. Akjouj, B. Sylla, L. Dobrzyński, Introduction à la théorie des systèmes composites: exemples simples de matériaux lamellaires, Ann. Phys. Fr. 18 (1993) 363.

[11] L. Dobrzyński, D.L. Mills, Effect of a surface field on the properties of the Heisenberg ferromagnet, Phys. Rev. 186 (1969) 538.

[12] T. Holstein, H. Primakoff, Field Dependence of the Intrinsic Domain Magnetization of a Ferromagnet, Phys. Rev. 58 (1940) 1098.

[13] T. Jarlborg, A.J. Freeman, Magnetism of metallic superlattices, J. Appl. Phys. 53 (1982) 8041.

[14] T. Shinjo, K. Kawaguchi, R. Yamamoto, N. Hosaito, T. Takada, FE-Mg multilayered films with artificial superstructure, J. Phys. (Paris) Colloq. 45 (1984) CS-367.

[15] N.K. Flevaris, J.B. Ketterson, J.E. Hilliard, Magnetic properties of compositionally modulated thin films, J. Appl. Phys. 53 (1982) 8046.

[16] R.E. Camley, T.S. Rahman, D.L. Mills, Magnetic excitations in layered media: Spin waves and the light-scattering spectrum, Phys. Rev. B 27 (1983) 261.

[17] F. Herman, P. Lambin, D. Jepsen, Electronic and magnetic structure of ultrathin cobalt-chromium superlattices, Phys. Rev. B 31 (1985) 4394.

[18] K. Mika, P. Grunberg, Dipolar spin-wave modes of a ferromagnetic multilayer with alternating directions of magnetization, Phys. Rev. B 31 (1985) 4465.

[19] L. Dobrzyński, B. Djafari-Rouhani, H. Puszkarski, Theory of bulk and surface magnons in Heisenberg ferromagnetic superlattices, Phys. Rev. B 33 (1986) 3251.

[20] M.N. Baibich, J.M. Broto, A. Fert, F. Nguyen Van Dau, F. Petroff, P. Etienne, G. Creuzet, A. Friederich, J. Chazelas, Giant Magnetoresistance of (001)Fe/(001)Cr Magnetic Superlattices, Phys. Rev. Lett. 61 (1988) 2472.

[21] B. Djafari-Rouhani, L. Dobrzyński, Quelques propriétés vibrationnelles et magnétiques d'une interface et d'un défaut plan dans l'approximation élastique, J. Phys. C 10 (1977) 2321.

[22] M. Dupuis, R. Mazo, L. Onsager, Surface Specific Heat of an Isotropic Solid at Low Temperatures, J. Chem. Phys. 33 (1960) 1452.

[23] A.A. Maradudin, R.F. Wallis, Theory of Surface Effects on the Thermal Diffuse Scattering of X Rays or Electrons from Crystal Lattices, Phys. Rev. 148 (1966) 962.

[24] R.F. Wallis, A.A. Maradudin, I.P. Ipatova, A.A. Klochikhin, Surface spin waves, Solid State Commun. 5 (1967) 89.

[25] B.N. Filippov, Soviet Phys. Solid State 9 (1967) 1098.

[26] H. Puszkarski, Surface Spin Waves in Hexagonal Cobalt Thin Films, Phys. Status Solid 22 (1967) 355.

[27] D.L. Mills, W.M. Saslow, Surface Effects in the Heisenberg Antiferromagnet, Phys. Rev. 171 (1968) 488.

[28] D.L. Mills, A.A. Maradudin, Some thermodynamic properties of a semi-infinite Heisenberg ferromagnet, J. Phys. Chem. Solids 28 (1967) 1855.

[29] A.J. Henderson, D.G. Onn, H. Meyer, J.P. Remeika, Calorimetric Study of Yttrium and Rare-Earth Iron Garnets between 0.4 and 4.5° K, Phys. Rev. 185 (1969) 1218.

[30] E.R. Jones, J.T. McKinney, M.B. Webb, Surface Lattice Dynamics of Silver. I. Low-Energy Electron Debye-Waller Factor, Phys. Rev. 151 (1966) 476.

[31] L. Dobrzynski, D.L. Mills, Vibrational properties of an adsorbed surface layer on a simple model crystal, J. Phys. Chem. Solids 30 (1969) 1043.

[32] C. Kittel, Excitation of Spin Waves in a Ferromagnet by a Uniform rf Field, Phys. Rev. 110 (1958) 1295.

[33] L. Rayleigh, On waves propagated along the plane surface of an elastic solid, Proc. Lond. Math. Soc. 17 (1887) 4.

[34] M. Abramowitz, I. Stegun (Eds.), Handbook of Mathematical Functions, US Department of Commerce, Washington, DC, 1964, p. 806.

[35] A.A. Maradudin, D.L. Mills, Some thermodynamic properties of a semi-infinite Heisenberg ferromagnet, J. Phys. Chem. Solids 30 (1969) 784(E).

[36] L. Dobrzynski, A. Akjouj, E. El Boudouti, et al., Interface Response Theory, in: Interface Transmission Tutorial Book Series: Phononics, Elsevier, Amsterdam, 2017, p. 1.

Chapter 6

One-Dimensional Magnonic Crystals

Leonard Dobrzyński*, Abdellatif Akjouj*, Housni Al-Wahsh[†] and Bahram Djafari-Rouhani*

*Department of Physics, Faculty of Sciences and Technologies, Institute of Electronics, Microelectronics and Nanotechnology, UMR CNRS 8520, Lille University, Villeneuve d'Ascq Cedex, France †Faculty of Engineering, Benha University, Cairo, Egypt

Chapter Outline

6.1 Introduction   221
6.2 Bulk Magnetic Response Function for a Two-Slab 1D Crystal   223
6.3 Surface Ferromagnetic Response Function for a Two-Slab 1D Crystal   226
6.4 Bulk and Surface Magnons in a Two-Slab 1D Crystal   227
6.5 Bulk and Surface Magnons in a Three-Slab 1D Crystal   229
6.6 Discussion   231
References   231

6.1 INTRODUCTION

An exciting development in materials science is the appearance of new samples of alternating thin layers of two different materials, with the thickness and composition of each element subject to precise control. The resulting entity may possess new physical properties. Most particularly, by means of sputtering technique, specimens were prepared from two metals, each of which is present as a layer with thickness from a few angstroms to several hundred angstroms [1–6]. The nature of the spin-wave spectrum of such systems started to be studied theoretically [2, 3, 7–11] and experimentally by light scattering [12]. Since these early works on one-dimensional (1D) magnonic crystals called also magnetic superlattices, very exciting developments followed. Our aim here is not to review them all, as this was done before, see, for example, Cottam [13]. Are cited here those who lead to giant magneto resistance and to Albert Fert and Peter Grunberg Nobel prize [14, 15]. Experimental developments [16–21]

Magnonics. https://doi.org/10.1016/B978-0-12-813366-8.00006-6

enabling to master the scattering and transmission of surface spin waves start paving the ways for the use of magnons for circuits.

A large number of papers address the problem of magnon band structure of 1D magnetic composites. Most of these papers focus on the existence of band gaps in the spin-wave spectra. Dobrzynski et al. [10] and Albuquerque et al. [22] calculated the dispersion equation for spin waves propagating in a general direction of an infinite 1D crystal made of two alternating ferromagnetic layers. They showed that in a certain frequency domain the 1D crystal dispersion curves exhibit broad pass bands and narrow stop bands. Dobrzynski et al. [10] investigated the existence of surface-localized magnons in the spin-wave spectra of semiinfinite ferromagnetic 1D crystals. Barnas [23] analyzed theoretically the spin-wave spectra of infinite, semiinfinite, and finite ferromagnetic 1D crystals in the exchange dominated region. Several authors [24, 25] have carried out the study of magnetic properties of 1D crystals constructed by alternating slabs of ferromagnetic and antiferromagnetic layers. Samples have also been prepared in which one of the two materials is a ferromagnetic metal and the second is a nonmagnetic one [12, 26]. The nature of the spin-wave spectrum of such a system was studied theoretically and experimentally by light scattering.

The present chapter addresses first a 1D crystal made from slabs of two different ferromagnetic crystalline materials. Then, it considers the case of three alternating slabs. It studies this 1D magnonic material within an atomic model, the Heisenberg model, including exchange effects between first-nearest neighbors and neglecting dipolar and Zeeman energies. Of course, a more complete study will also have to include these effects. But here, for a tutorial approach of this type of a ferromagnetic system, use is made of only the simplest Heisenberg model.

Note that the present model of a ferromagnetic 1D crystal is, from a mathematical point of view, an easy transposition of a model previously used for the study of surface phonons in 1D crystal [27]. The 1D crystal under study here is built up from alternating L_1 and L_2 (001) atomic planes of two different simple-cubic lattices having the same lattice parameter a_0 and characterized by their Heisenberg exchange interactions (J_1 and J_2) between first nearest-neighbor atoms. The alternating thin layers is bound together by an exchange interaction J between the interface atoms. This simple model enables to obtain in closed form the bulk and (001) surface response function for these 1D crystals. The analytic knowledge of these functions enables to study easily all the bulk and surface magnetic properties of this 1D magnonic crystal.

Are given here the analytic expressions obtained for: the response functions, the folded bulk magnon dispersion curves, and the surface localized modes, which may appear within the extra gaps that exist between the folded bulk bands. These expressions enable to discuss easily the effects of the physical parameters defined previously. These surface magnons also depend on the kind of layer (1 or 2) being near the (001) surface of the 1D crystal.

Then is considered also the case of a 1D crystal made out of three alternating different ferromagnetic slabs.

6.2 BULK MAGNETIC RESPONSE FUNCTION FOR A TWO-SLAB 1D CRYSTAL

Start from an infinite simple-cubic lattice described by the Heisenberg Hamiltonian

$$H_{01} = - \sum_{l,l'} J_1 \mathbf{S}_l \cdot \mathbf{S}_{l'}, \tag{6.1}$$

where are retained only the exchange interactions J_1 between the spins $\mathbf{S}_l$ and $\mathbf{S}_{l'}$ situated on nearest-neighbor atoms. The linearized Holstein-Primakoff transformation enables to rewrite this Hamiltonian in the following form

$$H_{01} = -S \sum_{l,l'} J_1 (b_l^\dagger b_{l'} + b_{l'}^\dagger b_l + b_l^\dagger b_l + b_{l'}^\dagger b_{l'}) + Cte, \tag{6.2}$$

where $b_l^\dagger$ and b_l are the usual creation and annihilation operators, and S is the spin amplitude.

The equations of motion of the $b_l^\dagger$ operators can be written in the following matrix form

$$(\omega \mathbf{I} - \mathbf{H}_{01})\mathbf{u} = 0, \tag{6.3}$$

where $\mathbf{u}$ is a column vector having as many rows as atoms in the crystal and representing the ensemble of the operators $b_l^\dagger$. So the Heisenberg Hamiltonian $\mathbf{H}_{01}$ in this matrix form has matrix elements only between nearest-neighbor atoms.

The diagonalization of this Hamiltonian provides the bulk magnon dispersion relation

$$\omega = 4SJ_1(3 - \cos k_1 a_0 - \cos k_2 a_0 - \cos k_3 a_0), \tag{6.4}$$

where $\mathbf{k}$ is the propagation vector and a_0 the distance between first nearest-neighbor atoms.

Now construct out of this lattice a slab of L_1 layers bounded by a pair of (001) free surfaces. Each (001) atomic plane of this slab is labeled by $1 \leq l_3 \leq L_1$.

The equation of motion of the $b_l^\dagger$ operators of this slab can be written as

$$(\omega \mathbf{I} - \mathbf{H}_1) \cdot \mathbf{u} = 0. \tag{6.5}$$

A corresponding response function $\mathbf{U}_1$ can be defined as

$$(\omega \mathbf{I} - \mathbf{H}_1) \cdot \mathbf{U}_1 = \mathbf{I}, \tag{6.6}$$

where $\mathbf{I}$ is the unit matrix.

The advantage of this model is that this slab response function $\mathbf{U}_1$ can be worked out in closed form, once the corresponding surface response function is known [27]. The mathematical procedure is very similar to the one used for a similar vibrational model of surfaces [28, 29] and slabs [30].

Taking advantage of the periodicity of the slab in directions parallel to the surfaces, introduce a 2D position vector

$$\mathbf{x}_\|(l) = a_0(l_1\hat{x}_1 + l_2\hat{x}_2), \tag{6.7}$$

a 2D wave vector parallel to the surfaces

$$\mathbf{k}_\| = k_1\hat{x}_1 + k_2\hat{x}_2, \tag{6.8}$$

and a Fourier transformation of the response function

$$\mathbf{U}_1(l, l'; \omega) = \frac{1}{N^2} \sum_{\mathbf{k}_\|} \mathbf{U}_1(\mathbf{k}_\|, \omega; l_3, l_3') \exp\{i\mathbf{k}_\| \cdot [\mathbf{x}_\|(l) - \mathbf{x}_\|(l')]\}, \tag{6.9}$$

where N^2 is the number of atoms in a (001) plane.

The explicit expression of $\mathbf{U}_1(\mathbf{k}_\|, \omega; l_3, l_3')$ is calculated as a function of

$$\zeta_1 = 3 - \cos(k_1 a_0) - \cos(k_2 a_0) - \frac{\omega}{4SJ_1} \tag{6.10}$$

and

$$t_1 = \begin{cases} \zeta_1 - (\zeta_1^2 - 1)^{1/2}, & \zeta_1 > 1 \\ \zeta_1 + j(1 - \zeta_1^2)^{1/2}, & -1 < \zeta_1 < 1 \\ \zeta_1 + (\zeta_1^2 - 1)^{1/2}, & \zeta_1 < -1 \end{cases}, \tag{6.11}$$

and is

$$\mathbf{U}_1(\mathbf{k}_\|, \omega; l_3, l_3') = \frac{1}{2SJ_1} \frac{t_1^{|l_3 - l_3'|+1}}{t_1^2 - 1} + \frac{1}{2SJ_1} \frac{t_1^{l_3 + l_3'}}{t_1^2 - 1}$$
$$+ \frac{1}{2SJ_1} \frac{t_1}{t_1^2 - 1} \frac{t_1^{2L_1}}{1 - t_1^{2L_1}} \left(t_1^{-l_3 - l_3'+1} + t_1^{-l_3 + l_3'} + t_1^{l_3 - l_3'} + t_1^{l_3 + l_3'-1} \right). \tag{6.12}$$

In the same manner, construct another slab of $L_2(001)$ layers. In order to distinguish these two slabs one from the other, use an index $\kappa = 1$ or 2. The corresponding response function $\mathbf{U}_2(\mathbf{k}_\|, \omega; l_3, l_3')$ can be obtained from the previous equations by changing all indices 1–2. Remark that for this $\kappa = 2$ slab one has $1 \le l_3 \le L_2$.

This $\kappa = 2$ slab is in epitaxy with the $\kappa = 1$ slab. Characterize this double slab by another integer n. An infinite repetition $-\infty < n < +\infty$ of this double slab gives the starting point for this model of a ferromagnetic 1D crystal. Couple all these alternating $\kappa = 1$ and $\kappa = 2$ slabs by exchange interactions J between the interface atoms facing each other. This creates the infinite 1D crystal. In the same manner as for the slab (Eq. 6.6), define a response function $\mathbf{D}$ for this 1D crystal. Its elements can be worked out explicitly and are given as functions of the q_κ defined in terms of the $t_\kappa = 1$ or 2, of Eq. (6.11) by

$$t_\kappa = e^{q_\kappa} \tag{6.13}$$

and a new variable t defined by

$$t = \begin{cases} \eta - (\eta^2 - 1)^{1/2}, & \eta > 1 \\ \eta + j(1 - \eta^2)^{1/2}, & -1 < \eta < 1 \\ \eta + (\eta^2 - 1)^{1/2}, & \eta < -1 \end{cases} \tag{6.14}$$

with

$$\eta = B_1 B_2 + \frac{1}{2J^2} C_1 C_2 + \frac{1}{2}(B_1 C_2 + B_2 C_1) + \frac{1}{2}(A_1 C_2 + A_2 C_1), \tag{6.15}$$

where

$$A_\kappa = \frac{1}{J_\kappa} \frac{sh[(L_\kappa - 1)(q_\kappa)]}{sh(q_\kappa)}, \tag{6.16}$$

$$B_\kappa = \frac{ch[(L_\kappa - 1/2)q_\kappa]}{ch(q_\kappa/2)}, \tag{6.17}$$

and

$$C_\kappa = 2J_\kappa th(q_\kappa/2)sh(L_\kappa q_\kappa). \tag{6.18}$$

Note that A_κ is function of $(L_\kappa - 1)$, B_κ is function of $(L_\kappa - 1/2)$, and C_κ is function of L_κ. In what follows, one sees that the 1D crystal bulk and surface modes may be expressed in compact forms with the help of these entities. However, in the analytical calculation process, it is also helpful to use the following entities:

$$A_\kappa(l_3) = \frac{1}{J_\kappa} \frac{sh[(l_3)(q_\kappa)]}{sh(q_\kappa)}, \tag{6.19}$$

$$B_\kappa(l_3) = \frac{ch[(l_3)q_\kappa]}{ch(q_\kappa/2)}, \tag{6.20}$$

and

$$C_\kappa(l_3) = 2J_\kappa th(q_\kappa/2)sh[(l)q_\kappa], \tag{6.21}$$

where l_3 may indicate any layer position difference within the κ slab of the 1D crystal.

The magnetic response function of a 1D crystal is expressed here below with the above six entities. Are given now the explicit expressions of the elements $D(k_\parallel, \omega | n, \kappa, l_3; n', \kappa', l_3')$ of the 1D crystal response function $\mathbf{D}$. The elements of $\mathbf{D}$ between different $\kappa = 1$ and $\kappa = 2$ slabs are

$D(n, 1, l_3; n', 2, l_3')$

$$= \frac{t}{t^2 - 1}[K_{12}(l_3, l_3')t^{|n-n'|} + K_{12}(L_1 - l_3 + 1, L_2 - l_3' + 1)t^{|n-n'-1|}], \tag{6.22}$$

where

$$K_{12}(l_3, l_3') = \frac{1}{2JS} B_1 \left(l_3 - \frac{1}{2} \right) B_2 \left(L_2 - l_3' + \frac{1}{2} \right)$$
$$+ \frac{1}{2S} \left[A_1(l_3 - 1)B_2 \left(L_2 - l_3' + \frac{1}{2} \right) + A_2(L_2 - l_3')B_1 \left(l_3 - \frac{1}{2} \right) \right]$$

$$(6.23)$$

and

$$D(n, 2, l_3; n', 1, l_3') = D(n, 1, l_3'; n', 2, l_3). \qquad (6.24)$$

The elements of **D** between the same κ slabs are

$$D(n, 1, l_3; n', 1, l_3') = \frac{1}{2S} A_1(l_3 - l_3')t^{|n-n'|} sgn[I_1(n - n') + l_3 - l_3']$$
$$+ \frac{1}{2J_1 S} K_{11}(l_3, l_3') \frac{t^{|n-n'|+1}}{t^2 - 1}, \qquad (6.25)$$

where

$$K_{11}(l_3, l_3') = \frac{J_1}{4\cosh(q_1/2)} [B_1(L_1 + l_3 - l_3') + B_1(L_1 - l_3 + l_3')$$
$$+ 2B_1(L_1 - l_3 - l_3' + 1)] \times \left[\frac{C_2(L_2)}{J^2} + \frac{2}{J} B_2(L_2 - 1/2) + A_2(L_2 - 1) \right]$$
$$+ \frac{1}{2\sinh(q_1)} \left[\frac{C_2(L_2)\cosh(q_1/2)}{2J_1 \sinh(q_1)} [B_1(L_1 + l_3 - l_3' - 1) + B_1(L_1 - l_3 + l_3' - 1) \right.$$
$$- 2B_1(L_1 - l_3 - l_3' + 1)] + 2J_1 \sinh(q_1)[A_1(L_1 - 1/2)B_1(l_3 - l_3')$$
$$\left. -A_1(1/2)B_1(L_1 - l_3 - l_3' + 1)] \times [C_2(L_2)/J + B_2(L_2 - 1/2)] \right] \qquad (6.26)$$

and $D(n, 2, l_3; n', 2, l_3')$ can be obtained from $D(n, 1, l_3; n', 1, l_3')$ by interchanging in J_κ and L_κ all the indices $\kappa = 1$ and 2. Now proceed to use these results for the calculation of the surface response functions of this magnonic 1D crystal.

6.3   SURFACE FERROMAGNETIC RESPONSE FUNCTION FOR A TWO-SLAB 1D CRYSTAL

Consider a surface slab with same width as corresponding bulk slabs. So one can create two free surfaces by equating to zero all interactions between atoms in the plane ($n = 0$, $\kappa = 2$, $l_3 = L_2$) and atoms in the plane ($n = 1$, $\kappa = 1$, $l_3 = 1$). Define a corresponding surface response function **G**. The relation between **G** and the one of the infinite 1D crystal **D** enables to find [27] for n and $n' \geq 1$:

$$G(n, \kappa, l_3; n', \kappa', l_3') = D(n, \kappa, l_3; n', \kappa', l_3')$$
$$+ \frac{J}{\Delta_{S1}} D(1, 1, 1; n', \kappa', l_3')[D(n, \kappa, l_3; 0, 2, L_2) - D(n, \kappa, l_3; 1, 1, 1)], \quad (6.27)$$

where

$$\Delta_{S1} = \frac{tA_S - 1}{t^2 - 1} \tag{6.28}$$

with

$$A_S = B_1 B_2 + B_1 C_2/J + C_2 A_1. \tag{6.29}$$

It is also possible to calculate in closed form the surface response function for a surface slab with width smaller than corresponding bulk slabs, see Dobrzynski et al. [10].

Now proceed to use the earlier results for the determination of the bulk and surface magnons in a 1D crystal.

6.4 BULK AND SURFACE MAGNONS IN A TWO-SLAB 1D CRYSTAL

The bulk magnons of the 1D crystal made out of two different alternating layers can be obtained from the knowledge of the bulk response function (Eqs. 6.22–6.26). First, recall that for the infinite simple-cubic lattice described earlier, the bulk magnon dispersion relation (Eq. 6.4), can be obtained from Eq. (6.10) and is given by

$$\zeta_1 = \cos(k_3 a_0), \quad -\pi < k_3 a_0 < +\pi. \tag{6.30}$$

In the same manner, Dobrzynski et al. [27] for the infinite 1D crystal, one obtains the bulk magnons from

$$\eta = \cos[k_3(L_1 + L_2)a_0], \tag{6.31}$$

where η is given by Eq. (6.15). Because the periodicity in the direction x_3 is now given by $(L_1 + L_2)a_0$, one has

$$-\pi < k_3(L_1 + L_2)a_0 < +\pi. \tag{6.32}$$

More explicitly, the bulk magnons of this 1D crystal are given by

$$\cos[k_3(L_1 + L_2)a_0] = B_1 B_2 + \frac{1}{2J^2} C_1 C_2$$
$$+ \frac{1}{2}(B_1 C_2 + B_2 C_1) + \frac{1}{2}(A_1 C_2 + A_2 C_1). \tag{6.33}$$

As a consequence of this larger periodicity in the direction x_3, one has a folding of the magnon dispersion curves in a reduced Brillouin zone specified by Eq. (6.32) and an opening of new gaps between these folded dispersion curves (see Fig. 6.1).

In the gaps, new surface magnons may appear. They can be found from the new poles in the surface response functions (Eqs. 6.27, 6.28) due to the creation of the free surface.

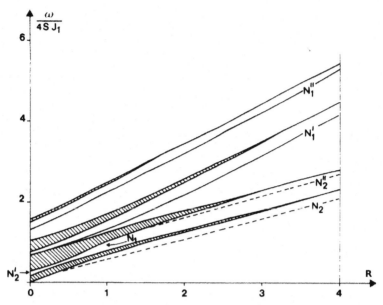

FIG. 6.1 Bulk (*shaded areas*) and surface (*solid and dashed lines*) magnons of a 1D crystal with two atomic planes in each slab in function of $R = 2 - \cos k_1 a_0 - \cos k_2 a_0$. The slab labeled by the index $\kappa = 1$ is at the surface. The N'_1, N''_1 surface magnons are obtained for $J_1 = 2J_2$, and the N'_2, N''_2 surface magnons are obtained for $J_1 = \frac{J_2}{2}$.

In particular, in this case for which the surface slab has the same width as the corresponding bulk slabs, the diagonal element of the surface response function **G** on the surface plane is

$$G(1,1,1;1,1,1) = \frac{(t - A_S)}{2SD(\omega)}, \tag{6.34}$$

where

$$D(\omega) = \frac{C_1 C_2}{J} + C_2 B_1 + C_1 B_2. \tag{6.35}$$

From the poles of this response function, one finds the frequencies ω_S of the surface modes localized at the free surface ($n = 1$, $\kappa = 1$, $l_3 = 1$) and decaying inside the bulk situated at n and $l_3 > 1$. They are given by

$$D(\omega_s) = 0, \tag{6.36}$$

together with the condition coming from Eq. (6.28)

$$\left| \frac{C_2}{C_1} \right| > 1, \tag{6.37}$$

which ensures that these modes decay inside the bulk [27].

In Fig. 6.1 are presented the bulk and surface magnons of a 1D crystal with two atomic planes in each slab ($L_1 = L_2 = 2$) in function of the parameter

$R = 2 - \cos(k_1 a_0) - \cos(k_2 a_0)$. One assumes here $J_1 = 2J_2$ or $J_1 = J_2/2$. The exchange at the interface is assumed to be $J = (J_1 + J_2)/2$. The slab labeled by the index $\kappa = 1$ is at the surface. The shaded areas represent the bulk magnons (four here for each value of $k_\parallel$). The N_1', N_1'' surface magnons (solid line) refer to $J_1 = 2J_2$, the N_2', N_2'' surface magnons (dashed line) to $J_1 = J_2/2$.

6.5 BULK AND SURFACE MAGNONS IN A THREE-SLAB 1D CRYSTAL

Consider now three different ferromagnetic slabs and construct out of them a three-slab 1D crystal. Call now $J_{\kappa,\kappa'}$ the first nearest-neighbor exchange interactions at the interfaces between the κ and κ' slabs. An analytic derivation, similar to the one given above for a two-layer 1D crystal may be done. In this section, are just given its results.

The bulk magnons of this three slab 1D crystal are given by

$$2\cos[k_3(L_1 + L_2 + L_3)a_0] = 2B_1B_2B_3 + \frac{C_1C_2C_3}{J_{12}J_{23}J_{31}}$$
$$+ \epsilon_{ijk}\left[B_iA_jC_k + \frac{1}{2}\left(\frac{1}{J_{12}} + \frac{1}{J_{23}} + \frac{1}{J_{31}} \right) B_iB_jC_k \right.$$
$$+ \left. \frac{1}{2J_{jk}}\left[A_iC_jC_k + \left(\frac{1}{J_{ki}} + \frac{1}{J_{ij}} \right) B_iC_jC_k \right] \right],$$

(6.38)

where are used the notations defined by Eqs. (6.16)–(6.18). ϵ_{ijk} is the Levi-Civita symbol. It is equal to 1 if $ijk = 123$, 231, or 312. It is equal to -1 if $ijk = 321$, 132, or 213. It is equal to 0 if $i = j$ or $j = k$ or $k = i$. The previous equation uses the Einstein convention, which means summation on the repeated indexes.

Fig. 6.2 is presented the magnon band structure of the 1D crystal for $L_1 = L_2 = L_3 = 2$. The bulk bands are represented by the hatched areas. Between them exist gaps where may appear surface magnons discussed here after.

In order to obtain a 1D crystal with a free surface, one proceeds as explained earlier for the 1D crystal made out of two different alternating slabs. One works out the surface element of the three-slab 1D crystal response function $\mathbf{G}'$. For a semiinfinite 1D crystal with the slab 1 at the surface, followed by slab 2, slab 3, and so on, one obtains

$$G'(1, 1, 1; 1, 1, 1) = \frac{(t - A_S')}{2SD'(\omega)},$$

(6.39)

where

$$A_S' = B_1B_2B_3 + A_1C_2B_3 + A_1B_2C_3 + B_1A_2C_3 + \frac{B_1C_2C_3}{J_{12}J_{23}}$$
$$+ \frac{1}{J_{23}}(A_1C_2C_3 + B_1B_2C_3) + \frac{1}{J_{12}}(B_1C_2B_3 + B_1B_2C_3)$$

(6.40)

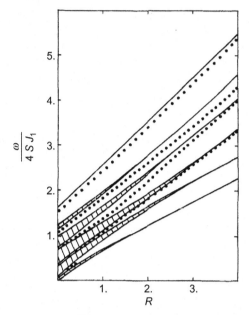

FIG. 6.2 Magnon band structure $\frac{\omega}{4S_1 J_1}$ for a three-slab 1D lattice, with $L_1 = L_2 = L_3 = 2$, as a function of $R = 2 - \cos(k_1 a_0) - \cos(k_2 a_0)$. The exchange interaction values are $J_3/J_1 = 0.5$, $J_2/J_1 = J_{31}/J_1 = 0.75$, $J_{12}/J_1 = 0.875$, $J_{23}/J_1 = 0.675$, and $S_2 = S_3 = S_1 = S$. The *hatched areas* show the bulk bands. The *dotted lines* are for the surface magnons when the slab $i = 1$ is at the surface.

and

$$D'(\omega) = B_1 B_2 C_3 + B_1 C_2 B_3 + C_1 B_2 B_3 + C_1 A_2 C_3 + \frac{C_1 C_2 C_3}{J_{12} J_{23}}$$

$$+ \frac{1}{J_{23}}(C_1 B_2 C_3 + B_1 C_2 C_3) + \frac{1}{J_{12}}(C_1 C_2 B_3 + C_1 B_2 C_3). \quad (6.41)$$

The frequencies of the surface modes are obtained from

$$D'(\omega_S) = 0, \quad (6.42)$$

together with the condition

$$|A'_S(\omega_S)| > 1, \quad (6.43)$$

which comes from the fact that the amplitude of these surface modes decay from the surface to the bulk of the 1D crystal. These surface modes are function of the nature of the surface slab, of its width, and also of the arrangement of the different slabs starting from the surface. For example, the succession 1/2/3, ... gives different surface magnons than the succession 1/3/2, ...

Fig. 6.2 shows the surface magnons for the succession 1/2/3, ... For this case four surface magnon branches appear in most of the Brillouin zone.

6.6 DISCUSSION

This chapter presents examples of surface and bulk magnons on a simple 3D atomic Heisenberg model of a ferromagnetic 1D crystal. The simplicity of this model enables to derive in closed form the bulk and surface magnetic response functions for this 1D crystals. From the poles of these response functions, one obtains analytic expressions for the bulk and surface magnons of this 1D crystal. The surface magnons given here are a special feature of this 1D crystal, as this model has neither surface nor interface magnons [31] localized on each of the two homogenous materials.

REFERENCES

[1] Z.Q. Zheng, C.M. Falco, J.B. Ketterson, I.K. Schuller, Magnetization of compositionally modulated CuNi films, Appl. Phys. Lett. 38 (1981) 424.

[2] T. Jarlborg, A.J. Freeman, Magnetism of metallic superlattices, J. Appl. Phys. 53 (1982) 8041.

[3] K. Flevaris, J.B. Ketterson, J.E. Hilliard, Magnetic properties of compositionally modulated thin films, J. Appl. Phys. 53 (1982) 8046.

[4] T. Shinjo, K. Kawaguchi, R. Yamamoto, N. Hosaito, T. Takada, Fe-Mg Multilayered films with artificial superstructure, J. Phys. (Paris) Colloq. 45 (1984) C5-367.

[5] A.J. Freeman, J.-H. Xu, S. Ohnishi, T. Jarlborg, Electronic structure and magnetism of interfaces: sandwiches and modulated structures, J. Phys. (Paris) Colloq. 45 (1984) C5–369.

[6] R. Krishnan, W. Jantz, Ferromagnetic resonance studies in compositionally modulated cobalt-niobium films, Solid State Commun. 50 (1984) 533.

[7] R.E. Camley, T.S. Rahman, D.L. Mills, Magnetic excitations in layered media: spin waves and the light-scattering spectrum, Phys. Rev. B 27 (1983) 261.

[8] F. Herman, P. Lambin, O. Jepsen, Electronic and magnetic structure of ultrathin cobalt-chromium superlattices, Phys. Rev. B 31 (1985) 4394.

[9] K. Mika, P. Grunberg, Dipolar spin-wave modes of a ferromagnetic multilayer with alternating directions of magnetization, Phys. Rev. B 31 (1985) 4465.

[10] L. Dobrzynski, B. Djafari-Rouhani, H. Puszkarski, Theory of bulk and surface magnons in Heisenberg ferromagnetic superlattices, Phys. Rev. B 33 (1986) 3251.

[11] A. Akjouj, L. Dobrzynski, B. Djafari-Rouhani, Magnons dans les super réseaux ferromagnétiques à 3-couches, J. Phys. (Paris) IV Colloq. 2 (1992) C3-191.

[12] A. Kueny, M.R. Khan, I.K. Schuller, M. Grimsditch, Phys. Rev. B 29 (1984) 2879.

[13] M.G. Cottam (Ed.), Linear and Nonlinear Spin Waves in Magnetic Films and Superlattices, World Scientific, Singapore, 1994.

[14] M.N. Baibich, J.M. Broto, A. Fert, F. Nguyen Van Dau, F. Petroff, P. Etienne, G. Creuzet, A. Friederich, J. Chazelas, Giant Magnetoresistance of (001)Fe/(001)Cr Magnetic Superlattices, Phys. Rev. Lett. 61 (1988) 2472.

[15] G. Binasch, P. Grunberg, F. Saurenbach, W. Zinn, Enhanced magnetoresistance in layered magnetic structures with antiferromagnetic interlayer exchange, Phys. Rev. B 39 (1989) 4828.

[16] A.V. Chumak, A.A. Serga, B. Hillebrands, M.P. Kostylev, Scattering of backward spin waves in a one-dimensional magnonic crystal, Appl. Phys. Lett. 93 (2008) 022508.

[17] A.V. Chumak, P. Pirro, A.A. Serga, M.P. Kostylev, R.L. Stamps, H. Schultheiss, K. Vogt, S.J. Hermsdoerfer, B. Laegel, P.A. Beck, B. Hillebrands, Spin-wave propagation in a microstructured magnonic crystal, Appl. Phys. Lett. 95 (2009) 262508.

[18] A.V. Chumak, A.A. Serga, S. Wolff, B. Hillebrands, M.P. Kostylev, Design and optimization of one-dimensional ferrite-film based magnonic crystals, J. Appl. Phys. 105 (2009) 083906.

[19] A.V. Chumak, A.A. Serga, S. Wolff, B. Hillebrands, M.P. Kostylev, Scattering of surface and volume spin waves in a magnonic crystal, Appl. Phys. Lett. 94 (2009) 172511.

[20] A.D. Karenowska, A.V. Chumak, A.A. Serga, J.F. Gregg, B. Hillebrands, Magnonic crystal based forced dominant wavenumber selection in a spin-wave active ring, Appl. Phys. Lett. 96 (2010) 082505.

[21] Z.K. Wang, V.L. Zhang, H.S. Lim, S.C. Ng, M.H. Kuok, S. Jain, A.O. Adeyeye, Nanostructured magnonic crystals with size-tunable bandgaps, ACSNANO 4 (2010) 643.

[22] E.L. Albuquerque, P. Fulco, E.F. Sarmento, D.R. Tilley, Spin waves in a magnetic superlattice, Solid State Commun. 58 (1986) 41.

[23] J. Barnas, Exchange modes in ferromagnetic superlattices, Phys. Rev. B 45 (10) (1992) 427.

[24] L.L. Hinchey, D.J. Mills, Magnetic properties of superlattices formed from ferromagnetic and antiferromagnetic materials, Phys. Rev. B 33 (1986) 3329.

[25] L.L. Hinchey, D.J. Mills, Magnetic properties of ferromagnet-antiferromagnet superlattice structures with mixed-spin antiferromagnetic sheets, Phys. Rev. B 34 (1986) 1689.

[26] M. Grimsditch, M.R. Khan, A. Kueny, I.K. Schuller, Collective behavior of magnons in superlattices, Phys. Rev. Lett. 51 (1983) 498.

[27] L. Dobrzynski, B. Djafari-Rouhani, O. Hardouin Duparc, Theory of surface phonons in superlattices, Phys. Rev. B 29 (1984) 3138.

[28] L. Dobrzynski, D.L. Mills, Effect of a a surface field on the properties of the Heisenberg ferromagnet, Phys. Rev. B 186 (1969) 538.

[29] L. Dobrzynski, Quelques propriétés vibrationnelles et magnétiques des surfaces cristallines, Ann. Phys. (Paris) 4 (1969) 637.

[30] P. Mazur, A.A. Maradudin, Mean-square displacements of atoms in thin crystal films, Phys. Rev. B 24 (1981) 2996.

[31] B. Djafari-Rouhani, L. Dobrzynski, Magnons et surstructures d'interfaces, J. Phys. (Paris) 36 (1975) 835.

Chapter 7

Two-Dimensional Magnonic Crystals

Leonard Dobrzyński*, Housni Al-Wahsh†, Abdellatif Akjouj*, Yan Pennec* and Bahram Djafari-Rouhani*

*Department of Physics, Faculty of Sciences and Technologies, Institute of Electronics, Microelectronics and Nanotechnology, UMR CNRS 8520, Lille University, Villeneuve d'Ascq Cedex, France †Faculty of Engineering, Benha University, Cairo, Egypt

Chapter Outline

7.1	Introduction	233	7.4	Summary	245
7.2	Method of Calculation	234		References	247
7.3	Magnon Band Structures	239			

7.1 INTRODUCTION

This chapter presents a tutorial introduction to the theoretical investigations of two-dimensional (2D) inhomogeneous magnetic media. Two-dimensional crystals are introduced by Yablonovitch [1] in Photonics. Sigalas and Economou [2] and Kushwaha et al. [3] start independently 2D Phononics. Dobrzynski et al. [4] introduce 2D Electronics. Vasseur et al. [5] initiate then 2D magnonics. Magnon band structure calculations may be undertaken for 2D periodic bi-material composites such as square arrays of infinite cylinders embedded in a host material. Many investigations are published, see, for example, Refs. [5–16]. The very important contributions of this exponentially growing research field are not fully reviewed here. The aim of this chapter is mostly tutorial. The readers, who would like to try and contribute to this fast-developing field, are advised to do their own bibliography in connection with their own interests, not only in the magnonic field but also in the photonic and phononic ones, see, for example, Refs. [17, 18] as this usually provides many possible transpositions.

Magnonics. https://doi.org/10.1016/B978-0-12-813366-8.00007-8

7.2 METHOD OF CALCULATION

Consider a model system composed of an array of infinite cylinders of circular cross-section made of a ferromagnetic material A embedded in an infinite ferromagnetic matrix B. The cylinders are assumed to be parallel to the x_3-axis of the Cartesian coordinate system. Consequently, intersections of their axis with the transverse (x_1, x_2) plane form a 2D periodic structure (square lattice, see Fig. 7.1).

The square lattice parameter is a and the filling fractions are f and $(1 - f)$ for the materials A and B, respectively. In the theoretical model, taking the dipolar interactions and the exchange coupling into account [19], both ferromagnetic materials A and B are described by their spontaneous magnetization M_{S_A} and M_{S_B} and exchange constants A_A and A_B. Thus the spontaneous magnetization and the exchange constant in the composite system are space dependent with respect to the position vector $\mathbf{X} = (x_1, x_2)$ in the transverse plane and can be written as

$$M_S = M_{S_A} \delta_m + M_{S_B}(1 - \delta_m) \tag{7.1}$$

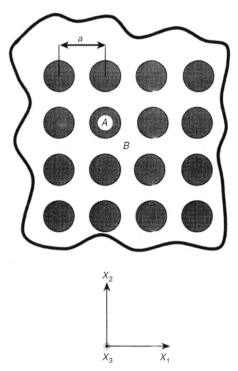

FIG. 7.1 A transverse cross-section of the binary composite system: a square array of infinite cylinders (A) periodically distributed in an infinite matrix (B). The lattice parameter is a.

and

$$A = A_A \delta_m + A_B(1 - \delta_m),\tag{7.2}$$

where δ_m equals 1 inside the inclusions and 0 outside. A static magnetic field H_0 is applied along the x_3-axis and both ferromagnetic materials are assumed to be magnetized parallel to H_0.

The equation of motion in the composite material is

$$\frac{\partial}{\partial t}\mathbf{M}(\mathbf{r}, t) = \gamma \mu_0 \mathbf{M}(\mathbf{r}, t) \times \mathbf{H}_{eff}(\mathbf{r}, t),\tag{7.3}$$

where γ is the gyromagnetic ratio ($\gamma < 0$), assumed to be the same in materials A and B and $\mathbf{H}_{eff}(\mathbf{r}, t)$ is the effective field acting on the magnetization $\mathbf{M}(\mathbf{r}, t)$, $\mathbf{r}$ being the 3D position vector. Eq. (7.3) is the Landau-Lifshitz equation without a damping term [19, 20].

Neglecting an anisotropy field, $\mathbf{H}_{eff}(\mathbf{r}, t)$ can be written for an inhomogeneous material

$$\mathbf{H}_{eff}(\mathbf{r}, t) = H_0 \, \mathbf{e}_3 + \mathbf{h}(\mathbf{r}, t) + \frac{2}{\mu_0 M_S}\left(\nabla \cdot \frac{A}{M_S}\nabla\right)\mathbf{M}(\mathbf{r}, t),\tag{7.4}$$

where

$$\nabla = \mathbf{e}_1(\partial/\partial x_1) + \mathbf{e}_2(\partial/\partial x_2) + \mathbf{e}_3(\partial/\partial x_3),\tag{7.5}$$

($\mathbf{e}_1$, $\mathbf{e}_2$, $\mathbf{e}_3$ are unit vectors along the x_1, x_2, x_3 axes, respectively), $\mathbf{h}(\mathbf{r}, t)$ is the dynamic dipolar field, and the last term describes the exchange field. The difference of this equation from the corresponding one for homogeneous media is in the $\frac{A}{M_S}$ scalar put in between the two ∇, in order to satisfy automatically the boundary conditions [19, 21] at the internal interfaces.

The boundary conditions imply the continuity at the internal interfaces of $(A/M_S)(\partial/\partial n)\mathbf{M}$ where $(\partial/\partial n)$ is an interface normal derivative. Such boundary conditions include the macroscopic variations of A and M_S (Eqs. 7.1, 7.2).

Different boundary conditions [6, 7] are sometimes adopted for an inhomogeneous material with a different form for $\mathbf{H}_{eff}(\mathbf{r}, t)$ in our Eq. (7.4)

$$\mathbf{H}_{eff}(\mathbf{r}, t) = H_0 \, \mathbf{e}_3 + \mathbf{h}(\mathbf{r}, t) + \frac{2}{\mu_0}\left(\nabla \cdot \frac{A}{M_S^2}\nabla\right)\mathbf{M}(\mathbf{r}, t).\tag{7.6}$$

Such particular boundary conditions imply the continuity at the internal interfaces of $(A/M_S^2)(\partial/\partial n)\mathbf{M}$ rather than that of the usual $(A/M_S)(\partial/\partial n)\mathbf{M}$, where $(\partial/\partial n)$ is an interface normal derivative.

The microscopic modifications of the exchange constant A at these interfaces (so-called interface exchange coupling) are not taken into account. It is justified to neglect such modifications since such microscopic variations of the exchange have negligible effects on the bulk magnons studied in this chapter. The interface exchange coupling can lead to nonnegligible effects only when the thicknesses of the materials out of which the composite material is build out are of the order of a few inter atomic distances [22, 23].

In order to resolve Eq. (7.3), write

$$\mathbf{M}(\mathbf{r}, t) = M_S\, \mathbf{e}_3 + \mathbf{m}(\mathbf{r}, t), \tag{7.7}$$

where $\mathbf{m}(\mathbf{r}, t)$ is the dynamic component of the magnetization. Suppose that

$$\mathbf{m}(\mathbf{r}, t) = \mathbf{m}(\mathbf{r})e^{-i\omega t}, \tag{7.8}$$

where ω is the wave circular frequency and

$$\mathbf{h}(\mathbf{r}, t) = \mathbf{h}(\mathbf{r})e^{-i\omega t}, \tag{7.9}$$

with $\mathbf{h}(\mathbf{r}) = -\nabla\Psi(\mathbf{r})$, where $\Psi(\mathbf{r})$ is a magnetostatic potential. This potential obeys the following equation:

$$\nabla^2\Psi(\mathbf{r}) - \lambda\left(\frac{\partial m_1(\mathbf{r})}{\partial x_1} + \frac{\partial m_2(\mathbf{r})}{\partial x_2} + \frac{\partial m_3(\mathbf{r})}{\partial x_3}\right) = 0 \tag{7.10}$$

($\lambda = 4\pi$ in Gaussian units and $\lambda = 1$ in SI [Système International] units used throughout this chapter), since

$$\nabla(\mathbf{h}(\mathbf{r}) + \lambda\mathbf{m}(\mathbf{r})) = 0. \tag{7.11}$$

Use the usual linear-magnon approximation of neglecting, in Eq. (7.3), the small terms of second order in $\mathbf{m}(\mathbf{r})$ and $\mathbf{h}(\mathbf{r})$ [24]. This approximation is equivalent to setting $\mathbf{m}(\mathbf{r}) \cdot \mathbf{e}_3 = 0$ [19].

On substituting Eqs. (7.4), (7.7)–(7.10) into Eq. (7.3), one obtains

$$i\Omega m_1(\mathbf{r}) + (\nabla \cdot Q\nabla)m_2(\mathbf{r}) - m_2(\mathbf{r}) - \frac{M_S}{H_0}\frac{\partial\Psi(\mathbf{r})}{\partial x_2} = 0, \tag{7.12}$$

$$i\Omega m_2(\mathbf{r}) - (\nabla \cdot Q\nabla)m_1(\mathbf{r}) + m_1(\mathbf{r}) + \frac{M_S}{H_0}\frac{\partial\Psi(\mathbf{r})}{\partial x_1} = 0, \tag{7.13}$$

where

$$\Omega = \frac{\omega}{|\gamma|\mu_0 H_0} = B\omega \tag{7.14}$$

and

$$Q = \frac{2A}{M_S\mu_0 H_0}. \tag{7.15}$$

Considering the double periodicity in the (x_1, x_2) plane, expand Q and M_S in Fourier series

$$Q(\mathbf{X}) = Q(x_1, x_2) = \sum_{\mathbf{G}} Q(\mathbf{G})e^{i\mathbf{G}\cdot\mathbf{X}}, \tag{7.16}$$

$$M_S(\mathbf{X}) = M_S(x_1, x_2) = \sum_{\mathbf{G}} M_S(\mathbf{G})e^{i\mathbf{G}\cdot\mathbf{X}}, \tag{7.17}$$

where $\mathbf{G}$ are the reciprocal lattice vectors in the (x_1, x_2) plane of components (G_1, G_2), and $\mathbf{X}$ is the 2D position vector of components (x_1, x_2). The Fourier coefficients in Eq. (7.16) take the form

$$Q(\mathbf{G}) = \frac{1}{S} \iint d^2 \mathbf{X} Q(\mathbf{X}) e^{-i\mathbf{G}\cdot\mathbf{X}}, \tag{7.18}$$

where the integration is performed over the unit cell surface $S = a^2$.

For $\mathbf{G} = 0$, Eq. (7.18) gives the average Q, $\overline{Q}$

$$Q(\mathbf{G} = 0) = \overline{Q} = Q_A f + Q_B(1 - f). \tag{7.19}$$

For $\mathbf{G} \neq 0$, Eq. (7.18) may be written as

$$Q(\mathbf{G} \neq 0) = (Q_A - Q_B)F(\mathbf{G}) = \Delta Q F(\mathbf{G}), \tag{7.20}$$

where $F(\mathbf{G})$ is the structure factor

$$F(\mathbf{G}) = \frac{1}{S} \iint d^2 \mathbf{X} \, e^{-i\mathbf{G}\cdot\mathbf{X}} = 2f \frac{J_1(GR)}{GR}, \tag{7.21}$$

where $f = \pi(R^2/a^2)$ is the filling fraction, J_1 the Bessel function of the first kind, and R the radius of the cylinders. In Eq. (7.21), the integration is performed only on material A.

In an entirely similar way, Eq. (7.18) gives

$$M_S(\mathbf{G} = 0) = \overline{M}_S = M_{S_A} f + M_{S_B}(1 - f), \tag{7.22}$$

$$M_S(\mathbf{G} \neq 0) = (M_{S_A} - M_{S_B})F(\mathbf{G}) = \Delta M_S F(\mathbf{G}). \tag{7.23}$$

On the other hand, from the property of translational invariance in the x_3-direction, it follows that $m_1(\mathbf{r})$, $m_2(\mathbf{r})$, and $\Psi(\mathbf{r})$ must be of the form

$$m_1(\mathbf{r}) = m_1(\mathbf{X}) \, e^{iK_3 x_3}, \tag{7.24}$$

$$m_2(\mathbf{r}) = m_2(\mathbf{X}) \, e^{iK_3 x_3}, \tag{7.25}$$

$$\Psi(\mathbf{r}) = \Psi(\mathbf{X}) \, e^{iK_3 x_3}, \tag{7.26}$$

where K_3 is the x_3 component of the 3D wave vector (K_1, K_2, K_3).

For spin-wave propagation in the (x_1, x_2) plane (which means $K_3 = 0$), one can consider the 2D wave vector $\mathbf{K}(K_1, K_2)$ and use the Bloch theorem to write

$$m_1(\mathbf{X}) = e^{i\mathbf{K}\cdot\mathbf{X}} \sum_{\mathbf{G}} m_{1_K}(\mathbf{G}) e^{i\mathbf{G}\cdot\mathbf{X}}, \tag{7.27}$$

$$m_2(\mathbf{X}) = e^{i\mathbf{K}\cdot\mathbf{X}} \sum_{\mathbf{G}} m_{2_K}(\mathbf{G}) e^{i\mathbf{G}\cdot\mathbf{X}}, \tag{7.28}$$

$$\Psi(\mathbf{X}) = e^{i\mathbf{K}\cdot\mathbf{X}} \sum_{\mathbf{G}} \Psi_K(\mathbf{G}) e^{i\mathbf{G}\cdot\mathbf{X}}. \tag{7.29}$$

One can notice that for $K_3 = 0$, $\Psi(\mathbf{r})$ does not depend anymore on x_3. Therefore, $h_3(\mathbf{r}) = -(\partial\Psi(\mathbf{r})/\partial x_3) = 0$. One observes that for spin-wave propagation in the (x_1, x_2) plane and in the linear-magnon approximation, the vectors $\mathbf{m}(\mathbf{r})$ and $\mathbf{h}(\mathbf{r})$ are perpendicular to the x_3-axis.

After some algebra and considering the dimensionless vectors $\mathbf{k} = (a/2\pi)\mathbf{K}$ and $\mathbf{g} = (a/2\pi)\mathbf{G}$, the equations of motion can be rewritten as

$$
i\omega B m_{1_k}(\mathbf{g}) = \frac{\lambda \overline{M}_S}{H_0} \frac{(k_1 + g_1)(k_2 + g_2)}{|\mathbf{k} + \mathbf{g}|^2} m_{1_k}(\mathbf{g}) + \left\{ 1 + \left(\frac{2\pi}{a} \right)^2 |\mathbf{k} + \mathbf{g}|^2 \overline{Q} \right.
$$
$$
+ \frac{\lambda \overline{M}_S}{H_0} \frac{(k_2 + g_2)^2}{|\mathbf{k} + \mathbf{g}|^2} \left. \right\} m_{2_k}(\mathbf{g}) + \sum_{\mathbf{g} \neq \mathbf{g}'} \left\{ F \left(\frac{2\pi}{a} (\mathbf{g} - \mathbf{g}') \right) \right.
$$
$$
\times \left(\frac{\lambda \, \triangle \, M_S}{H_0} \frac{(k_1 + g_1')(k_2 + g_2')}{(\mathbf{k} + \mathbf{g}')^2} m_{1_k}(\mathbf{g}') \right.
$$
$$
+ \left[\left(\frac{2\pi}{a} \right)^2 \triangle Q(\mathbf{k} + \mathbf{g}')(\mathbf{k} + \mathbf{g}) + \frac{\lambda \, \triangle \, M_S}{H_0} \frac{(k_2 + g_2')^2}{(\mathbf{k} + \mathbf{g}')^2} \right] m_{2_k}(\mathbf{g}') \left. \right) \left. \right\},
$$

$$
\tag{7.30}
$$

$$
i\omega B m_{2_k}(\mathbf{g}) = - \left\{ 1 + \left(\frac{2\pi}{a} \right)^2 |\mathbf{k} + \mathbf{g}|^2 \overline{Q} + \frac{\lambda \overline{M}_S}{H_0} \frac{(k_1 + g_1)^2}{|\mathbf{k} + \mathbf{g}|^2} \right\} m_{1_k}(\mathbf{g})
$$
$$
- \frac{\lambda \overline{M}_S}{H_0} \frac{(k_1 + g_1)(k_2 + g_2)}{|\mathbf{k} + \mathbf{g}|^2} m_{2_k}(\mathbf{g}) - \sum_{\mathbf{g} \neq \mathbf{g}'} \left\{ F \left(\frac{2\pi}{a} (\mathbf{g} - \mathbf{g}') \right) \right.
$$
$$
\times \left(\left[\left(\frac{2\pi}{a} \right)^2 \triangle Q(\mathbf{k} + \mathbf{g}')(\mathbf{k} + \mathbf{g}) + \frac{\lambda \, \triangle \, M_S}{H_0} \frac{(k_1 + g_1')^2}{(\mathbf{k} + \mathbf{g}')^2} \right] m_{1_k}(\mathbf{g}') \right.
$$
$$
+ \frac{\lambda \, \triangle \, M_S}{H_0} \frac{(k_1 + g_1')(k_2 + g_2')}{(\mathbf{k} + \mathbf{g}')^2} m_{2_k}(\mathbf{g}') \left. \right) \left. \right\}.
$$

$$
\tag{7.31}
$$

In going from Eq. (7.12), (7.13) to Eqs. (7.30), (7.31), $\Psi(\mathbf{r})$ has been eliminated using Eqs. (7.10), (7.27)–(7.29) with $m_3 = 0$ and $h_3 = 0$.

One notes that there appear in those equations two types of terms: exchange terms depending on Q and $\triangle Q$ and dipolar interactions terms depending on M_s/H_0 and $\triangle M_s/H_0$. Eqs. (7.30), (7.31) correspond to an infinite set of linear equations where the unknowns are the Fourier components of the magnetization. In practice, obviously, only a finite number of $\mathbf{g}$ vectors are taken into account for the numerical calculation. The determinant of this system of equations must vanish, which conditions yield the magnon band structure $\omega_n(k)$. Despite the fact that Eqs. (7.30), (7.31) involve complex imaginary terms, their solutions $\omega_n(k)$ are real. One can also notice that in Eqs. (7.30), (7.31) there is an explicit dependence of the frequency ω with the lattice parameter a. In the case of photonic [1] and phononic band structures [2, 3, 25] such a dependence is implicit in the sense that the band structures are given in terms of a reduced frequency depending on the lattice parameter ($\Omega = \omega a/2\pi c$, where c is a velocity) versus the reduced wave vector. Regarding Eqs. (7.30), (7.31), one can observe that, due to the existence of the dipolar interaction terms, it is not enough to define a reduced frequency to take fully into account the effect of the lattice parameter a on the magnon band structure.

The calculation method presented in this section is not the only one. The interested readers may find other possibilities in Ref. [18].

7.3 MAGNON BAND STRUCTURES

In this section, are presented magnon band structures calculated for square arrays of Fe cylinders in a EuO background as well as for square array of Co cylinders in a permalloy matrix. The inverse situations, that is, EuO cylinders in an Fe matrix, and permalloy fibers in a Co matrix, have also been investigated. Fe, EuO, permalloy, and Co are ferromagnetic materials. Their physical parameters M_S and A are listed in Table 7.1.

In the case of the 2D periodic system, Fe (cylinders)/EuO (background), the influence of the Fe filling fraction and the effect of the lattice parameter on the band structure are also studied. In the course of the numerical calculations, the dimensionless reciprocal lattice vectors $\mathbf{g}$ are given as $\mathbf{g} = n_1 e_1 + n_2 e_2$, where n_1 and n_2 are two integers limited to the interval $-N \leq n_1, n_2 \leq +N$. All the results sketched here are obtained with $N = 6$. However, some of the dispersion curves were also calculated with $N = 10$ and confirmed the good accuracy of the results for $N = 6$. The difference in the eigenvalues calculated with $N = 10$ and 6 is small. We choose $N = 6$ which is a good compromise between accuracy and computing time.

Fig. 7.2 shows the first nine magnon bands for the square array of Fe cylinders in a EuO matrix, the filling fraction f being equal to 0.5. This magnon band structure is shown in the three principal directions of the first 2D Brillouin zone ΓXM (see the inset in Fig. 7.2). The plots are given in terms of the reduced frequency $\Omega = B\omega = \omega/|\gamma|\mu_0 H_0$ versus the reduced Bloch wave vector $\mathbf{k}$. The reduced frequency Ω is defined here as being independent of the lattice parameter a. The lattice parameter a is equal to 100 Å (the radius R of the cylinders is then equal to 40 Å) and $\mu_0 H_0 = 0.1$ T.

In the range of frequency of Fig. 7.2, four band gaps appear, respectively, between the first and the second band, between the second and the third band, between the fourth and the fifth band, and between the fifth and the sixth band. The widths of the first two gaps are of the order of 2 and 4 reduced units, respectively, which correspond to frequencies of the order of 5.5 and 11 GHz.

TABLE 7.1 Values of the Exchange Constant A and Spontaneous Magnetization M_S for Fe [19, 20], EuO [20, 26], and [27], Co [19], and Permalloy [19]

	$A(10^{-11}$ J/m)	$M_S(10^6$ A/m)
Fe	2.1	1.752
EuO	0.1	1.910
Co	2.8	1.390
Permalloy	0.7	0.810

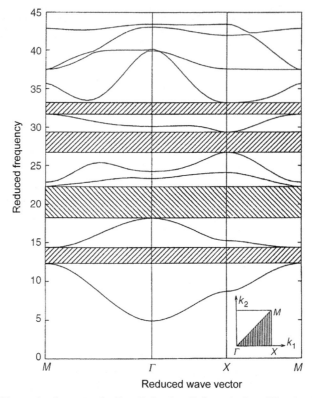

FIG. 7.2 Magnon band structure for Fe cylinders in a EuO matrix for a filling fraction $f = 0.5$, $a = 100$ Å, and $\mu_0 H_0 = 0.1$ T. The band structure is plotted in the three high-symmetry directions $\Gamma X M$ of the Brillouin zone (see *inset*). One can notice four gaps.

Fig. 7.3 shows the magnon band structure in the inverse situation, that is, the square array of EuO cylinders in a Fe matrix. In this case, the magnitudes of the first two gaps (close to 2 GHz) are smaller than those obtained in the previous case. However, the width of the third gap (appearing between the fourth and the fifth band) is of the same order of magnitude as in the former case.

In Fig. 7.4, the magnon band structure for a square array of Co cylinders in a permalloy matrix is drawn for $f = 0.5$. The other parameters used are the same as in Figs. 7.2 and 7.3. There is no gap in this case, neither in the inverse situation (permalloy cylinders in a Co matrix). One can notice that for the two binary composite systems Fe (cylinders)/EuO (matrix) and Co (fibers)/permalloy (background), the inclusions and the matrix have more or less the same spontaneous magnetization but different exchange constants (see Table 7.1).

Consider also a hypothetic binary composite system where the cylinders and the matrix have the same exchange constant but different spontaneous

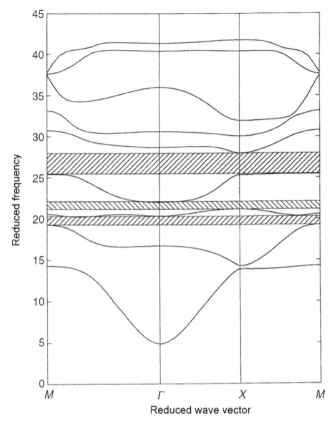

FIG. 7.3 Magnon band structure for EuO cylinders in an Fe matrix for a filling fraction $f = 0.5$, $a = 100$ Å, and $\mu_0 H_0 = 0.1$ T. There are three gaps. The width of the first two gaps is lower compared with Fig. 7.2. The width of the third gap is of the same order of magnitude in both cases.

magnetizations. In that case, gaps appear in the band structure only for the ratio $M_{S_A} M_{S_B}^{-1}$ greater than 10 or less than 0.1, that is, for a big difference in spontaneous magnetizations. Most of the usual ferromagnetic materials [20] (Fe, Co, Ni, Gd, EuO, permalloy) have spontaneous magnetizations of the same order of magnitude (from 0.5×10^6 to 2×10^6 A/m).

From these results, one can think that the existence of large gaps in the magnon band structure of square arrays of ferromagnetic cylinders embedded in a ferromagnetic background is associated with a very strong exchange contrast. The choice of an inclusion component of greater exchange constant than the matrix is more favorable for the opening of gaps in the magnon band structure. In Co and permalloy, the exchange constants are of the same order of magnitude whereas the Fe exchange constant is 21 times greater than 1 of EuO. One can also notice that the necessity of a strong contrast between the physical

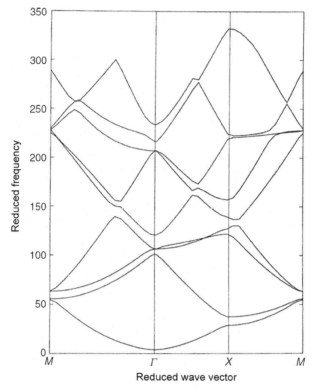

FIG. 7.4 Magnon band structure for Co cylinders in a permalloy matrix for a filling fraction $f = 0.5$, $a = 100$ Å, and $\mu_0 H_0 = 0.1$ T. There is no gap in this case either in the opposite situation (i.e., permalloy cylinders in a Co matrix).

parameters of the inclusions and the matrix has already been observed in previous works on elastic waves [3] and electrons [4].

Study now the influence of the inclusion filling fraction on the magnon band structure of 2D periodic ferromagnetic systems. In Fig. 7.5, the widths of the first three gaps in the magnon band structure of the square array of Fe cylinders in an EuO background are given as a function of the inclusion filling fraction f for $a = 100$ Å. Note the opening of gaps over a large range of the filling fraction, namely, $0.15 < f < 0.75$. The maximum gap width is obtained for $f = 0.65$. However, only two gaps appear for this value of f.

Investigate the effect of the lattice parameter on the magnon band structures. First of all, one can notice, looking at the equations of motion (Eqs. 7.30, 7.31), that the reduced frequency Ω depends on exchange terms and dipolar interactions terms, exchange terms being multiplied by the factor $(2\pi/a)^2$. As a result, for very low a, the magnon frequencies are not affected by the dipolar interactions. By contrast, for very large a, the first few magnon bands presented in our illustrations are strongly affected by the dipolar interactions. In Figs. 7.6

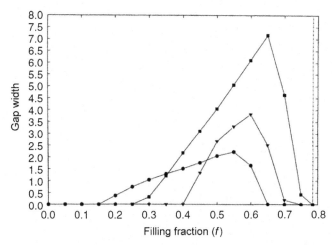

FIG. 7.5 The width of the first three band gaps in the magnon band structure of the square array of Fe cylinders in a EuO matrix for $a = 100$ Å and $\mu_0 H_0 = 0.1$ T as a function of the filling fraction f. The *vertical dashed line* corresponds to the close-packing value of f ($f = \pi/4$) for which one cylinder contacts another one. *Filled circles*: first gap (between the first and the second band). *Filled squares*: second gap (between the second and the third band). *Filled triangles*: third gap (between the fourth and the fifth band).

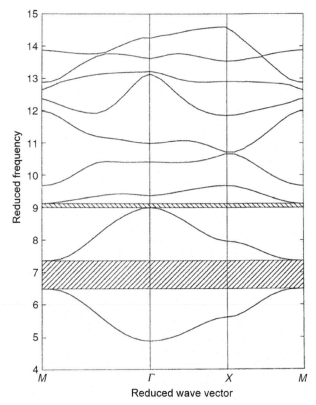

FIG. 7.6 Magnon band structure for Fe cylinders in a EuO matrix with the same parameters as in Fig. 7.2 but $a = 250$ Å.

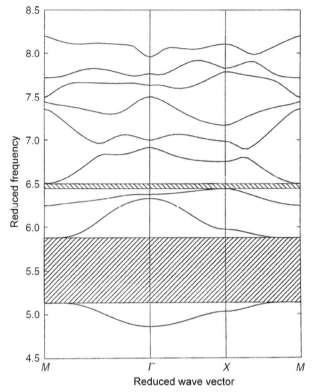

FIG. 7.7 Magnon band structure for Fe cylinders in a EuO matrix with the same parameters as in Fig. 7.2 but $a = 500$ Å.

and 7.7, the magnon band structures for the square array of Fe cylinders in a EuO matrix are plotted for two different values of the lattice parameter, $a = 250$ and 500 Å, respectively. Comparing Figs. 7.2, 7.6, and 7.7, one observes that the bottom of the first band (Γ point) still appears at, approximately, the same reduced frequency ($\Omega \approx 4.9$). On the other hand, the first gap width (as well as its location on the frequency scale) decreases with increasing lattice parameter. Moreover, in Figs. 7.6 and 7.7, the gaps of upper frequency domains (see Fig. 7.2) are very narrow and some have disappeared.

The opening of gaps in the magnon band structure of periodic square arrays of ferromagnetic cylinders embedded in a ferromagnetic matrix appears to be favored for arrays of low lattice parameter ($a \approx 100$ Å). One can notice that for $f = 0.5$ and $a = 100$ Å, the radius of the cylinders is $R = 40$ Å. This value corresponds approximately to the lower limit of the radius of the Co and Cu cylinders manufactured experimentally [28, 29].

7.4 SUMMARY

This chapter presents a few simple theoretical examples of the existence of band gaps in the magnon band structures of 2D composite systems composed of periodic arrays of infinite ferromagnetic cylinders embedded in a ferromagnetic matrix. For the periodic systems, Fe (cylinders)/EuO (matrix) and EuO (cylinders)/Fe (matrix), exist absolute band gaps for which the spin wave propagation in directions perpendicular to the cylinders axis is forbidden. The gap widths are in the range of 2–11 GHz, which is well above the experimentally available frequency resolution [20, 30]. The influence of the inclusion filling fraction and the effect of the lattice parameter on the band structure are also discussed. The existence of a strong contrast between the exchange constants of the inclusions and the matrix appears to be a necessary condition to obtain gaps in the magnon band structure of ferromagnetic composite systems. The width of these gaps and their frequency domains are strongly affected by the composition of the composite system and by the lattice parameter.

The first experiments on 2D magnonic crystals [31] are aimed at studying the processes of magnetization [32, 33] and magnetostatic wave transmission [34] in these structures. The periodic composite materials used in these studies are fabricated by drilling regularly distributed holes in a ferromagnetic material. Two-dimensional magnonic crystals can also be realized by systems of ferromagnetic rods forming a 2D crystallographic lattice and embedded in a magnetic or nonmagnetic matrix, or by a periodic antidot structure formed in a magnetic material [35–39]. The properties of a 2D magnonic crystal spin-wave spectrum depend on a number of factors. In an magnonic crystal composed of two different magnetic materials that form a periodic structure with lattice constant in the order of tens to hundreds of nanometers, exchange interactions (both within and between the constituent materials) are predominant, but the significance of dipolar interactions grows as the lattice constant of the magnonic structure increases. The situation is similar in magnonic crystals composed of a ferromagnet and a nonmagnetic material. In the three coupling ranges (exchange, dipolar exchange, and magnetostatic), spin waves in an magnonic crystal show qualitatively different dispersions. Studies of ferromagnetic layers with periodically modulated surfaces exist. Nikitov et al. [40] study an yttrium iron garnet (YIG) layer of 5–16 μm thickness with a lattice of holes with depth of 1–2 μm from the surface. He found that the surface periodicity has an impact on the spectrum of magnetostatic waves and on the magnonic gap opening.

It is worthy at this level to review some results concerning the 3D magnonic crystals. Actually, Krawczyk and Puszkarski [8] used the plane-wave method to determine spin-wave spectra of 3D magnonic crystals (the magnetic counterpart of photonic crystals) composed of two different ferromagnetic materials. The scattering centers in the magnonic crystal considered are ferromagnetic spheroids (spheres being a special case) distributed in sites of a cubic (sc, fcc, or bcc) lattice embedded in a matrix of a different ferromagnetic material. They demonstrate that magnonic gaps in such structures occur at spontaneous magnetization contrast and/or exchange contrast values above a certain critical level, which depends on the lattice type. Optimum conditions for magnonic gaps to open are offered by the structure in which the scattering centers are the most densely packed (the fcc lattice). They also show that in all three lattice types considered the reduced width of the gap (i.e., the width referred to the gap center) is, in good approximation, a linear function of both the exchange contrast and the magnetization contrast. Also, the gap width proves sensitive to scattering center deformation, and its maximum value corresponds to a scattering center shape close to a sphere. Moreover, the numerical results seem to indicate that dipolar interactions in general result in an effective reduction of the gap width, but their impact only becomes of importance when the lattice constant of the cubic magnonic structure is greater than the ferromagnetic exchange length of the matrix material.

On the other hand, a survey of fabrication methods of 3D periodic dielectric structures can be found in Refs. [41, 42]. As regard magnetic structure fabrication, the so far best developed methods are those of fabricating 1D systems (magnetic multilayers). A very promising method is ion implantation applied to ferromagnetic thin films [43, 44]. Widely used in submicron semiconductor technology [45], ion implantation has been applied for spatial modulation of anisotropy [46, 47], damping coefficient [48], and effective gyromagnetic ratio [43]. Ion implantation (with lithographic mask or directed ion beam) allows fabrication of magnonic crystals with either 2D or 3D periodicity [48]. Modeling magnetic properties by light or ion irradiation is discussed in detail in a survey by Fassbender et al. [49]. There is also an interesting possibility of "optical" fabrication of 3D magnonic crystals, based on the effect of local crystallization of Co_2MnSi (due to ferromagnetic phase formation) induced by femtosecond laser pulse. In general, interference of two laser beams can also result in a periodic magnetic structure reproducing the interference pattern [50].

Two-dimensional periodic structures are fabricated by lithographic methods, such as e-beam lithography [51] or ultraviolet lithography [52]. A promising technique using porous alumina templates allows fabrication of large regular lattice-based systems of ferromagnetic rods of length $0.2-200$ μm with lattice constant ranging from 50 to 500 nm [53, 54]. Block copolymer lithography is used as well, allowing fabrication of systems composed of a few nanolayers of periodically distributed magnetic particles with period of the order of ~56 nm [55]. The lithographic methods have an advantage of giving an almost free choice of the crystallographic structure of the fabricated 2D magnonic crystal, with a variety of dot shapes and a wide range of available filling fraction and lattice constant values [56, 57]. An excellent survey of fabrication methods of 2D ordered magnetic structures can be found in a paper by Terris and Thomson [58]. Methods to obtain 2D ordered magnetic nanostructures are also discussed in a survey by Martiń et al. [59]. Worthy of notice are also the self-assembled methods, allowing fabrication of 2D and 3D lattices of Co or Fe nanoparticles [60–63].

An important incentive to study magnonic crystals now is their prospective application to the construction of logic systems using the wave nature of spin excitations. Many papers on this subject discuss possible practical applications including wave front reversal (phase conjugation) of surface magnetostatic waves [64], spin-wave interference [65–67], or the possibilities of controlling spin-wave phase (an analog of the Aharonov-Bohm effect) [68]. Other prospective applications are based on negative magnetic refraction coefficient (left-handed spin waves) [69] or magnetostatic wave focusing [70, 71].

Magnonic circuits can by designed with the help of all magnetic crystals. One kind of magnetic materials of great interest is the man-made 2D magnonic crystals. The investigations of these artificial materials expand fast and in particular toward applications, see, for example, Refs. [9, 10, 13–16]. The simple tutorial examples presented in this chapter are already complemented by more realistic numerical simulations and by experiments for some long-wavelength spin waves. A lot remains to be done for microwaves, especially those used in modern telecommunications. Let us just mention that investigations on nonlinear effects in such structures exist also; see, for example, Zielinski et al. [72].

REFERENCES

[1] E. Yablonovitch, Inhibited Spontaneous Emission in Solid-State Physics and Electronics, Phys. Rev. Lett. 58 (1987) 2059.

[2] M.M. Sigalas, E.N. Economou, Band structure of elastic waves in two dimensional systems, Solid State Commun. 86 (1993) 141.

[3] M.S. Kushwaha, P. Halevi, L. Dobrzynski, B. Djafari-Rouhani, Acoustic band structure of periodic elastic composites, Phys. Rev. Lett. 71 (1993) 2022.

[4] L. Dobrzynski, B. Djafari-Rouhani, J. Vasseur, R. Kucharczyk, M. Steslicka, Electronic structure of some mesoscopic systems: electronic composites, Prog. Surf. Sci. 48 (1995) 213.

[5] J.O. Vasseur, L. Dobrzynski, B. Djafari-Rouhani, H. Puszkarski, Magnon band structure of periodic composites, Phys. Rev. B 54 (1996) 1043.

[6] M. Krawczyk, H. Puszkarski, Magnonic spectra of ferromagnetic composites versus magnetization contrast, Acta Phys. Pol. A 93 (1998) 805.

[7] H. Puszkarski, M. Krawczyk, Magnonic Crystals – the Magnetic Counterpart of Photonic Crystals, Solid State Phenom. 94 (2003) 125.

[8] M. Krawczyk, H. Puszkarski, Plane-wave theory of three-dimensional magnonic crystals, Phys. Rev. B 77 (2008) 054437.

[9] V.V. Kruglyak, S.O. Demokritov, D. Grundler, Magnonics, J. Phys. D Appl. Phys. 43 (2010) 264001.

[10] A.A. Serga, A.V. Chumak, B. Hillebrands, YIG magnonics, J. Phys. D Appl. Phys. 43 (2010) 264002.

[11] H. Al-Wahsh, A. Akjouj, B. Djafari-Rouhani, L. Dobrzynski, Magnonic circuits and crystals, Surf. Sci. Rep. 66 (2011) 29.

[12] B. Lenk, H. Ulrichs, F. Garbs, M. Munzenberg, The building blocks of magnonics, Phys. Rep. 507 (2011) 107.

[13] S.O. Demokritov, A.N. Slavin (Eds.), Magnonics, From Fundamentals to Applications, Springer, 2013.

[14] S.O. Demokritov (Ed.), Spin Wave Confinement, Pan Stanford and World Scientific Pub., 2009.

[15] M. Krawczyk, D. Grundler, Review and prospects of magnonic crystals and devices with reprogrammable band structure, J. Phys. Condens. Matter. 26 (2014) 123202.

[16] A.D. Karenowska, A.V. Chumak, A.A. Serga, B. Hillebrands, Magnon Spintronics, in: Y. Xu, D.D. Awschalom, J. Nitta (Eds.), Handbook of Spintronics, Springer, 2015, pp. 1505–1549.

[17] J.D. Joannopoulos, S.G. Johnson, J.N. Winn, R.D. Meade, Photonic Crystals Molding the Flow of Light, Princeton University Press, Princeton, NJ, 2008.

[18] Y. Pennec, B. Djafari-Rouhani, A. Akjouj, et al., in: L. Dobrzynski (Ed.), Transmission in 2D Phononic Crystals and Acoustic Metamaterials, Interface Transmission Tutorial Book Series, Elsevier, 2017, p. 271.

[19] M. Vohl, J. Barnas, P. Grünberg, Effect of interlayer exchange coupling on spin-wave spectra in magnetic double layers: Theory and experiment, Phys. Rev. B 39 (1989) 12003.

[20] F. Keffer, Spin Waves, Handbuch der Physik, vol. XVIII/2, Springer, Berlin, 1966.

[21] G.T. Rado, J.R. Weertman, Spin-wave resonance in a ferromagnetic metal, J. Phys. Chem. Solids 11 (1959) 315.

[22] H. Puszkarski, M. Cottam, Interface effective pinning and dynamical coupling of spin waves in bilayer ferromagnetic films with canted magnetizations, Acta Phys. Polon. A 79 (1991) 549.

[23] L. Dobrzynski, B. Djafari-Rouhani, H. Puszkarski, Theory of bulk and surface magnons in Heisenberg ferromagnetic superlattices, Phys. Rev. B 33 (1986) 3251.

[24] M.G. Cottam, D.R. Tilley, Introduction to Surface and Superlattice Excitations, Cambridge University Press, Cambridge, England, 1989.

[25] M.S. Kushwaha, P. Halevi, G. Martinez-Montes, L. Dobrzynski, B. Djafari-Rouhani, Theory of acoustic band structure of periodic elastic composites, Phys. Rev. B 49 (1994) 2313.

[26] B.T. Matthias, R.M. Bozorth, J.H. Van Vleck, Ferromagnetic Interaction in EuO, Phys. Rev. Lett. 7 (1961) 160.

[27] J.F. Dillon, C.E. Olsen, Ferromagnetic Resonance of EuO, Phys. Rev. 135 (1964) A434.

[28] A. Blondel, J.P. Meier, B. Doudin, J.P. Ansermet, Giant magnetoresistance of nanowires of multilayers, Appl. Phys. Lett. 65 (1994) 3019.

[29] J. de la Figuera, M.A. Huerta-Garnica, J.E. Prieto, C. Ocal, R. Miranda, Fabrication of magnetic quantum wires by step-flow growth of cobalt on copper surfaces, Appl. Phys. Lett. 66 (1995) 1006.

[30] M.G. Cottam, O.J. Lockwood, Light Scattering in Magnetic Solids, Wiley, New York, NY, 1987.

[31] Y.V. Gulyaev, S.A. Nikitov, L.V. Zhivotovskii, A.A. Klimov, P. Tailhades, L. Presmanes, C. Bonningue, C.S. Tsai, S.L. Vysotskii, Y.A. Filimonov, Ferromagnetic films with magnon bandgap periodic structures: Magnon crystals, JETP Lett. 77 (2003) 567.

[32] A.V. Butko, S.A. Nikitov, Y.A. Filimonov, Studying magnetization processes in magnon crystals using the meridional Kerr effect, J. Commun. Technol. Electron. 50 (2005) 88.

[33] A.V. Butko, A.A. Klimov, S.A. Nikitov, Y.A. Filimonov, Using the magneto-optical Kerr effect to study magnetization processes in two-dimensional magnonic crystals based on YIG thin films, J. Commun. Technol. Electron. 51 (2006) 944.

[34] S.L. Vysotskiĭ, S.A. Nikitov, Y.A. Filimonov, Magnetostatic spin waves in two-dimensional periodic structures (magnetophoton crystals), J. Exp. Theor. Phys. 101 (2005) 547.

[35] L. Torres, L. Lopez-Diaz, O. Alejos, J. Iniguez, Micromagnetic study of lithographically defined non-magnetic periodic nanostructures in magnetic thin films, Physica B 275 (2000) 59.

[36] C.C. Wang, A.O. Adeyeye, Y.H. Wu, Magnetic properties of asymmetric antirectangular Ni80Fe20 arrays, J. Appl. Phys. 94 (2003) 6644.

[37] L.J. Heyderman, H.H. Solak, F. Nolting, C. Quitmann, Fabrication of nanoscale antidot arrays and magnetic observations using x-ray photoemission electron microscopy, J. Appl. Phys. 95 (2004) 6651.

[38] C. Yu, M.J. Pechan, W.A. Burgei, G.J. Mankey, Lateral standing spin waves in permalloy antidot arrays, J. Appl. Phys. 95 (2004) 6648.

[39] S. McPhail, C.M. Gürtler, J.M. Shilton, N.J. Curson, J.A.C. Bland, Coupling of spin-wave modes in extended ferromagnetic thin film antidot arrays, Phys. Rev. B 72 (2005) 094414.

[40] S.A. Nikitov, Y.A. Filimonov, P. Tailhades, Magneto-photonic and Magnonic Crystals Based on Ferrite Films-New Types of Magnetic Functional Materials, Adv. Sci. Technol. (Faenza, Italy) 45 (2006) 1355.

[41] J.F. Galisteo, F. Garcia-Santamaria, D. Golmayo, B.H. Juarez, C. Lopez, E. Palacios, Self-assembly approach to optical metamaterials, J. Opt. A Pure Appl. Opt. 7 (2005) S244.

[42] C. Lopez, Three-dimensional photonic bandgap materials: semiconductors for light, J. Opt. A Pure Appl. Opt. 8 (2006) R1.

[43] S.G. Reidy, L. Cheng, W.E. Bailey, Dopants for independent control of precessional frequency and damping in Ni81Fe19 (50 nm) thin films, Appl. Phys. Lett. 82 (2003) 1254.

[44] J. Fassbender, J. McCord, Control of saturation magnetization, anisotropy, and damping due to Ni81Fe19 implantation in thin layers, Appl. Phys. Lett. 88 (2006) 252501.

[45] J. Melngailis, Focused ion beam technology and applications, J. Vac. Sci. Technol. B 5 (1987) 469.

[46] J. Fassbender, et al., Ion irradiation of exchange bias systems for magnetic sensor applications, Appl. Phys. A Mater. Sci. Process. 77 (2003) 51.

[47] J. McCord, T. Gemming, L. Schultz, J. Fassbender, M.O. Liedke, M. Frommberger, E. Quandt, Magnetic anisotropy and domain patterning of amorphous films by He-ion irradiation, Appl. Phys. Lett. 86 (2005) 162502.

[48] V. Dasgupta, N. Litombe, W.E. Bailey, H. Bakhru, Ion implantation of rare-earth dopants in ferromagnetic thin films, J. Appl. Phys. 99 (2006) 08G312.

[49] J. Fassbender, D. Ravelosona, Y. Samson, Tailoring magnetism by light-ion irradiation, J. Phys. D 37 (2004) R179.

[50] J.H. Kim, J. Kim, S.U. Lim, C.K. Kim, C.S. Yoon, G.J. Lee, Y.P. Lee, Spatially periodic magnetic structure produced by femtosecond laser-interference crystallization of amorphous Co2MnSi thin film, J. Appl. Phys. 99 (2006) 08G311.

[51] M.J. Pechan, C. Yu, D. Owen, J. Katine, L. Folks, M. Carey, Vortex magnetodynamics: Ferromagnetic resonance in permalloy dot arrays, J. Appl. Phys. 99 (2006) 08C702.

[52] J. Wang, A.O. Adeyeye, N. Singh, Magnetostatic Interactions in Mesoscopic Ni80Fe20 ring arrays, Appl. Phys. Lett. 87 (2005) 262508.

[53] K. Nielsh, R.B. Wehrspohn, J. Barthel, J. Kirschner, U. Gsele, S. Fischer, H. Kronmüller, Hexagonally ordered 100 nm period nickel nanowire arrays, Appl. Phys. Lett. 79 (2001) 1360.

[54] Y. Ikezawa, K. Nishimura, H. Uchida, M. Inoue, J. Magn. Magn. Mater. 272 (2004) 1690.

[55] J.Y. Cheng, W. Jung, C.A. Ross, Phys. Rev. B 70 (2004) 064417.

[56] C.A. Ross, et al., Magnetic nanostructures from block copolymer lithography: Hysteresis, thermal stability, and magnetoresistance, Phys. Rev. B 65 (2002) 144417.

[57] C.A. Ross, et al., Micromagnetic behavior of electrodeposited cylinder arrays, J. Appl. Phys. 91 (2002) 6848.

[58] D.B. Terris, T. Thomson, Magnetic behavior of lithographically patterned particle arrays, J. Phys. D 38 (2005) R199.

[59] J.I. Martín, J. Nogúes, K. Liu, J.L. Vicent, I.K. Schuller, Ordered magnetic nanostructures: fabrication and properties, J. Magn. Magn. Mater. 256 (2003) 449.

[60] M. Sachan, N.D. Walrath, S.A. Majetich, K. Krycka, C.-C. Kao, Interaction effects within Langmuir layers and three-dimensional arrays of ε-Co nanoparticles, J. Appl. Phys. 99 (2006) 08C302.

[61] D. Farrell, Y. Ding, S.A. Majetich, C. Sanchez-Hanke, C.C. Kao, Structural ordering effects in Fe nanoparticle two- and three-dimensional arrays, J. Appl. Phys. 95 (2004) 6636.

[62] S. Sun, C.B. Murray, Synthesis of monodisperse cobalt nanocrystals and their assembly into magnetic superlattices, J. Appl. Phys. 85 (1999) 4325.

[63] S. Sun, C.B. Murray, D. Weller, L. Folks, A. Moser, Monodisperse FePt nanoparticles and ferromagnetic FePt nanocrystal superlattices, Science 287 (2000) 1989.

[64] G.A. Melkov, V.I. Vasyuchka, A.V. Chumak, V.S. Tiberkevich, A.N. Slavin, Wave front reversal of nonreciprocal surface dipolar spin waves, J. Appl. Phys. 99 (2006) 08P513.

[65] S. Choi, K.-S. Lee, S.-K. Kim, Spin-wave interference, Appl. Phys. Lett. 89 (2006) 062501.

[66] J. Podbielski, F. Giesen, D. Grundler, Spin-Wave Interference in Microscopic Rings, Phys. Rev. Lett. 96 (2006) 167207.

[67] S.V. Vasiliev, V.V. Kruglyak, M.L. Sokolovskii, A.N. Kuchko, Spin wave interferometer employing a local nonuniformity of the effective magnetic field, J. Appl. Phys. 101 (2007) 113919.

[68] R. Hertel, W. Wulfhekel, J. Kirschner, Domain-Wall Induced Phase Shifts in Spin Waves, Phys. Rev. Lett. 93 (2004) 257202.

[69] D.D. Stancil, B.E. Henty, A.G. Cepni, J.P. Van't Hof, Observation of an inverse Doppler shift from left-handed dipolar spin waves, Phys. Rev. B 74 (2006) 60404.

[70] M. Bauer, C. Mathieu, S.O. Demokritov, B. Hillebrands, P.A. Kolodin, S. Sure, H. Dotsch, V. Grimalsky, Y. Rapoport, A.N. Slavin, Direct observation of two-dimensional self-focusing of spin waves in magnetic films, Phys. Rev. B 56 (1997) R8483.

[71] V. Veerakumar, R.E. Camley, Magnon focusing in thin ferromagnetic films, Phys. Rev. B 74 (2006) 214401.

[72] P. Zielinski, A. Kulak, L. Dobrzynski, B. Djafari-Rouhani, Propagation of waves and chaos in transmission line with strongly anharmonic dangling resonator, Eur. Phys. J. B 32 (2003) 73.

Index

Note: Page numbers followed by f indicate figures and t indicate tables.

A

Acoustic crystals, 151–152
Aharonov-Bohm systems, 55, 97, 137–138, 156
Asymmetric serial loop structures (ASLS), 131–138
Atomic model, 192, 222

B

Bessel function, 237
Bloch theorem, 151–152, 237
Block copolymer lithography, 246
Bose Einstein occupation-number factor, 208, 216
Bound in continuum states
    continuous systems, 48–49
    definition, 47–48
    discrete-continuous systems, 49
    discrete systems, 48
    fano and induced transparency resonances, 49–50
    general theorem, 49
Brillouin light scattering technics, 144–145
Brillouin zone, 186–187, 189f, 191f, 202, 227, 230, 239, 240f
Bulk Heisenberg model
Bulk response function, 187–188

C

Cartesian coordinate system, 59–61
Centered system phonons
    continuous phonon model, 44–47
    discrete phonon model, 42–44
Comb structures
    defect modes, 72–75, 73f, 74f, 76f
    fano resonances
        theoretical discussion, 84–86, 84f
        transmission gaps, 86–94
    magnonic stop bands, 68–72, 69f, 70f, 71f, 72f

1D waveguide/backbone, 64–66, 64f
pinning fields
    dispersion relations, 78–79
    huge gap results, 82–83
    results, 79–82, 80f, 83f
    transmission coefficients, 78–79
remarks, 76–77, 77f
single-grafted segment, 62–64, 62f
transmission coefficient, 66–68, 66f
transmission spectrum, 68–72, 69f, 70f, 71f, 72f
Confined slab magnons
    dipolar and Zeeman energies, 192
    the localized magnons, 198–199
    response function
        the ferromagnetic quantum well, 193–196
        particular limits, 196–197
    surface response operators, 192–193
Continuous phonon model, 44–47
Continuous systems, 48–49
Continuous theory, 58–59
Coupling infinite linear chain
    calculations, 148–154
    dipole-dipole interactions, 148, 152–153
    geometrical/magnetic parameters, 148, 150f
    model, 148–154
    nearest-neighbor dipolar systems, 150
    results, 155–160
    selecting/rejecting magnon filter, 148

D

Dangling resonators, 131–132
Discrete-continuous systems, 49
Discrete phonon model, 42–44
Discrete systems, 48
Discrete theory, 57–58
Dobrzynski's interface response theory, 113

E

Electromagnetic induced transparency (EIT)
  analytical results, 101–102
  motivations, 96–97, 98*f*
  transmission and reflection coefficients, 98–101
Electromagnetic waves, 54–55, 61
EuO cylinders, 239–240, 241*f*

F

Fano resonances, 55
  theoretical discussion, 84–86, 84*f*
  transmission gaps, 86–94
Ferromagnetic system, 222
Finite loop structures
  segment linking, 116–118
  two semiinfinite leads, 118
Finite surface-pinning field
  density of state variation, 216
  specific heat variation, 216–217
  the state phase shift, 214–216
Fourier analysis, 116
Fourier coefficients, 236–237
Free surface, 57, 59, 61–62, 78, 84

G

Green function method, 113
Green's function
  continuum approximation, 59
  diagonal and off-diagonal elements, 78
  finite wire, 61–62
  gyromagnetic ratio, 59–60
  infinite ferromagnetic medium, 59–61
  medium 1, 62, 64–67
  medium 2, 62, 64–66, 78
  semiinfinite medium, 61

H

Heisenberg exchange approach
  long-wavelength approximation
    five interacting objects, 33–37
    four interacting objects, 29–33
    N objects ($N \geq 2$), 37–40
    system responses, 40–41
    three interacting objects, 25–29
    two interacting objects, 22–25
Heisenberg exchange interaction model
  five interacting objects
    determinants, 14

eigenvalues and eigenvectors values, 14–16
    forced trapping and isolation, 17–18
    the inverse problem, 17
    matrix, 14
    resonant responses, 17
    response matrix, 16
    system responses, 16–17
  four interacting objects
    determinants, 11
    eigenvalues and eigenvectors values, 11–12
    forced trapping and isolation, 13
    the inverse problem, 13
    matrix, 10 11
    resonant responses, 13
    response matrix, 12
    system responses, 12–13
  N objects ($N \geq 2$)
    determinants, 18
    eigenvalues and eigenvectors values, 18–20
    matrix, 18
    response matrix, 20
    system responses, 20–21
  three interacting objects
    determinants, 8
    eigenvalues and eigenvectors values, 8–9
    forced trapping and isolation, 10
    the inverse problem, 9–10
    matrix, 7
    resonant responses, 10
    response matrix, 9
    system responses, 9
  two interacting objects
    determinants, 4
    dynamical matrix, 4
    eigenvalues and eigenvectors values, 4–5
    forced trapping and isolation, 6
    the inverse problem, 5–6
    resonant responses, 6
    response matrix, 5
    system responses, 5
Heisenberg ferromagnet, 187, 189, 204–206
Heisenberg model, 222, 231
Holstein-Primakoff transformation, 186, 223
Hypothetic binary composite system, 240–241

I

Infinite periodic system, 114*f*, 116
Interface magnons, 190–191
Interface response theory, 5, 9, 12, 16,
     22–23, 35, 40, 144–147
   continuous theory, 58–59
   discrete theory, 57–58

K

Kronecker symbol, 214

L

Landau-Lifshitz equation, 235
Larmor equation, 150–151
Levi-Civita symbol, 229
Localized magnons, 185–186, 190–191,
     191*f*, 198–199, 200*f*
Long-wavelength approximation
   five interacting objects
      complete eigenfunctions, 36
      determinants, 33
      eigenvalues and eigenvector values,
         34–35
      forced trapping and isolation, 37
      the inverse problem, 36
      matrix, 33
      resonant responses, 36–37
      response matrix, 35
      state conservation, 33
      system responses, 36
   four interacting objects
      complete eigenfunctions, 31–32
      determinants, 29
      eigenvalues and eigenvector values,
         30–31
      forced trapping and isolation,
         32–33
      the inverse problem, 32
      matrix, 29
      resonant responses, 32
      response matrix, 31
      state conservation, 29–30
      system responses, 32
   N objects (*N* ≥ 2), 37–40
      complete eigenfunctions, 40
      determinants, 37
      eigenvalue and eigenvector values,
         38–39
      response matrix, 39–40
      state conservation, 37–38

   system responses, 40–41
   three interacting objects
      complete eigenfunctions, 27–28
      determinants, 25
      eigenvalues and eigenvector values,
         26–27
      forced trapping, 29
      the inverse problem, 28
      matrix, 25
      resonant responses, 28
      response matrix, 27
      state conservation, 26
      system responses, 28
   two interacting objects
      complete eigenfunctions, 24
      determinants, 22
      eigenvalue and eigenvector values,
         22–23
      forced responses and trapping, 25
      the inverse problem, 24–25
      matrix, 22
      resonant responses, 25
      response function, 23–24
      system responses, 24
Low-dimensional spin systems, 54, 112
Low-lying modes, 208*f*, 211–212, 212*f*

M

Magnetic force microscopy,
     144–145
Magnetic induced transparency (MIT), 56
Magnetic memory, 150
Magnetic quantum well, 192
Magneto-optic Kerr effects, 144–145
Magnon band structures
   periodic square arrays, 244
   physical parameters, 239, 239*t*
   two binary composite systems, 240
Magnonic band gaps, 112, 119–123, 122*f*,
     137, 144
Magnonic crystals, 54–55, 76–77, 96–97,
     112, 137–138, 144, 156
Magnonic systems, 54–55, 97
Magnon nanometric multiplexer
   applications, 171–177
   calculations, 169–171
   results, 171–177
   telecommunication router devices,
     167–168
Medium 1, 62, 64–67
Medium 2, 62, 64–66, 78

MIT. *See* Magnetic induced transparency (MIT)
Modern telecommunications, 138
Molecular magnets, 150
Multiplexers, 102–103

N

Nanofabrication technology, 54, 112, 143–146, 177
Nanostructure, 54
Nearest-neighbor dipolar systems, 150
Neighboring magnetic moments, 59–60

O

One-dimensional (1D) crystals
    bulk and surface magnons, 228–229, 228*f*
    ferromagnetic and antiferromagnetic layers, 222
    magnetic superlattices, 221–222
    magnon band structure, 229, 230*f*
1D propagation vector, 115–116

P

Phenomenological damping factor, 59–60
Photonic crystals, 144
Pinning fields
    dispersion relations, 78–79
    huge gap results, 82–83
    results, 79–82, 80*f*, 83*f*
    transmission coefficients, 78–79
Planar defect magnons, 188

Q

Quantum computing, 150
Quantum well, particular limits
    an adsorbed ferromagnetic slab on another semiinfinite ferromagnet, 197
    an interface between two semiinfinite ferromagnets, 197
    free surface ferromagnetic slab, 197
    free surface of a semiinfinite ferromagnet, 197
Quasibox structures
    applications, 163–167
    calculations, 160–163
    model, 160–163
    results, 163–167

S

Serial loop structure (SLS), 113
Single-cell spectral gaps
    application, 137–138
    dangling resonators, 131–134
    magnonic band structure, 135–136, 136*f*
SLS. *See* Serial loop structure (SLS)
Soft magnon, 200–201
Spin systems, 144
Spin-transport devices, 112
Surface magnons
    ferromagnetic and antiferromagnetic stability, 202, 203*f*
    first nearest-neighbor interactions, 189–190
    reconstruction, 200–203
    second nearest-neighbor interactions, 190
    soft magnon, 200–203
Surface-pinning fields
    elementary considerations
        with an infinite surface-pinning field, 211–213
        without surface-pinning fields, 207–211
    motivations, 203–205
    specific heat reduction factor, 217, 218*f*
Symmetric serial loop structures
    analytical and numerical results, 127–128
    defect modes, 123–126
    magnonic band gaps, 119–123
    selective transmission, 123–126
    theoretical model
        finite loop structures, 116–118
        1D infinite serial loop, 114–116, 114*f*
        structure with a defect, 118–119
    transmission spectra, 119–123

T

Telecommunication router devices, 167–168
Three-slab 1D crystal, bulk and surface magnons, 229–230
Two-dimensional (2D) crystals
    composite system, 245
    magnetostatic wave focusing, 247
    magnon band structure, 240, 242*f*
    microscopic modifications, 235
    plane-wave method, 245–246
    theoretical model, 234–235
2D periodic system, 239

Two-slab 1D crystal
    advantage, 224
    bulk and surface magnons, 227–229
    bulk magnetic response function,
        223–226
    surface ferromagnetic response function,
        226–227

V
Variation of the density of states (VDOS),
        155f, 156

Y
Yttrium iron garnet (YIG), 245

Printed in the United States
By Bookmasters